Bhanu Prasad (Ed.)

Soft Computing Applications in Business

Studies in Fuzziness and Soft Computing, Volume 230

Editor-in-Chief

Prof. Janusz Kacprzyk
Systems Research Institute
Polish Academy of Sciences
ul. Newelska 6
01-447 Warsaw
Poland
E-mail: kacprzyk@ibspan.waw.pl

Further volumes of this series can be found on our homepage: springer.com

Bhanu Prasad (Ed.)

Soft Computing
Applications in Business

 Springer

Editor

Dr. Bhanu Prasad
Department of Computer and Information Sciences
Florida A&M University
Tallahassee, FL 32307
USA
Email: bhanu.prasad@famu.edu

Associate Editors

Dr. Pawan Lingras
Department of Math &
Computing Science,
Saint Mary's University, Halifax,
Nova Scotia, Canada

Dr. M.A. Karim Sadiq
Applied Research Group,
Satyam Computer Services Ltd.,
Indian Institute of Science Campus,
Bangalore, India

Dr. T.N. Nagabhushan
Department of Information Science &
Engineering,
S J College of Engineering,
Mysore - 570 006, India

Dr. Shailendra Singh
Samsung India Software Center,
Noida -201301, India

Dr. Yenumula B. Reddy
Department of Mathematics and
Computer Science,
Grambling State University, Grambling,
Louisiana 71245, USA

ISBN 978-3-642-09783-6   e-ISBN 978-3-540-79005-1

DOI 10.1007/978-3-540-79005-1

Studies in Fuzziness and Soft Computing   ISSN 1434-9922

Printed in acid-free paper

9 8 7 6 5 4 3 2 1

springer.com

Preface

Welcome to the book "Soft Computing Applications in Business". This book consists of 15 scientific papers. Each paper was independently reviewed by 2 experts in the topical area. The papers include the following:

Ensembles of Classifiers in Arrears Management by Chris Matthews and Esther Scheurmann from Latrobe University, Victoria, Australia

Bicluster Analysis of Currency Exchange Rates by Haizhou Li and Hong Yan from City University of Hong Kong, Hong Kong and the University of Sydney, Sydney, Australia

Predicting the Effects of Alternative Pricing Strategies in an Artificial Society Undergoing Technology Adoption by Judy K. Frels, Debra Heisler and James A. Reggia from the University of Maryland, College Park, USA

An Evolutionary Programming based Knowledge Ensemble Model for Business Risk Identification by Lean Yu, Kin Keung Lai and Shouyang Wang from the Chinese Academy of Sciences, Beijing, China and the City University of Hong Kong, Hong Kong

The application of Fuzzy Decision Trees in Company Audit Fee Evaluation: A Sensitivity Analysis by Malcolm J. Beynon from Cardiff University, Cardiff, UK

An Exposition of NCaRBS: Analysis of US Banks and Moody's Bank Financial Strength Rating by Malcolm J. Beynon from Cardiff University, Cardiff, UK

A Clustering Analysis for Target Group Identification by Locality in Motor Insurance Industry by Xiaozhe Wang and Eamonn Keogh from the University of Melbourne, Melbourne, Australia and the University of California, Riverside, USA

Probabilistic Sales Forecasting for Small and Medium-Size Business Operations by Randall E. Duran from Catena Technologies Pte Ltd, Singapore

User Relevance Feedback Analysis in Text Information Retrieval: A Rough Set Approach by Shailendra Singh and Bhanu Prasad from Samsung India Software Center, Noida, India and Florida A&M University, Tallahassee, USA

Opportunistic Scheduling and Pricing Strategies for Automated Contracting in Supply Chains by Sabyasachi Saha and Sandip Sen from University of Tulsa, Tulsa, USA

Soft Computing in Intelligent Tutoring Systems and Educational Assessment by Rodney D. Nielsen, Wayne Ward, James H. Martin from University of Colorado, Boulder, USA

A Decision Making System for the Treatment of Dental Caries by Vijay Kumar Mago, Bhanu Prasad, Ajay Bhatia and Anjali Mago from DAV College, Jalandhar, India; Florida A&M University, Tallahassee, USA; ICFAI, Hyderabad, India and Mago Dental Clinic, Jalandhar, India

CBR Based Engine for Business Internal Control by M. L. Borrajo, E. S. Corchado, M. A. Pellicer and J. M. Corchado from University of Vigo, Vigo, Spain; University of Burgos, Burgos, Spain and University of Salamanca, Salamanca, Spain

An EMD-based Neural Network Ensemble Learning Model for World Crude Oil Spot Price Forecasting by Lean Yu, Shouyang Wang and Kin Keung Lai from Chinese Academy of Sciences, Beijing, China and City University of Hong Kong, Hong Kong and

Structured Hidden Markov Models: A General Tool for Modeling Agent Behaviors by Ugo Galassi, Attilio Giordana and Lorenza Saitta from Università del Piemonte Orientale, Alessandria, Italy.

Sincerely
Bhanu Prasad
Editor
Department of Computer and Information Sciences,
Florida A&M University, Tallahassee, FL 32307, USA

Contents

Ensembles of Classifiers in Arrears Management

Chris Matthews and Esther Scheurmann

Faculty of Science, Technology & Engineering Latrobe University
P.O. Box 199 Bendigo 3552, Victoria, Australia
Tel.: +61 3 54447557; Fax: +61 3 54447998
c.matthews@latrobe.edu.au, ess.scheurmann@gmail.com

Abstract. The literature suggests that an ensemble of classifiers outperforms a single classifier across a range of classification problems. This chapter provides a brief background on issues related ensemble construction and data set imbalance. It describes the application of ensembles of neural network classifiers and rule based classifiers to the prediction of potential defaults for a set of personal loan accounts drawn from a medium sized Australian financial institution. The imbalanced nature of the data sets necessitated the implementation of strategies to avoid under learning of the minority class and two such approaches (minority over-sampling and majority under-sampling) were adopted here. The ensembles outperformed the single classifiers, irrespective of the strategy that was used. The results suggest that an ensemble approach has the potential to provide a high rate of classification accuracy for problem domains of this type.

Keywords: neural network and rule based ensembles, data set imbalance, loan default, arrears management.

1 Introduction

Authorised Deposit-Taking Institutions (ADIs) are corporations that are authorised under the Australian Banking Act (1959) to invest and lend money. ADIs include banks, building societies and credit unions. ADIs generate a large part of their revenue through new lending or extension of existing credit facilities as well as investment activities. The work described here focuses on lending, in particular the creation and management of customer personal loan accounts. Many financial institutions use credit scoring models to assist in loan approval. Traditionally these have been statistically based and are built using techniques such as logistic regression, linear regression and discriminant analysis (Lewis 1992, Mays 2001). More recently the use of computational intelligence approaches, including artificial neural networks; rule based and fuzzy approaches in credit scoring have attracted some research interest.

The use of artificial neural networks (ANNs) in business applications is well established in the literature (for example, see the survey of application areas in Vellido, Lisboa & Vaughan (1999) and McNelis (2005)). This indicates that ANNs show promise in various areas where non-linear relationships are believed to exist in data and where statistical approaches are deficient. One of the earliest applications of ANNs to business loan approval was by Srivastava (1992). In this work a multilayer perceptron forms a module in a larger loan approval system. The work uses simulated data and claims that the system can successfully emulate human decision-making. Several comparative studies have been undertaken to compare the performance of

B. Prasad (Ed.): Soft Computing Applications in Business, STUDFUZZ 230, pp. 1–18, 2008.
springerlink.com   © Springer-Verlag Berlin Heidelberg 2008

ANN credit scoring models with more traditional approaches. For example, Desai, Crook & Overstreet (1995) found that ANN models compared favourably with those constructed using logistic regression techniques, particularly in the classification of bad loans. A more recent study (West 2000) compared a variety of ANN approaches (such as multilayer perceptrons, mixture of experts and radial basis functions) with traditional statistical techniques and non-parametric approaches. This work was an investigation into the accuracy of quantitative models used by the credit industry. The comparison was conducted across two datasets: a German credit scoring dataset and an Australian one. This research indicates that ANNs can improve credit-scoring accuracy, but that logistic regression is still a good alternative. However the study showed that k-nearest neighbour and linear discriminant analysis did not produce as encouraging results across the two datasets.

Hybrid approaches have also attracted attention, in particular researchers have attempted to combine ANNs with genetic algorithms and fuzzy systems. Hoffman, Baesens, Martens, Put & Vanthienen (2002) compared two hybrid combination classifiers for credit scoring. One was a genetic fuzzy classifier while the other was based on the NEFCLASS neuro-fuzzy classifier (Nauck & Kruse 1995). The genetic fuzzy classifier evolved a fuzzy system and infered approximate fuzzy rules where each rule had its own definition of the membership functions. These rules were generated iteratively where examples matching the newly generated rule were removed from the training dataset. The rule generation process continued until all training examples are sufficiently covered. The neuro-fuzzy classifier learnt descriptive fuzzy rules where all fuzzy rules shared a common, interpretable definition for the membership functions. The real life credit scoring data used in this work came from a Benelux financial institution and C4.5 rules were generated as a benchmark. The results indicated that the genetic fuzzy classifier yielded the best classification accuracy at the expense of intuitive and humanly understandable rules. The neuro-fuzzy classifier showed a lesser classification performance, although the fuzzy rules generated were more comprehensible for a human to understand. Another hybrid approach is to combine a supervised and an unsupervised artificial neural network learning approach (Huysmans, Baesens, Vanthienen & Van Getsel 2006). Initially the powerful data visualisation capabilities of a self organising map[1] are used to distinguish between good and bad credit risk examples. Once the map is built it can be clustered and then labeled. The map now functions as a classifier for unseen examples. To improve the classification accuracy of the poorer performing neurons in the map, a classifier is built for each using a supervised approach and a subset of the training set data projected onto the neuron. A second approach is to train the supervised classifiers on the entire training set prior to building the map. A weighted output of these classifiers, together with original input variables is used to build the map. As with the neurofuzzy and the genetic fuzzy hybrid approaches the real world data sets used in this work were obtained from a Benelux financial institution.

---

[1] A self organising map uses an unsupervised learning algorithm i.e. there is no target but instead the neurons in the feature layer (map) are grouped according to their closeness or similarity to the training set examples. In other words the map takes on the characteristics found in the training data.

2 Arrears Management

Substantial amounts of money are spent on the recovery of defaulted loans, which could be significantly decreased by having the option of tracking the high default risk borrowers' repayment performance. This process is sometimes referred to as *arrears* or *collections management*. A typical collections strategy is shown in figure 1. The details may vary from financial institution to financial institution, but the process is initiated once a customer fails to meet their repayment obligations.

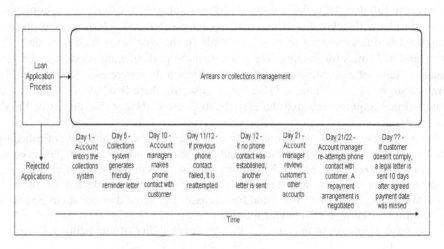

Fig. 1. An Arrears or collection strategy

This is essentially a classification problem. Loan accounts could be classified as high or low risk depending on the risk of the customer not meeting these obligations. The predictors could be a mixture of applicant data collected at the time of loan approval and that obtained during the life of the account. If one was also interested in assessing the effectiveness of the credit scoring models used to assist in the loan approval process then the predictor attributes would be reduced to only those collected at the point of application. The success of computational intelligence approaches such as artificial neural networks in credit scoring suggests that they may also be of use in the identification of risky loan accounts. Not surprisingly neural network models have been built to predict both loan default and early loan repayment (Baesens, Van Gestel, Stepanova & Vanthienen 2003). Again real world rather than simulated data sets (obtained from a U.K. financial institution) were used. The results were promising and suggested that the neural network approach outperformed the more traditional approaches, particularly for the prediction of early repayment.

The work described here focuses on loan default and applies an ensemble as well as a single classifier approach. As well as using ensembles of neural network classifiers we also use rule based approaches. One advantage of a rule based classifier is that it is possible to seek an explanation for a particular classification through an examination of the clauses in the rule sets. This is not possible when using a numerical learning approach such as an artificial neural network. Instead the classifier is treated as a 'black box' with no explanation provided for the classification result.

3 Classifiers and Ensembles

In simplest terms, a classifier divides examples into a number of categories. Classifiers may be built using a training data set and then tested on unseen data to determine their generalisation capabilities. Typically training uses a supervised learning approach i.e the target class is known for both the training and testing data. To reduce any bias that may arise in the selection of examples for training and testing *N-fold* cross validation has been suggested as a suitable method to assess generalisation. In this case the full data set is divided into *N* disjoint subsets. *N-1* of these are combined and used to build the classifier. The remaining set is reserved for testing. This process is repeated *N* times ensuring that each example in the data set is used only once for testing and *N-1* times for training. The generalisation performance is given as the percentage of data set examples correctly classified when they were used for testing. The literature suggests 10 as suitable value for *N*, ensuring there is a balance between the computational requirements and the classification error (Hastie, Tibshirani & Friedman 2003).

One alternative to cross validation is the *bootstrap*. Here a training set is formed by sampling *with replacement* from the original data set. The number of samples is equal to the data set size. So, although the training set has the same number of examples as the original data set, there are duplicates. Those examples not selected are then reserved for testing. It can be shown that for a reasonably sized data set about 36.8% of the examples will not be selected for training.

If we have a data set of *m* examples then the probability of one being selected for training is $\frac{1}{m}$ and of not being selected is $(1-\frac{1}{m})$. For a total of *m* such selections the probability (*P*) of an example not being selected for training is

$$P = \left(1 - \frac{1}{m}\right)^m, where$$

$$P \approx e^{-1} = 0.368 \ (provided \ m \ is \ large).$$

So the ratio of unique examples in the training to testing set is approximately 0.632:0.368. The procedure could be repeated several times and an average value taken as an indication of generalisation (Witten & Frank 2005).

An *ensemble* consists of a series of individual classifiers whose results are combined in some way to give a single classification. There are sound theoretical reasons suggesting that the use of an ensemble, rather than a single classifier, can significantly improve classification performance (Hansen & Salamon 1990, Dietterich 2000, Wu & Arribas 2003). This assumes that each ensemble member is *accurate* and that *diversity* exists between the ensemble members. Accuracy is related to the performance of an individual classifier on the data set while diversity is related to a difference in classification performance of the ensemble members on a given data set example. It is expected that the probability of correct classification of each ensemble member is greater than 0.5 i.e. better than a random guess.

3.1 Diversity Measures

The development of suitable diversity measures and the correlation between these and ensemble accuracy is an area of ongoing research (Kucheva & Whitaker 2003, Ko, Sabourin & Britto 2006). Broadly speaking a diversity measure can be classified into one of two categories: *pairwise* measures and *non-pairwise* measures. Pairwise measures use the concept of agreement and disagreement between pairs of classifiers and estimates an overall diversity based on the average of the measures for all possible pairs in the ensemble. To illustrate this consider a data set of N examples and two classifiers C_a and C_b which can classify the examples either correctly or incorrectly. There are four possibilities:

N^{11}: the number of examples that both C_a and C_b classify correctly
N^{00}: the number of examples that both C_a and C_b classify incorrectly
N^{10}: the number of examples that C_a classifies correctly and
C_b incorrectly
N^{01}: the number of examples that C_a classifies incorrectly and
C_b correctly

where $N = N^{11} + N^{00} + N^{01} + N^{10}$
   The two classifiers disagree in $N^{10} + N^{01}$ cases and agree in $N^{11} + N^{00}$ cases.
   Several simple measures have been proposed using these. For example, a simple disagreement measure(D_{ab}) would be the proportion of the data set where the classifiers disagreed i.e.

$$D_{ab} = \frac{N^{10} + N^{01}}{N^{11} + N^{00} + N^{10} + N^{01}} .$$

The measure ranges from 0 (no disagreement) to 1 (total disagreement).
   Another approach is to use a measure that assesses the similarity of the classifiers. For example, the Q statistic[2] has been applied

$$Q_{ab} = \frac{N^{11}N^{00} - N^{01}N^{10}}{N^{11}N^{00} + N^{01}N^{10}} .$$

If the two classifiers disagree more than they agree (i.e N^{01}, N^{10} are greater than N^{11}, N^{00}) then the measure will tend to be negative. In the case where there is no agreement (i.e. N^{11} and N^{00} are both equal to zero) then the measure is -1. In the opposite case the measure is 1.
   Non-pairwise measures focus on the voting patterns of the ensemble members across the data set examples. A simple example is the entropy based diversity measure(E). An entropy measure can be considered as an indication of disorder or impurity. In this context it relates to the extent to which the ensemble members agree (or disagree) when classifying a particular example. If all members agree then the measure is 0, if there is complete disagreement (i.e half the ensemble vote one way and the other half the other way) then measure is 1. The measure can be defined as follows:

---

[2] The Q statistic is a well established co-efficient of association used in statistics to estimate the association between attributes (see Yule & Kendall (1950) for further details).

For an ensemble of L classifiers, $[L/2]$ represents the integer part of $L/2$. If, for a particular example z, there are X classifiers that correctly classify it then $(L-X)$ classifiers have incorrectly classified it. The diversity is.

$$E_z = \frac{1}{L - [L/2]} \times \min(X, (L - X)).$$

For a data set of N examples the overall entropy based diversity is the average across all examples i.e.

$$E(N) = \frac{1}{N} \sum_{z=1}^{N} E_z.$$

For a more complete discussion of these and other diversity measures the reader is directed to Kucheva & Whitaker (2003).

3.2 Ensemble Construction and Combination Methods

Ensembles are particularly useful for classification problems involving large data sets (Chawla, Hall, Bowyer & Kegelmeyer 2004, Bernardini, Monard & Prati 2005) and can be constructed in various ways (Dietterich 2000, Granitto, Verdes & Ceccatto 2005). Each ensemble member could be trained and tested on a subset of the total data set. This approach works well for unstable learning algorithms such as those used by artificial neural networks and in rule induction[3]. There are three possible approaches: *bagging (bootstrap aggregation)*, *cross-validated committees* and *boosting*. In the first two cases the ensembles members are constructed independently of each other whereas in the boosting approach they are constructed sequentially i.e. the construction of a given ensemble member is dependant on that built before.

The bagging approach is illustrated in figure 2. The ensemble is constructed from a data set of m examples. Each member is trained and tested using the bootstrap approach. Each ensemble member focuses on differing subsets of the original data set with some overlap between them.

The cross-validated committee approach (figure 3) is based on cross validation as previously described. For an ensemble of L classifiers the data set is subdivided into L disjoint subsets. Each classifier is trained on *(L-1)* of these with a different one left out each time. The left out subset can be used for testing if necessary Again the ensemble members focus on different subsets of the full data set with some overlap.

Boosting algorithms build the ensemble members by focusing on previously mis-classified examples (figure 4). Initially each example in the data set is equally weighted. A classifier is built using randomly selected training data and is then tested on the data set. Misclassified examples are more heavily weighted. A second classi-fier is then built. The more heavily weighted examples i.e. those miss-classified by the first classifier, have a greater chance of selection for training than before. In this way the second classifier focuses on correctly classifying those examples

---

[3] An unstable learning algorithm is one where the classifier output undergoes significant changes in response to small changes in the training data (Dietterich 2000).

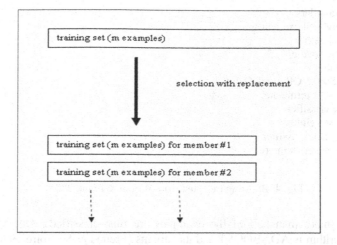

Fig. 2. Bagging (bootstrap aggregation)

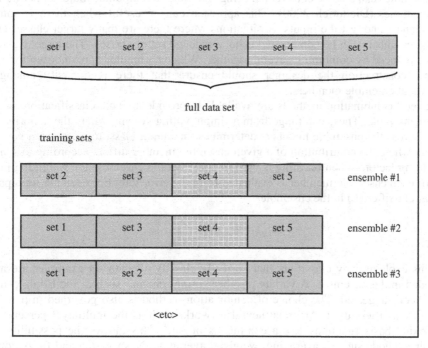

Fig. 3. A five (5) member ensemble constructed using a cross-validated committee approach

miss-classified by the first classifier. The second classifier is then tested on the data set and the weights are adjusted again. This process is repeated until the desired number of classifiers are built. The process may terminate earlier when a classifier is built that classifies all examples correctly (i.e. the error rate is zero) or when the error rate

```
Let N = #classifiers
Equally weight all examples
Build new classifier
Evaluate error rate(e)
FOR N iterations:
        IF e >= 0.5 OR e = 0 THEN
                terminate
        Store classifier
        Adjust weights
        Build new classifier
        Evaluate error rate(e)
END FOR
```

Fig. 4. Boosting (adapted from Witten & Frank 2005)

exceeds 0.5 (more than half of the examples are miss-classified). One well known boosting algorithm is ADABOOST and its variants (Freund & Schapire 1996, Kucheva 2003).

Ensembles can also be constructed using subsets of the input attributes rather training examples (Dietterich 2000). This approach can be useful when there is some redundancy amongst the inputs. In situations where there are many target classes ensemble members can be constructed using a reduced set of classes. The target classes can be reduced by combining several together. Whatever methods are choosen for ensemble construction the designer should ensure that there is diversity amongst individual ensemble members.

Several combination methods are available to provide a single classification result for an ensemble. These can range from a simple voting system, where the majority of votes across the ensemble members determines the output class, to a weighted voting system where the contribution of a given ensemble member differs according to some weighting factor. When boosting is used to construct the ensemble the weighting factor for each ensemble member is evaluated from the error rate. For example for a particular classifier (L) in the ensemble

$$weight_L = -\log\left(\frac{e_L}{1 - e_L}\right)$$

This explains why classifiers that correctly classify all data set examples are not included in the ensemble. A variety of other voting schemes such as the Borda count have been suggested. The choice of combination method is also governed in part by type of classifiers used. Artificial neural networks such as the multilayer perceptron and radial basis function have a continuous output[4]. This allows the possibility of further methods such as averaging, weighted averaging, Nash voting and fuzzy combination. For discussion and evaluation of these and other approaches the reader is directed to Wanas & Kamel (2001) and Torres-Sospedra, Fernandez-Redono & Hernandez-Espinosa (2005).

---

[4] Although the output is continuous artificial neural networks can solve classification problems. The target classes are coded using discrete values from the network output interval. The network is trained to approximate these and hence recognise the classes

4 The Application Area and Classification Approaches

4.1 The Application Area

The data used in this study was real life data sourced from a medium sized Australian bank and was subject to ethics approval and commercial confidentiality. It included a low proportion of bad accounts. The classifier models were developed using personal loan accounts created in May 2003. The observation point was 12 months later i.e May 2004. This was considered sufficient time before a realistic assessment of their performance could be made (figure 5). The data set totaled 1534 accounts consisting of 1471 `good' examples and 63 `bad' examples. The imbalanced nature of the data set was typical across the unsecured loan accounts of the financial institution involved. Personal loan accounts created in June, November and December 2003 and whose performance was assessed twelve months later (i.e. in 2004) were used for testing. Twenty-three input attributes were available, of which twenty-two were collected at the time of loan approval and one during the observation period, reflecting loan per-formance. Of these, seventeen were continuous and six discrete. There was no signifi-cant correlation between any of the input attributes collected at the time of loan approval and the target class. However there was some correlation between the one attribute collected during the observation period and the target. This was expected as it related to repayment behavior and was seen as a reasonable predictor for loan default.

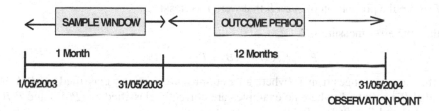

Fig. 5. Data Selection

The study was guided by two questions:

1. Is it possible to use machine learning approaches such as artificial neural net-works and rule induction to predict whether a loan account is likely to default at some point in the future?
2. If so, can we identify those factors which might indicate whether an account is at risk of defaulting i.e. what are the strong predictors of default?

A preliminary study which used similar datasets suggested that ensembles of clas-sifiers out performed single classifiers, particularly when classifying the minority class (Scheurmann & Matthews 2005). However only neural network ensembles were used and all available attributes were included when building the models. Due to the non-symbolic nature of a neural network approach only the first of the two questions could be addressed. The work described here extends that study and uses rule based as well as neural network ensembles. As indicated earlier an advantage of a symbolic approach such as rule induction is that a model which can provide an explanation for the classification is created. In this way some insight into the factors affecting loan default can be gained. In other words it may be possible to seek an answer to the second question.

The highly imbalanced nature of the data sets presents a challenge. The literature suggests that classifiers trained on imbalanced data sets of the type used here tend to learn the majority class at the expense of the minority one (Guo & Viktor 2004). In arrears management it is important that classifiers predict well the minority class (i.e. the `bad' accounts) and so strategies need to be developed to encourage this. Both over-sampling of the minority class and under-sampling of the majority have been suggested as suitable approaches (Guo & Viktor 2004, Cohen, Hilario, Sax, Hugonnet & Geissbuhler 2006). The greater importance of correct classification of the minority class also influences the way in which the accuracy of a single classifier or ensemble is accessed. For a reasonably balanced data set a simple measure of correct (or incorrect) classifications is sufficient. However in this case a weighted error measure that rewards correct classification of the minority class at the expense of the majority class was used when comparing the performance of individual classifiers and ensembles. It is based on the error rate measure given in Drummond & Holte (2005). The more severe the imbalance the greater the weighting given to the minority class. For two classes A and B let

$P(A)$ = the proportion of examples of class A in the data set
$P(B)$ = the proportion of examples of class B in the data set
$P(A^\sim)$ = the proportion of class A that were misclassified
$P(B^\sim)$ = the proportion of class B that were misclassified

then the error measure is

$$E = P(A^\sim) \times P(B) + P(B^\sim) \times P(A)$$

The measure ranges from 0 (where all examples are correctly classified i.e. $P(A^\sim) = 0$ and $P(B^\sim) = 0$) to 1 (where no examples are correctly classified i.e.($P(A^\sim)$ and $P(B^\sim)$ are both 1 and $E = P(B) + P(A)$).

In the case where a classifier successfully classifies all of class A ($P(A^\sim) = 0$) but none of class B ($P(B^\sim) = 1$) then the error measure is $P(A)$, or in the opposite case $P(B)$. In a data set where there is equal representation of both classes ($P(A) = 0.5$, $P(B) = 0.5$) the measure is same as the overall error. For data set of size N where x examples of class A and y examples of class B are misclassified, the overall error is $\frac{(x+y)}{N}$. There are $0.5N$ examples in each class so the measure becomes

$$E = \frac{x}{0.5N} \times 0.5 + \frac{y}{0.5N} \times 0.5 = \frac{x+y}{N}$$

4.2 Classification Approaches

Two basic classification approaches were used in this work: the multi-layer perceptron (MLP) and rule induction. A multi-layer perceptron is a numeric learning system that can be considered a non-linear function approximator. They are feed forward

```
PROCEDURE BuildTree (ExampleSubset)
NumberOf Classes = calculate the number of classes in the example subset
IF NumberOfClasses = 0 THEN Null leaf
ELSE
     IF NumberOfClasses = 1 THEN
         store the output class as a leaf in the tree
     ELSE
         DecisionNodeInput = Determine the best attribute
         IF the DecisionNodeInput = 0 THEN
           Error: more than one class has all the same attributes
         ELSE
           Create a decision node for the DecisionNodeInput
           FOR all values of the DecisionNodeInput
               Determine the NewExampleSubset for this input value.
               BuildTree(NewExampleSubset)
         ENDFOR
       ENDIF
     ENDIF
   ENDIF
```

Fig. 6. A simple decision tree algorithm

neural networks consisting of a collection of weighted interconnected parallel processing units (called neurons) arranged in layers. These units process weighted numerical inputs to derive an output value. Each processing unit accepts as input a
weighted sum of the outputs of those neurons connected to it. A simple architecture
consists of three components: a layer of input neurons connected to a layer of hidden
neurons, which in turn is connected to a layer of output neurons. Depending on the
number of hidden layers, the connected units may be input layer neurons or neurons
from a previous layer. The processing units then apply a transformation function and
output a numeric value. The inputs to a multi-layer perceptron must be numeric and
each neuron is responsible for the output of a single piece of data. Multi-layer perceptrons use supervised learning i.e. by being presented with training examples where the
target is known. During training examples are presented and an output evaluated. An
error is evaluated and this propagated back through the network and used to adjust the
weighted connections to reduce the error. The process is iterative and training finishes
when the error is minimised or a set number of training cycles have been reached[5].

On the other hand rule induction is a symbolic approach that builds a set of production rules from a decision tree based on the continual partitioning of a data set into
smaller and smaller subsets. The method used here is based on Quinlan's C4.5 decision
tree algorithm (Quinlan 1993). This is a recursive algorithm that uses an entropy based
measure to select an appropriate attribute on which to split the data set (figure 6). Continuous attributes are discretised prior to construction of the tree and a pruning
algorithm can be applied to simplify 'bushy' trees.

---

[5] There are many basic references outlining the theory and the application of artificial neural
networks to classification problems. For example, see Hagan, Demuth & Beale 1996 and Karray & de Silva C 2004.

All of the experimental work was conducted using the Waikato Environment for Knowledge Acquisition (*Weka*) development platform (Witten & Frank 2005). The advantage of this platform is that a series of symbolic learning approaches (including rule induction and decision tree building) as well as numerical approaches such as the multilayer perceptron are available. The software also provides data analysis tools and implements boosting and bagging algorithms which may be useful in ensemble construction.

5   Experimental Results and Discussion

A series of preliminary experiments using single multi-layer preceptron (MLP) and rule based classifiers were conducted. These used both the original training data set and a larger data set where the minority class was over-sampled. The results are shown in table 1. The ratio of majority to minority class examples is given for each model. The over-sampled data set was formed by replicating the minority class examples twenty five times to produce an enlarged set with almost equal representation of each class. Two error measures are given for model performance. The overall error is simply the fraction of the data set misclassified whereas the weighted error uses the class proportions to modify this as described above. High values for the weighted error measure and low values for the overall error indicate poor classification performance on the minority class and good classification on the majority class. For each data set, the final row in the table gives the weighted error and the average error that would

Table 1. Single classifier models using all attributes

Model (majority: minority)	May (training) weighted error (overall error)	June (testing) weighted error (overall error)	November (testing) weighted error (overall error)	December (testing) weighted error (overall error)
MLP 1 ($\approx 0.96:0.04$)	0.244 (0.014)	0.709 (0.052)	0.672 (0.067)	0.782 (0.066)
MLP.2 ($\approx 0.96:0.04$)	0.244 (0.013)	0.790 (0.040)	0.755 (0.062)	0.805 (0.058)
MLP 3 ($\approx 0.96:0.04$)	0.213 (0.013)	0.729 (0.042)	0.707 (0.058)	0.748 (0.062)
MLP oversampled ($\approx 0.48:0.52$)	0.001 (0.001)	0.790 (0.050)	0.839 (0.068)	0.817 (0.067)
Single rule base ($\approx.0.96:0.04$)	0.171 (0.010)	0.630 (0.018)	0.562 (0.067)	0.537 (0.034)
Single rule base oversampled ($\approx.0.48:0.52$)	0.004 (0.004)	0.466 (0.018)	0.493 (0.071)	0.494 (0.032)
$P(A^-) = 1$ $P(B^-) = 0$	0.959 (0.041)	0.972 (0.028)	0.957 (0.043)	0.953 (0.047)

result if all minority class examples were misclassified (i.e. $P(A^-) = 1$) and all majority class examples were classified correctly (i.e. $P(B^-) = 0$).

The three multilayer perceptron models represent networks of differing architectures. Several comments can be made.

- As expected classifiers built using the data set without over-sampling tended to learn the majority class (the `good' accounts) at the expense of the minority class (the `bad' accounts). This can be seen by noting the higher value of the weighted error with respect to that of the overall error on the training data. The testing performance of these classifiers was relatively poor on the minority class, again evidenced by the high weighted error values.
- The classifiers built using the minority oversampled training data had low training set error measures but still had difficulty classifying the minority class in the testing data.
- The rule based classifiers performed better than the multilayer perceptron classifiers on the testing data.

To obtain a better measure of generalisation the data sets were combined and 10-fold cross validation was performed using both a rule based and a multilayer perceptron approach. In both cases the overall error measure was very low (0.044 for the MLP and 0.036 for the rule based approach), however the weighted error measures (0.855 for the MPL and 0.456 for the rule based approach) suggested that the rule based approach generalised far better on the minority class. This is consistent with the relative testing set performances mentioned above.

Ensembles of nine classifiers were then built. Again the training sets for each were taken from the May 2003-2004 loan accounts and the June, November and December data sets were used for testing. All minority class examples (i.e the 'bad' accounts) were used in each training set. The majority class (i.e. the 'good' accounts) was under-sampled with disjoint subsets of this class used for the training sets. The proportion of 'good' accounts to 'bad' accounts was approximately 2.5:1 in each case. Ensembles of both multi-layer perceptrons and of rule based classifiers were built. Table 2 shows the training performance of each ensemble member.

The weighted error and overall error measures for each were very low indicating that the majority class under-sampling combined with all minority class examples resulted in the construction of accurate classifiers that were able to classify both classes well. Three diversity measures are also provided, two pairwise (the Q statistic and the disagreement measure) and an entropy based measure. They calculated using the training set data. Taken together the error and diversity measures show that in all cases the MLP ensemble members are more accurate than the corresponding rule base members, and that the MLP ensemble is more diverse.

A second pair of ensembles were constructed using data containing only those attributes collected at the time of loan approval. This is of interest as it might be possible to test the effectiveness of the loan approval process, including the credit scoring models used. However the classification problem becomes more difficult as the attribute relating to loan performance is removed. The results are shown in table 3.

Table 2. Ensemble members built using all attributes

Ensemble member training set	MLP weighted error (overall error)	Rule base weighted error (overall error)
#1	0.000 (0.000)	0.055 (0.026)
#2	0.011 (0.004)	0.114 (0.044)
#3	0.023 (0.009)	0.100 (0.044)
#4	0.023 (0.009)	0.130 (0.053)
#5	0.016 (0.009)	0.100 (0.058)
#6	0.011 (0.004)	0.169 (0.071)
#7	0.000 (0.000)	0.139 (0.062)
#8	0.023 (0.009)	0.093 (0.044)
#9	0.035 (0.013)	0.130 (0.066)
average	0.016 (0.006)	0.115 (0.052)
sd	0.011 (0.004)	0.031 (0.013)
Diversity(Disagreement)	0.202	0.104
Diversity(Q-statistic)	0.600	0.820
Diversity(Entropy)	0.231	0.113

Table 3. Ensemble members using only those attributes collected at the time of loan application

Ensemble member training set	MLP weighted error (overall error)	Rule base weighted error (overall error)
#1	0.000 (0.000)	0.125 (0.062)
#2	0.011 (0.004)	0.141 (0.071)
#3	0.023 (0.009)	0.132 (0.062)
#4	0.011 (0.004)	0.198 (0.093)
#5	0.016 (0.009)	0.148 (0.071)
#6	0.011 (0.004)	0.107 (0.044)
#7	0.016 (0.009)	0.127 (0.071)
#8	0.011 (0.004)	0.141 (0.071)
#9	0.023 (0.009)	0.142 (0.057)
average	0.014 (0.006)	0.140 (0.067)
sd	0.005 (0.002)	0.024 (0.012)
Diversity(Disagreement)	0.260	0.321
Diversity(Q-statistic)	0.586	0.412
Diversity(Entropy)	0.302	0.390

A similar trend to that in table 2 was observed. The multi-layer perceptron ensemble members were more accurate than their rule base equivalents. However in this case they appear less diverse. It should also be noted that a positive value for the Q statistic suggests that the ensemble members are tending to agree with each other on the training data. The size of the ensemble rule sets depended very much on the attribute used for training. Those trained using all attributes (table 2) ranged in size from 11 to 18 rules (sd 1.8) whereas those trained only the attributes collected at the time of loan application were larger, varying from 28 to 33 rules (sd 1.4). This reflects the relative difficulty of the classification task across the two data sets.

The ensembles were then tested using the June, November and December data sets. A simple majority voting scheme was used to obtain a single classification result for each testing example. The results are shown in table 4. The best performance was that of the rule based ensemble built using all attributes. The low weighted and overall error measures suggest that it generalises well on both the majority and minority classes. The MLP ensemble built using the same data sets had some difficulty generalising on the minority class, particularly for the June and November data sets.

Table 4. Ensemble performance

Ensemble	June weighted error (overall error)	November weighted error (overall error)	December weighted error (overall error)
MLP (all attributes)	0.610 (0.121)	0.687 (0.142)	0.396 (0.135)
Rule based (all attributes)	0.102 (0.013)	0.018 (0.135)	0.001 (0.011)
MLP (attributes at time of loan application)	0.794 (0.169)	0.857 (0.196)	0.778 (0.207)
Rule based (attributes at time of loan application)	0.854 (0.155)	0.879 (0.170)	0.822 (0.185)

This is different from the observations in the previous study (Scheurmann & Matthews 2005) where the ensemble generalised very well across both classes in all testing sets. It should be noted that there were differences in the way the data was preprocessed and selected, and in the size of the ensemble. More significantly a variation on the basic back propagation algorithm, *quickprop* (Fahlman 1988) was used in the training of the ensemble members. This suggests that data pre-processing decisions, testing set selection and the learning algorithm are important considerations when constructing ensembles of this type. As expected the performance of the two ensembles built using only those attributes collected at the time of loan application was less than that of those built using all attributes.

Finally a series of MLP and rule based ensembles, using all atributes, were built the bagging and boosting techniques described earlier. The results are shown in

Table 5. Ensembles built using bagging and boosting

Ensemble	June weighted error (overall error)	November weighted error (overall error)	December weighted error (overall error)
MLP (bagging)	0.850 (0.026)	0.802 (0.041)	0.815(0.042)
Rule based (bagging)	0.648 (0.019	0.528 (0.058)	0.609 (0.030)
Rule based (boosting)	0.648 (0.019)	0.503 (0.041)	0.689 (0.034)

table 5. It should be noted that when boosting was applied to the MLP classifiers the algorithm terminated after only one iteration and consequently the results were effectively the same as those for the single classifiers (table 1).

Two further comments can be made regarding the ensemble results.

- In terms of generalisation the ensembles outperformed the single classifiers, even when over-sampling of the minority class was used.
- Those ensembles built using disjoint subsets of the minority class together with the entire minority class outperformed those constructed using boosting and bagging.

The rule based ensemble members were analysed to identify those attributes that classify the minority class. As expected the attribute related to loan repayment behaviour appears in many of the rules that classify `bad' accounts. When this attribute was removed further attributes such as occupation, residential status, number of past jobs and residential addresses (all collected at the time of loan application) also appeared frequently, suggesting that they also may have some influence on the identification of potential loan defaults.

6   Concluding Remarks

The use of ensembles of classifiers, particularly those containing rules, in an application area of this type is somewhat novel. When machine learning approaches such as artificial neural networks and/or rule induction have been applied to credit scoring or as in this case, to arrears management, typically single classification models have been built. Interest has normally been in comparing the approaches or in an evaluation of results against those obtained using the more traditional statistical methods such as logistic regression or discriminant analysis. Our interest has been in the use of machine learning approaches to develop an ensemble classification model capable of dealing with highly imbalanced real world data sets. As well as providing a prediction for loan default (in this case the minority class) we were also interested in identifying those factors which may be an indicator of the potential for default.

Our results are consistent with the literature in that the overall generalisation performance of the ensembles was superior to that of the individual members. The ensembles also outperformed the single classifiers, whether rule based or an artificial neural network, built using the entire training set. On the other hand, for those ensembles built using only attributes collected at the time of loan application, the generalisation performance on the minority class of the individuals was superior to that of the ensemble. This may simply reflect the fact that the credit scoring models used as part of the loan approval process are not resulting in the approval of risky loans. Thus the major predictors of potential default are those identified during the loan account's life rather than at time of approval.

References

Baesens, B., Van Gestel, T., Stepanova, M., Vanthienen, J.: Neural Network Survival Analy-sis for Personal Loan Data. In: Proceedings of the Eighth Conference on Credit Scoring and Credit Control (CSCC VIII 2003), Edinburgh, Scotland (2003)

Bernardini, F.C., Monard, M.C., Prati, R.C.: Constructing Ensembles of Symbolic Classifiers. In: Proceedings of the 5th International Conference on Hybrid Intelligent Systems (HIS 2005), Rio de Janeiro Brazil, pp. 315–322. IEEE Press, Los Alamitos (2005)

Chawla, N.V., Hall, L.O., Bowyer, K., Kegelmeyer, W.P.: Learning Ensembles from Bites: A Scalable and Accurate Approach. Journal of Machine Learning 5, 421–451 (2004)

Cohen, G., Hilario, M., Sax, H., Hugonnet, S., Geissbuhler, A.: Learning form imbalanced data in surveillance of nosocomial infection. Artificial Intelligence in Medicine 37, 7–18 (2006)

Desai, V.S., Crook, J.N., Overstreet, G.A.: A Comparison of Neural Networks and Linear Scoring Models in the credit union environment. European Journal of Operational Research 95, 24–39 (1995)

Dietterich, T.G.: Ensemble Methods in Machine Learning. In: Kittler, J., Roli, F. (eds.) MCS 2000. LNCS, vol. 1857, Springer, Heidelberg (2000)

Drummond, C., Holte, R.C.: Severe Class Imbalance: Why Better Algorithms Aren't the Answer. In: Gama, J., Camacho, R., Brazdil, P.B., Jorge, A.M., Torgo, L. (eds.) ECML 2005. LNCS (LNAI), vol. 3720, pp. 539–546. Springer, Heidelberg (2005)

Fahlman, S.E.: Faster-learning variations on back-propagation: An empirical study. In: Sejnowski, T.J., Hinton, G.E., Touretzky, D.S. (eds.) 1988 Connectionist Models Summer School, Morgan Kaufmann, San Mateo (1988)

Freund, Y., Schapire, R.E.: Experiments with a new boosting algorithm. In: Saitta, L. (ed.) Proceedings of the Thirteenth International Conference on Machine Learning, Bari, Italy, pp. 148–156. Morgan Kaufmann, San Francisco (1996)

Grannitto, P.M., Verdes, P.F., Ceccatto, H.A.: Neural network ensembles: evaluation of aggregation algorithms. Artificial Intelligence 163, 139–162 (2005)

Guo, H., Viktor, H.L.: Learning form Imbalanced Data Sets with Boosting and Data Generation: The DataBoost-IM Approach. ACM SIGKDD Explorations Newsletter: Special Issue on Learning from Imbalanced Datasets 6, 30–39 (2004)

Hagan, M.T., Demuth, H.B., Beale, M.: Neural Network Design. PWS Publishing, Boston (1996)

Hansen, L.K., Salamon, P.: Neural Network Ensembles. IEEE Transactions on Pattern Analysis and Machine Intelligence 12, 993–1001 (1990)

Hastie, T., Tibshirani, R., Friedman, F.: The Elements of Statistical Learning: Data Mining. In: Inference and Prediction, Springer, New York (2003)

Hoffman, F., Baesens, B., Martens, J., Put, F., Vanthienen, J.: Comparing a Genetic Fuzzy and a NeuroFuzzy Classifier for Credit Scoring. International Journal of Intelligent Systems 17, 1067–1083 (2002)

Huysmans, J., Baesens, B., Vanthienen, J., Van Getsel, T.: Failure Prediction with Self Organising Maps. Expert Systems with Applications 30, 479–487 (2006)

Karray, F.O., de Silva, C.: Soft Computing and Intelligent Systems Design, Harlow, England. Pearson Educa-tion Limited, London (2004)

Ko, A.H.-R., Sabourin, R., deS (jnr), B.A.: Combining Diversity and Classification Accu-racy for Ensemble Selection in Random Subspaces. In: Proceedings 2006 International Joint Conference on Neural Networks, Vancouver Canada, pp. 2144–2151 (2006)

Kucheva, L.I.: Error Bounds for Aggressive and Conservative Adaboost. In: Windeatt, T., Roli, F. (eds.) MCS 2003. LNCS, vol. 2709, pp. 25–34. Springer, Heidelberg (2003)

Kucheva, L.I., Whitaker, C.J.: Measures of Diversity in Classifiers Ensembles and Their Relationship with the Ensemble Accuracy. Machine Learning 51, 181–207 (2003)

Lewis, E.M.: An Introduction to Credit Scoring. The Athena Press, San Rafael, California (1992)

Mays, E.: Handbook of Credit Scoring. Glenlake Publishing, Chicago (2001)

McNelis, P.D.: Neural Networks in Finance: Gaining Predictive Edge in the Market. El-sevier Academic Press, Burlington, MA (2005)

Nauck, D., Kruse, R.: NEFCLASS - A Neuro-Fuzzy Approach for the Classification of Data. In: George, K.M., Carrol, J.H., Deaton, E., Oppenheim, D., Hightower, J. (eds.) Proceedings of the 1995 ACM Symposium on Applied Computing, Nashville, Tennessee, ACM Press New York, New York (1995)

Quinlan, J.R.: C4.5: Programs for Machine Learning. Morgan Kaufmann, San Mateo (1993)

Scheurmann, E., Matthews, C.: Neural Network Classifiers in Arrears Management. In: Duch, W., Kacprzyk, J., Oja, E., Zadrożny, S. (eds.) ICANN 2005. LNCS, vol. 3697, pp. 325–330. Springer, Heidelberg (2005)

Srivastra, R.P.: Automating judgemental decisions using neural networks: a model for processing business loan applications. In: Agrawal, J.P., Kumar, V., Wallentine (eds.) Proceedings of the 1992 ACM Conference on Communications, Kansas City, Missouri ACM Press, New York (1992)

Torres-Sospedra, J., Fernandez-Redono, M., Hernandez-Espinosa, C.: Combination Methods for Ensembles of MF. In: Duch, W., Kacprzyk, J., Oja, E., Zadrożny, S. (eds.) ICANN 2005. LNCS, vol. 3697, pp. 131–138. Springer, Heidelberg (2005)

Vellido, A., Lisboa, P.J.G., Vaughan, J.: Neural Networks in Business: a survey of applications (1992-1998). Expert Systems with Applications 17, 51–70 (1999)

Wanas, N.M., Kamel, M.S.: Decision Fusion in Neural Network Ensembles. In: Proceedings of the International Joint Conference on Neural Networks (IJCNN 2001), vol. 4, pp. 2952–2957. IEEE Press, Los Alamitos (2001)

West, D.: Neural network credit scoring models. Computers & Operations Research 27, 1131–1152 (2000)

Witten, I.H., Frank, E.: Data Mining: Practical Machine Learning Tools and Techniques, 2nd edn. Morgan Kaufmann, San Francisco (2005)

Wu, Y., Arribas, J.I.: Fusing Output Information in Neural Networks: Ensemble Performa Better. In: Proceedings of the 25th Annual Conference of IEEE EMBS, Cancum, Mexico (2003)

Yule, G.U., Kendall, M.G.: An Introduction to the Theory of Statistics, 14th edn., Griffin, London (1950)

Bicluster Analysis of Currency Exchange Rates

Haizhou Li[1] and Hong Yan[1,2]

[1] Department of Electronic Engineering, City University of Hong Kong, Kowloon, Hong Kong
[2] School of Electrical and Information Engineering, University of Sydney, NSW 2006, Australia

Abstract. In many business applications, it is important to analyze correlations between currency exchange rates. In this Chapter, we develop a technique for bicluster analysis of the rates. Our method is able to extract coherent cluster patterns from a subset of time points and a subset of currencies. Experimental results show that our method is very effective. The bicluster patterns are consistent with the underlying economic reasons.

1 Introduction

Analysis of the correlation between different currency exchange rates is of interest to many people in business (Belke and Göcke 2001; Herwartz and Reimers 2002; Herwartz and Weber 2005). This kind of correlation is beneficial for much economic research work, such as accurate financial forecasting. The US dollar is the most important currency today in the international monetary system and it is regarded as a reference for all other currencies in the world. Therefore, the exchange rate mentioned in this chapter is the one between a currency and the US dollar. There are several types of relationships between exchange rates and each type has unique characteristics. For example, a desire to maintain the competitiveness of Japanese exports to the United States with German exports to the United States leads the Bank of Japan to intervene to ensure a matching depreciation of the yen against the dollar whenever the Deutsche mark (DM) depreciates against the US dollar (Takagi 1996). On the other hand, a preference for price stability may lead the Bank of Japan to intervene to ensure a matching appreciation of the yen against the dollar, whenever the DM appreciates against the US dollar. The following is another example. Because of the fixed exchange rate regime strictly implemented in Argentina and Egypt in the 1990s, the Argentine peso versus US dollar and the Egyptian pound versus US dollar exchange rates experience almost no change.

The correlation between different currency exchange rates consists of three key factors. The first one is the type of correlation. The second one relates to which exchange rates are involved in the correlation. The third is at which time points these involved exchange rates have this type of correlation. This is a typical biclustering problem. The exchange rates of different currencies can be organized into a two dimensional matrix. The columns of this matrix correspond to the selected exchange rates, and the rows correspond to the selected time points. The exchange rate at a time point can be the real exchange rate value or the difference value. The biclustering algorithm can group a certain set of exchange rates together based on the fact that they have a certain type of defined correlation in a certain set of time points (Lazzeroni and Owen 2002).

Biclustering has recently become a very active research topic in bioinformatics. Now it is being used in many other research fields. For example, in document clustering (Dhillon 2001), biclustering algorithms are applied to the clustering of documents and words simultaneously. In this chapter, the biclustering algorithm is used to analyze the correlations among currency exchange rates.

Biclustering, also called bidimentional clustering, and subspace clustering, was first investigated by Hartigan (1972). It was first applied to gene expression matrices for simultaneous clustering of both genes and conditions by Cheng and Church (2000). Existing biclustering algorithms have recently been surveyed in (Madeira and Oliveira 2004; Tanay et al. 2006). The general strategy in these algorithms can be described as adding or deleting rows or columns in the data matrix to optimize a predefined cost or merit function.

A different viewpoint of biclustering is proposed in (Gan et al. 2005). It is called geometric biclustering. It is performed in terms of the spatial geometrical distribution of points in a multi-dimensional data space. Based on the concept of geometric biclustering, we develop a new method for currency exchange rate data analysis. In our algorithm, we choose the Fast Hough transform (FHT) (Li et al. 1986) to identify line patterns in two dimensional spaces, each containing a pair of exchange rates. Each line represents a sub-bicluster. An expansion algorithm is used to combine these sub-biclusters to form the complete biclusters based on comparison and merging.

The chapter is organized as follows. Section 2 demonstrates that different biclusters can be formulated using the linear relation in a column-pair space. Section 3 describes our biclustering algorithm based on the FHT algorithm and the expansion algorithm. In Section 4, we present results of several experiments to verify the performance of our biclustering algorithm for the extraction of correlations between different exchange rates. Finally, we conclude the chapter in Section 5.

2 Bicluster Patterns

In this chapter, we address five major classes of biclusters:

1. Biclusters with constant values.
2. Biclusters with constant rows.
3. Biclusters with constant columns.
4. Biclusters with additive coherent values.
5. Biclusters with multiplicative coherent values.

Table 1–5 show examples for the five classes of biclusters. A constant bicluster reveals a subset of exchange rates with similar expression values within a subset of time points. A constant row bicluster identifies a subset of time points with similar expression values across a subset of exchange rates, but allows the values to differ from time point to time point. Similarly, a constant column bicluster identifies a subset of exchange rates presenting similar expression values within a subset of time points, but the value may differ from exchange rate to exchange rate. A bicluster with additive coherent values identifies a subset of exchange rates having an additive relationship within a subset of time points. That is, a column can be obtained by adding a constant to another column. Similarly, a bicluster with multiplicative coherent values identifies a subset of exchange rates having multiplicative relationship within a subset of time points. Actually, constant biclusters are a special case of all other types of biclusters. Constant row and constant column biclusters are special cases of additive and multiplicative biclusters. Thus, we reduce the original five classes of biclusters to two classes, additive and multiplicative ones and only consider these two structures for currency exchange rate analysis presented in this chapter.

Table 1. A bicluster with constant values

Time Point	Expression value of ExRate A	Expression value of ExRate B	Expression value of ExRate C
month a	7.5%	7.5%	7.5%
month b	7.5%	7.5%	7.5%
month c	7.5%	7.5%	7.5%

ExRate means exchange rate.

Table 2. A bicluster with constant rows

Time Point	Expression value of ExRate A	Expression value of ExRate B	Expression value of ExRate C
month a	7.5%	7.5%	7.5%
month b	10%	10%	10%
month c	9.5%	9.5%	9.5%

Table 3. A bicluster with constant columns

Time Point	Expression value of ExRate A	Expression value of ExRate B	Expression value of ExRate C
month a	7.5%	10%	9.5%
month b	7.5%	10%	9.5%
month c	7.5%	10%	9.5%

Table 4. A bicluster with additive coherent values

Time Point	Expression value of ExRate A	Expression value of ExRate B	Expression value of ExRate C
month a	7.5%	10.5%	15%
month b	10%	13%	17.5%
month c	9.5%	12.5%	17%

Table 5. A bicluster with multiplicative coherent values

Time Point	Expression value of ExRate A	Expression value of ExRate B	Expression value of ExRate C
month a	7.5%	11.25%	15%
month b	10%	15%	20%
month c	9.5%	14.25%	19%

The two classes of biclusters both have an important property that any two columns (a column-pair) in the bicluster form a sub-bicluster of the same type. For example, in the additive bicluster shown above, columns 1 and 2 form a sub-additive bicluster, and so do columns 1 and 3, and columns 2 and 3. Let us represent a column variable as x_i for exchange rate i. Then additive and multiplicative biclusters can be represented using the linear equation

$$x_j = kx_i + c, \tag{1}$$

where k and c are constants. The equation represents an additive bicluster when k equals 1 and a multiplicative one when c equals 0. Therefore, we can find these linear relations in exchange rate pair spaces first and then combine similar sub-biclusters to identify a complete bicluster, instead of searching through all exchange rates at the same time using the strategy of the original geometric biclustering algorithm (Gan et al. 2005).

3 Geometric Biclustering Algorithm

We present a biclustering algorithm based on the concept of geometric biclustering. A similar method has already been used for microarray gene expression data analysis (Zhao et al. 2007). In the first phase of our biclustering algorithm, the lines are detected in the 2D space corresponding to each exchange rate pair. In the second phase, we use our expansion algorithm to form the complete biclusters. The FHT algorithm is chosen because of its robustness for line detection. The expansion algorithm consists of comparison and mergence. In this section, we first introduce the FHT and then propose the expansion algorithm.

3.1 The Fast Hough Transform for Line Detection

The FHT algorithm is a powerful technique for multi-dimensional pattern detection and parameter estimation. It has advantages when little is known about the pattern (Li et al. 1986). In this chapter, we use the FHT to detect the lines in the 2D spaces.

We use x-y to represent the 2D data space, and p-q to represent the corresponding 2D parameter space. The FHT algorithm will transform a point in the x-y space into a line in the p-q parameter space. If n points on a line in the x-y data space are known, the line obtained from each such point should intersect at the same point in the p-q

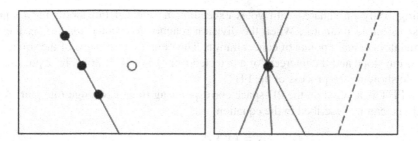

Fig. 1. Several points on a line to be detected and a random point in the original data space (left) and lines in the parameter space (right), corresponding to the points in the data space. We can detect the line in the data space by locating the intersection point of the lines in the parameter space.

space. The intersection point defines the parameters of the line in the x-y space. Fig. 1 shows the relation between line and point in their respective space.

In the FHT algorithm, points of concentration are detected by dividing the p-q parameter space into a list of accumulators, which are organized into a hierarchical data structure. The FHT algorithm recursively divides the p-q parameter space into squares from low to high resolutions and performs the subdivision and subsequent "vote

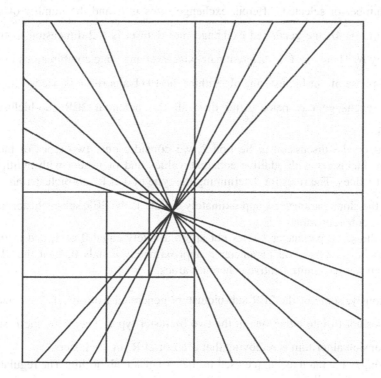

Fig. 2. Illustration of the FHT. The accumulators are organized into a hierarchical data structure to improve the computing speed.

counting" only on squares with votes exceeding a selected threshold. Each square represents an accumulator. When the division reaches a selected resolution, the left accumulators contain points of concentration. The center coordinates of the square are used as the slope and the intercept of a detected line (Li et al. 1986). Fig. 2 presents an example showing the process of the FHT.

The FHT is applied on the 2D space corresponding to an exchange rate pair. A detected line can be described in the equation

$$y = kx + c,$$
(2)

where k is the slope, and c takes the value of the point of intersection with the ordinate axis. Parameters k and c are the key factors that determine the correlation type between the exchange rates. The line detection process can be regarded as subbiclustering. Each detected line can be regarded as a sub-bicluster. Because the detected sub-bicluster in this step contains two exchange rates, it is called a 2ER subbicluster (2 exchange rates sub-biclusters). Each 2ER sub-bicluster comprises all the approximate points on the detected line.

3.2 The Expansion Algorithm

If the number of selected different exchange rates is N and the number of selected time points is M, the generated exchange rate dataset is a 2-dimension matrix with size M by N. There are C_N^2 different pair-wise exchange rate combinations in total. In the first phase of our biclustering algorithm, the FHT algorithm is needed for each of the C_N^2 exchange rate pairs. After that, all the potential 2ER sub-biclusters are detected.

Based on the discussion in Section 2, we consider only two types of biclusters. They are biclusters with additive coherent values and biclusters with multiplicative coherent values. The rules for determining the type of a 2ER sub-bicluster are:

1. If the slope parameter k approximately equals 1, the 2ER sub-bicluster has additive coherent values.
2. If the slope parameter k does not approximately equal 0 or 1, and at the same time the intersection parameter c approximately equals 0, then the 2ER subbicluster has multiplicative coherent values.

Obviously, some of the 2ER sub-biclusters generated from the C_N^2 exchange rate pairs does not fit into either one of the two bicluster types. Therefore, the first step in our expansion algorithm is removing that kind of 2ER sub-biclusters.

Merging is the main technique used in the expansion algorithm. The regulations for mergence are as follows:

1. The conditions of merging two 2ER sub-biclusters together are:

   - The two 2ER sub-biclusters have a common exchange rate.
   - The two 2ER sub-biclusters have the same type.
   - The two 2ER sub-biclusters have a reasonable degree of overlap among the time points they contain.

   Only when all these conditions are satisfied at the same time, can the two 2ER sub-biclusters be merged into a new 3ER sub-bicluster (3 exchange rates sub-biclusters). This new 3ER sub-bicluster comprises the three exchange rates contained in the two 2ER sub-biclusters. The time point set contained in this new 3ER sub-bicluster is the intersection of the time point sets of the two 2ER sub-biclusters. The type of this new 3ER sub-bicluster is the same as the type of those two 2ER sub-biclusters.

2. Every new nER sub-bicluster (n exchange rates sub-bicluster) is generated by combining a 2ER sub-bicluster and an (n-1)ER sub-bicluster. The conditions for merging a 2ER sub-bicluster and an (n-1)ER sub-bicluster are similar to those of merging two 2ER sub-biclusters:

   - The 2ER sub-bicluster and the (n-1)ER sub-bicluster have a common exchange rate.
   - They are of the same type.
   - They have a reasonable degree of overlapping between the time points they contain.

   In the same way, only when all these three conditions are fulfilled at the same time, can the 2ER sub-bicluster and the (n-1)ER sub-bicluster be merged into a new nER sub-bicluster. This new nER sub-bicluster is comprised of the n exchange rates in the 2ER sub-bicluster and the (n-1)ER sub-bicluster. The time point set contained in this new nER sub-bicluster is the intersection of the time point sets of the 2ER sub-bicluster and the (n-1)ER sub-bicluster. The type of this new nER sub-bicluster is the same as those of the two original sub-biclusters.

We put all the 2ER sub-biclusters detected in the first phase of our biclustering algorithm into set Ω. Set Ψ is used to contain the generated complete biclusters in the process of the expansion algorithm. The expansion algorithm consists of the following steps:

1. Remove the 2ER sub-biclusters that can not be classified into either one of the two bicluster types that we concern from Ω.
2. Select the first 2ER sub-bicluster b_1 in Ω as a temporary complete bicluster t and remove b_1 from Ω.
3. Search through all the 2ER sub-biclusters in Ω and find the first 2ER sub-bicluster b that can be merged with the temporary complete bicluster t. Merge b and t to generate the new temporary complete bicluster. Remove b from Ω.

4. Repeat Step 3 until no 2ER sub-bicluster in Ω can be merged with the temporary complete bicluster t. Here t will represent a final complete bicluster. Put t into Ψ. Search through all the 2ER sub-biclusters in Ω and remove all the ones that can be regarded as part of t.

5. Repeat Step 2 to Step 4 until Ω is empty. Here all the complete biclusters are recorded in Ψ.

4 Experiment Results

We use an exchange rate dataset to test the performance of our biclustering algorithm. All the historical exchange rate data used here are obtained from the internet. Seventeen currencies are selected for our experiments. They are ARS (Argentine Peso), AUD (Australian Dollar), BRL(Brazilian Real), CAD (Canadian Dollar), CHF (Swiss Franc), EGP (Egyptian Pound), GBP (British Pound), IDR (Indonesian Rupiah), INR (Indian Rupee), JPY (Japanese Yen), MXN (Mexican Peso), PHP (Philippine Peso), RUB (Russian Rouble), SGD (Singapore Dollar), THB (Thai Baht), TWD (Taiwan Dollar), and ZAR (South African Rand). The exchange rates of these 17 currencies are all valued against the USD (US Dollar). The date range is from January 1, 1996 to December 31, 2005, a total of ten years. However, there is nearly no change in the exchange rates if measured day by day. This does not match our purpose of analyzing the principle of evolution of the exchange rate. Therefore, the comparison is made on a month by month basis. The average exchange rate of a month is used to represent the exchange rate of that month. Totally there are 120 months, so the size of the dataset is 120 by 17. Because the evolution of the exchange rate is what we are interested in, a difference dataset is generated from the original exchange rate dataset. Every element of this difference dataset is the difference between the exchange rate in a month and that of the preceding month. A positive element means an increase, a negative element means a decrease, and zero means no change. The size of this difference dataset becomes 119 by 17.

4.1 Results on the Simplified Difference Dataset

In this experiment we work on a simplified difference dataset. A threshold α is defined to make the simplification more reasonable. We use the symbol a to represent an element of the difference dataset. The simplification is as follows:

1. If $a>0$ and $|a|>\alpha$, a is replaced with 1.
2. If $a<0$ and $|a|>\alpha$, a is replaced with -1.
3. If $|a|\leq\alpha$, a is replaced with 0

After simplification, the generated new dataset contains 1, -1 and 0 only. 1 means upward change, -1 means downward change and 0 means no change. Fig. 3 shows the simplified difference dataset.

Two detected biclusters are analyzed to verify the performance of our biclustering algorithm in the following.

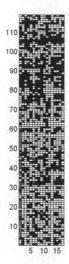

Fig. 3. The simplified difference dataset. White square represents 1, gray square represents 0, and black square represents −1.

Bicluster 1

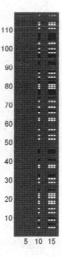

Fig. 4. One constant row bicluster extracted using the proposed algorithm. The gray squares are the background. The white squares and the black squares show the bicluster.

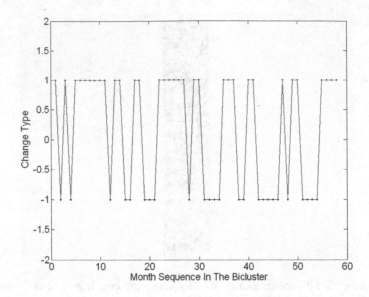

Fig. 5. The common evolution that the four exchange rates have in the 58 months. In the diagram, 1 means upward change, 0 means no change and −1 means downward change.

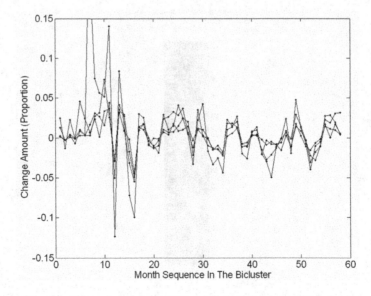

Fig. 6. The real amount of proportional changes of the four exchange rates in the 58 months. A value above 0 means increase, one equal to 0 means no change, and one below 0 means decrease.

Fig. 4 shows an extracted bicluster. This bicluster is a constant row bicluster, which contains four exchange rates and 58 months. Fig. 5 shows the common evolution that the four exchange rates have in the 58 months, and Fig. 6 presents the corresponding real amount of proportional changes. The four currencies contained in this bicluster are JPY, SGD, THB and TWD. The change types in the exchange rates for these four currencies are the same in 58 months. These discontinuous 58 months nearly equals five years, which is half of ten years. Therefore, this bicluster reflects that JPY, SGD, THB and TWD have a very close relationship. The conclusion accords with the fact in the real world. Japan, Singapore, Thailand and Taiwan are all in Far East. They are geographically very close to each other. Their economies are also related with each other very closely. Japan is the second strongest developed country in the world, and it is the most developed country in Asia. The other three countries or regions were all developing fast. There is plenty of cooperation between these four economies. Japan's financial policy has a certain degree of impact on the financial policies of all other countries or regions. Therefore this detected bicluster accurately reflects the close relationship between the four economies.

Bicluster 2

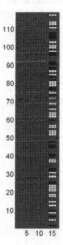

Fig. 7. Another constant row bicluster extracted using the proposed algorithm

Fig. 7 presents another extracted bicluster. This bicluster is also a constant row bicluster, which contains three exchange rates and 71 months. Fig. 8 shows the common evolution that the three exchange rates have in the 71 months, and Fig. 9 presents the corresponding real amount of proportional changes. The three currencies contained in this bicluster are SGD, THB, and TWD. The change types in the exchange rates for these three currencies are the same in 71 months, which are not continuous and occupy sixty percent of the total 119 months. Just as the analysis for Bicluster 1, this bicluster accurately reflects the close relationship in the economies of the three regions, Singapore, Thailand and Taiwan. However, if we observe the bicluster carefully, we

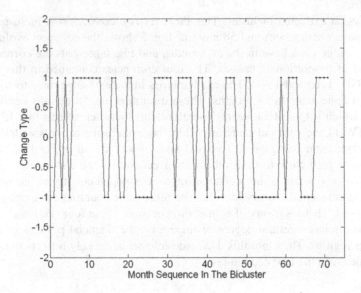

Fig. 8. The common evolution that the three exchange rates have in the 71 months. The value 1 means upward change, 0 means no change, and −1 means downward change.

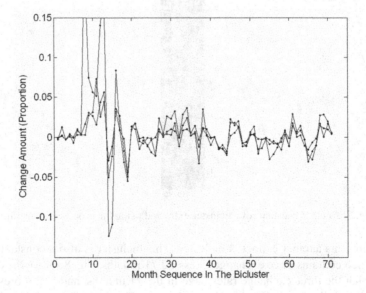

Fig. 9. The real amount of proportional changes of the three exchange rates in the 71 months. A value above 0 means increase, one equal to 0 means no change, and one below 0 means decrease.

see that the changes in the exchange rates for the three currencies are all up from the 8th month to the 13th month in the bicluster, and the increment is larger than the usual level. The period from the 8th month to the 13th month in the bicluster corresponds to the period from July 1997 to January 1998 without September 1997. What happened in that period? The answer is the finance crisis of Asia in 1990s. This crisis begins from July of 1997, and lasted almost two years. At the starting phase of this crisis, the exchange rates for THB, SGD and TWD increased quickly. After nearly half a year, the exchange rates for these currencies become steady. Therefore, this detected bicluster is significant and reflects the happenings in the real currency exchange world. The experiment proves the usefulness of the biclustering based technique proposed in this chapter.

4.2 Results on Real Difference Dataset

In this experiment, the real difference dataset is used directly to test the performance of the proposed biclustering algorithm. Fig. 10 shows the real amount of proportional changes in the exchange rates for three currencies ARS, BRL, and EGP in 31 non-continuous months. Similarly, Fig. 11 shows the amount changes in exchange rates for currencies ARS, EGP, and PHP in the 31 months. These two biclusters are analyzed in details as follows.

Bicluster 1

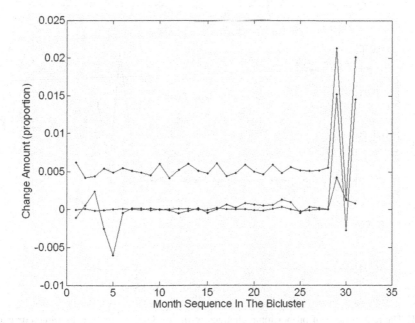

Fig. 10. The real amount of proportional changes in the exchange rates for the three currencies in the 31 non-continuous months

The first bicluster contains three exchange rates and 31 months. The three currencies contained in this bicluster are ARS, BRL, and EGP. Fig. 10 presents the real
amount of proportional changes that the three exchange rates have in the 31 months.
This bicluster belongs to the additive model, though the amount of changes in the exchange rates for the three currencies does not fulfill the additive relationship accurately in the last four months. The first 27 months of the total 31 months are all in the
period from February 1996 to July 1998. In addition, there are 30 months from
February 1996 to July 1998. That is to say, almost all of the months from February
1996 to July 1998 are contained in this bicluster. In this period, the amount of change
in the exchange rate for BRL can be approximately regarded as 0.5 percent, and the
amount of change in the exchange rates for ARS and EGP can be approximately regarded as 0. These are all the essential information contained in this bicluster. Now,
how about the fact that the exchange rates for the three currencies in the period from
February 1996 to July 1998? In this period, Argentina and Egypt both implemented a
strictly fixed exchange rate regime, which means that there is nearly no change in the
exchange rate. Brazil also implemented a fixed exchange rate regime, and it made the
exchange rate increase by 6 percent each year by design. Six percent in one year
means 0.5 percent in one month. We find that the essential information contained in
this bicluster accurately accord with the fact.

Bicluster 2

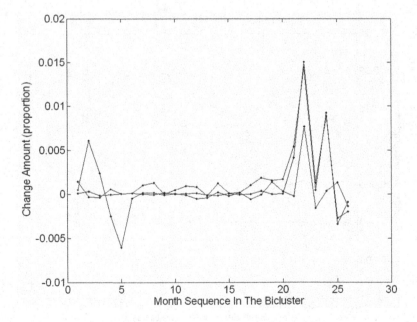

Fig. 11. The real amount of proportional changes in the exchange rates for the three currencies
in the 26 non-continuous months

The second bicluster contains three exchange rates and 26 months. The three currencies contained in the bicluster are ARS, EGP, and PHP. Fig. 11 presents the real amount of proportional changes that the three exchange rates have in the 26 months. This bicluster belongs to the constant row model. In this bicluster, the first 16 months of the total 26 months are in the period from March 1996 to June 1997. In this period, the exchange rates for the three currencies are almost unchanged. Just as the analysis for Bicluster 1, this essential accords with the fact that Argentina, Egypt and Philippine implemented a strictly fixed exchange rate regime in the period from March 1996 to June 1997. We also find that the amount of change in the exchange rates for the three currencies becomes larger from the 17th month of the total 26 months. The 17th month in the bicluster is June 1999. And the largest change happens in the 22nd month of the 26 months, which is September 2002. This essential can be explained with the following facts. The Philippines changed the fixed exchange rate regime to a flexible exchange rate regime in the second half of 1997 because of the finance crisis. Argentina also changed the fixed exchange rate regime to flexible exchange rate regime in August 2002. Though Egypt did not change the fixed exchange rate regime to a flexible exchange rate regime until January 2003 which is after September 2002, it already had a period in 2001 and 2002 when EGP depreciated sharply.

5 Conclusion

In this chapter, we have presented a biclustering based algorithm to analyze the correlation between different currency exchange rates. We use the FHT to detect sub-biclusters in the exchange rate pair spaces, and then use the expansion algorithm to combine the sub-biclusters to form the complete biclusters. We generate several exchange rate datasets to verify the performance of our method. Experiment results show that our method is effective. The bicluster patterns are consistent with the underlying economic reasons.

Acknowledgement. This work is supported by the Hong Kong Research Grant Council (Project CityU 122506).

References

Belke, A., Göcke, M.: Exchange rate uncertainty and employment: an algorithm describing 'play'. J Applied Stochastic Models in Business and Industry 17, 181–204 (2001)

Cheng, Y., Church, G.M.: Biclustering of expression data. In: Proc. of the Eighth International Conference on Intelligent Systems for Molecular Biology (ISMB), vol. 8, pp. 93–103 (2000)

Dhillon, I.S.: Co-clustering documents and words using bipartite spectral graph partitioning. In: Proc. of the Seventh ACM SIGKDD international conference on Knowledge discovery and data mining, pp. 269–274 (2001)

Gan, X.C., Liew, A.W.C., Yan, H.: Biclustering Gene Expression Data Based on a High Dimensional Geometric Method. In: Proc. of the Fourth International Conference on Machine Learning and Cybernetics., vol. 6, pp. 3388–3393 (2005)

Hartigan, J.A.: Direct Clustering of a Data Matrix. J American Statistical Association 67, 123–129 (1972)

Herwartz, H., Reimers, H.E.: Empirical modeling of the DEM/USD and DEM/JPY foreign exchange rate: Structural shifts in GARCH-models and their implications. J Applied Stochastic Models in Business and Industry 18, 3–22 (2002)

Herwartz, H., Weber, H.: Exchange rate uncertainty and trade growth – a comparison of linear and non-linear (forecasting) models. J Applied Stochastic Models in Business and Industry 21, 1–26 (2005)

Lazzeroni, L., Owen, A.: Plaid models for gene expression data. J Statistica Sinica 12, 61–86 (2002)

Li, H., Lavin, M.A., Le Master, R.J.: Fast Hough transform: a hierarchical approach. J Computer Vision, Graphics, and Image Processing 36, 139–161 (1986)

Madeira, S.C., Oliveira, A.L.: Biclustering Algorithms for Biological Data Analysis: A Survey. J IEEE/ACM Transactions on Computational Biology and Bioinformatics 1, 24–45 (2004)

Takagi, S.: The Yen and Its East Asian Neighbors, 1980-1995: Cooperation or Competition? NBER Working Paper No. W5720 (1996)

Tanay, A., Sharan, R., Shamir, R.: Biclustering Algorithms: A Survey. In: Aluru, S. (ed.) Handbook of Computational Molecular Biology. Computer and information Science Series. Chapman & Hall/CRC, Boca Raton,FL (2006)

Zhao, H.Y., Liew, A.W.C., Yan, H.: A new strategy of geometrical biclustering for microarray data analysis. In: Proc. of the Fifth Asia-Pacific Bioinformatics Conference, pp. 47–56 (2007)

Predicting the Effects of Alternative Pricing Strategies in an Artificial Society Undergoing Technology Adoption

Judy K. Frels[1], Debra Heisler[2], and James A. Reggia[3]

[1] Robert H. Smith School of Business, Van Munching Hall, University of Maryland,
College Park, MD 20742.
Tel.: 301-405-7607, Fax: 410-510-1830
jfrels@rhsmith.umd.edu
[2] Department of Computer Science at the University of Maryland
heisler@cs.umd.edu
[3] Department of Computer Science, A. V. Williams Bldg., University of Maryland,
College Park MD 20742
reggia@cs.umd.edu

Abstract. Artificial societies are computer models in which the collective behavior of a population of simulated human decision makers is observed over time. Here we describe an artificial society of "soft computing agents" (model consumers) making probabilistic purchasing decisions about new technological products that are introduced by competing firms. The model is studied under varying conditions to determine the relative success of these firms as they pursue different pricing strategies designed to increase their market share. We find that a critical factor in determining the success of different pricing strategies is the utility that an individual consumer gains from other consumers adopting the same technology. Further, financial success is uncoupled from market share under some conditions, so (for example) while an inferior technology may gain substantial market share by aggressive price-cutting, it is unlikely to gain financial rewards. These results add to growing evidence that artificial society models may prove useful in improving our understanding of collective decision making in complex sociological, economic and business management situations.

1 Technology Markets with Network Externalities

Expanding the scientific understanding of collective human behavior and decision making in the context of uncertainty is of interest in many fields, including psychology, sociology, economics, and business management. Most research in this area has been based on experimental data collection and analysis from populations of human subjects. However, it is increasingly being recognized that agent-based models of human populations provide a potentially powerful framework for investigating the collective behavior of people, in part due to their explicit representation of space, and their focus on concurrent local computations that lead to self-organization and emergent behaviors. Such *artificial societies* typically consist of a two dimensional cellular space in which simple computer-simulated agents (individuals representing people) exist, make decisions, and possibly move around. Given that these models often focus on human decision making in the face of uncertainty, and the simplifications needed to produce computationally-tractable models of human social interactions, it is not

B. Prasad (Ed.): Soft Computing Applications in Business, STUDFUZZ 230, pp. 35–55, 2008.
springerlink.com   © Springer-Verlag Berlin Heidelberg 2008

surprising that soft computing methods such as neural networks, evolutionary computation, swarm intelligence and probabilistic reasoning are playing an increasing role in these and related multi-agent systems (Loia 2003; Wagner et al, 2006; Lapizco-Encinas et al, 2005; Mazanec 2006; Libai, Muller and Peres 2005).

In the context of the complexities of human cognition and decision making, it might be expected that artificial societies would be unlikely to produce unexpected or interesting results. However, it has repeatedly been found that, even with relatively simple rules governing agent behaviors, these models can produce collective behaviors reminiscent of those seen in human social situations. For example, in one of the earliest and best known artificial societies involving a cellular space, individuals/agents occupying a fraction of the cells in a two dimensional lattice would count how many of their neighbors shared with them a single cultural feature (in this case the feature was called "race") (Schelling 1969). If this number was too low, the individual would move to an adjacent empty cell, causing initially randomly distributed individuals to form spatial clusters of agents having the same feature ("segregated neighborhoods") over time. A related model subsequently showed that if individuals have multiple cultural features (political affiliation, dress style, etc.), and similar individuals influenced each other to become more similar, multiple homogeneous "cultural regions" would emerge over time (Axelrod 1997). Most closely related to the work here are similar agent-based models that examine difficult issues in economics and marketing. For example, the influential *Sugarscape* artificial society (Epstein and Axtell 1996) has demonstrated that fairly simple rules can produce a skewed distribution of wealth similar to that seen in human societies, while a more recent model has been used to explain temporal patterns in consumer electronics markets' sales data (Goldenberg, Libai and Muller 2002). Cellular automata have been used to show that commonly accepted rules of market entry strategies may not be robust under differing levels of innovativeness and various response functions to advertising (Libai, Muller and Peres 2005). Other recent work has examined the effectiveness of market segmentation schemes using agents that have some degree of autonomy and limited learning capability (Mazanec 2006). Many other related models in economics and marketing are described in (Tesfatsion 2002). Other studies have used artificial societies to examine such issues as evolution of cooperative behavior (Hauert and Doebeli 2004), traffic flow (Gaylord and Nishidate 1996), urban growth patterns (Back 1996), and the emergence of communication (Reggia, Schulz, Wilkinson and Uriagereka 2001; Wagner, Reggia, Uriagereka, and Wilkinson 2003).

In the following we describe our research using an artificial society that we call *StandardScape* to investigate human decision making during the adoption of new technology and the development of de facto "standards". The core of *StandardScape* is a population of soft computing agents, each intended to be a simplified, probabilistic approximation to human decision making in the context of uncertainties about a product's future success and the weighting that people give to different factors during product adoption decisions. As part of model development, we collected and analyzed data from over 140 people that provided us with weight estimates used by the model's agents. Our goal in creating *StandardScape* was to determine whether such a model of technology adoption could be developed effectively, and if so, to use it to examine the relative success of technology firms adopting different pricing strategies to compete for market share. This examination is done under varied conditions, such as equal versus asymmetric intrinsic value of competing technologies, availability of local versus global information to agents about decisions made by others, and the importance of "network effects" on a technology's value. The latter factor, network influence, was of special interest to us because of its recognized importance in contemporary marketing theory, as follows.

In many markets, particularly information technology markets, consumers gain benefits from doing what other consumers do, i.e., from adopting compatible products. Such markets are called *network markets* (Besen and Farrell 1994; Frels, Shervani and Srivastava 2003) because the value that a consumer gets from a product depends on the network of other consumers associated with that product. For example, telephones and fax machines are valuable only when other people (a "network") have purchased such products. Network markets tend to involve positive feedback, and thus to be "tippy" or "winner-takes-all," meaning that as one competing product gets ahead, it becomes more attractive to the next adopter who, upon adopting it, makes it even more attractive to the next adopter (e.g., VHS vs. Betamax video recorders). Thus, one characteristic of such markets is often only one technical standard is ultimately likely to prevail, and because the value the network provides can overwhelm the value provided by the technology itself, it is possible that the prevailing standard will not be the most technologically advanced. In such markets, management may adopt strategies to ensure the market tips in their direction rather than their competitor's.

Strategies appropriate in network markets often run counter to recommended strategies in non-network markets because it can be critical for a firm to gain an early lead in adoption (Besen and Farrell 1994). Therefore, in such a market, it may be worthwhile to pursue strategies that are not immediately revenue enhancing under the assumption that greater revenue can be captured once the market has tipped in the firm's direction. Examples of strategies suggested to managers of network markets include penetration pricing, signaling, preannouncements, licensing of the technology, and support for development of complementary products (Besen and Farrell 1994; Hill 1997). Evidence of such strategies exist in standards battles such as the home video game console / high definition DVD wars. Sony's PlayStation3 was released at a penetration price estimated to be several hundred dollars below its actual manufacturing cost, and Microsoft's XBOX 360 was similarly introduced at a money-losing low price (*Financial Times* 2006). It is widely assumed that in both cases, each firm's goal has been to gain a dominant market share in the console industry, drive volume up (and hence, production prices down) and make up the lost revenue in game licenses. More importantly, it is also assumed that in these "console wars" there will be one dominant player who takes the lion's share of the market and the profits in the long-term, and penetration pricing has been the weapon of choice for key industry players in waging this war.

In the following, we first describe *StandardScape*, our artificial society in which agents make decisions about purchases of competing technologies. We then present the results of systematic computational experiments in which we examine how market penetration and share are influenced by different manufacturer pricing strategies, high versus low levels of network importance, and local versus global access of consumer agents to information about product adoptions by others. We conclude with a discussion of some of the implications of this work.

2 Methods

We study an artificial society model of human decision-making in network markets. Our primary goal is to better understand the effects of different pricing strategies that a business can take to promote adoption of its technology where decisions by one consumer affect those of others. Our research is conducted in an artificial world we call *StandardScape*, where active cells are viewed as agents representing adopters (individual consumers or firms) who choose between two incompatible technological products (A or B) and whose goal is to maximize their utility. We also model technology sponsors representing firms competing to achieve a dominant market share and then recoup their investment in this simulated market. The technology sponsors are external to but interact with and influence the cellular space model. Each technology sponsor firm has a cost of goods sold, and we track the revenue each firm obtains from the agents adopting its products. If, through the various strategies enacted, the technology sponsors deplete their capital, they go out of business. Their strategy is based on penetration pricing, the essence of which is explained next. When a firm has a market share it does not deem sustainable, it lowers its price to gain further market share. When a firm has a market share that it deems "dominant" and overall market penetration by all providers is significant, it raises its price to recoup earlier revenue foregone by price cutting and to take advantage of its near monopolistic situation (Farrell and Saloner 1986).

2.1 Consumers as Agents in a Cellular World

The current set of simulations was done in a 60x60 cellular space having 600 randomly located cells designated to be agents/consumers (see Fig. 1). Agent cells have multi-attribute states, i.e., states composed of multiple fields (Chou and Reggia 1997). For example, one field is Product-Adopted and can take on values N (neither product), A (product A adopted) or B (B adopted).

All non-agent cells are permanently quiescent. Each agent/consumer cell updates its state based on the state of other agents that are considered to be its neighbors. Two neighborhood sizes are used in different situations. In simulations using *incomplete information*, an agent can only see technology purchase decisions made by just those other agents (about 18 on average) in its local 11x11 neighborhood. In simulations using *complete information*, an agent can see the total numbers of each technology purchased by all other agents in the world. In each time period (conceptualized as a month in this setting), agents seek to adopt a product (A or B) that maximizes their utility. Adopting a product creates network externalities (as well as switching costs) that then influence agents' future decisions. We model a multi-period adoption scenario because repeated adoption is common in business settings (Frels, Shervani and Srivastava 2003); agents may revert from A or B to neither. Cell state changes are determined by a consumer/agent's utility function that is based on a multi-attribute model (Fishbein and Ajzen 1975) reflecting the stand-alone technological value of the product, the price of the product, the network of users associated with the product, the consumer's expectations of the future size of the network (Besen and Farrell 1994), and the agents' switching costs (Burnham, Frels and Mahajan 2003). In the equation

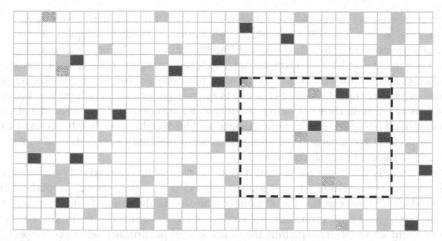

Fig. 1. A small portion of the cellular space, where quiescent cells are white, and agents adopting product A (gray), B (black), or neither (hatched) are shaded cells. The agent in the center of the box enclosed by dashed lines can only observe the state of other agents within its 11 x 11 neighborhood.

below, p designates a particular product, i designates an individual consumer agent, and weights $w1_i$—$w5_i$ are each positive and sum to 100. Agent i's utility U_{pi} is:

$$U_{pi} = w1_i(Tech_p) - w2_i(Price_p) + w3_i(Network_{pi}) + w4_i(Expect_{pi}) + \qquad (1)$$
$$w5_i(Investment_{pi})$$

where $Tech_p$ represents the performance rating of each of the technologies A and B, $Price_p$ represents the price of each technology (initially $100 for each), $Network_{pi}$ is calculated as ln(number of neighbors who have bought product p) / ln(total number of neighbors), $Expect_{pi}$ represents agent-specific expectations about the future size of the user network for products A and B (Besen and Farrell 1994), and $Investment_{pi}$ represents the investments an agent has made in product A or B to date or their switching costs. A logarithmic scale is used for $Network_{pi}$ to represent diminishing effects with larger numbers of product purchases.

At each time tick, which is taken to represent one month of real time, each simulated consumer agent attempts to purchase the technology with the greatest utility provided that the consumer agent has sufficient capital, and that the utility of the chosen product is greater than the reservation utility of threshold (the amount of utility above and beyond the price that a product must provide in order for the consumer to purchase a product rather than choosing "no choice" as a preferred option). Each time period consumer agents are endowed with an increment of capital ($100) that they are able to spend on technology A or B or save for future spending, depending on their utility function. In the simulations reported here, this generally means that agents have sufficient capital to make a purchase each month if they decide to do so. If an agent purchases a product, an amount of capital equal to the price of the product (set by the technology sponsor agents) is transferred to the technology sponsor agent as payment. The technology sponsor agents amass capital by a simple calculation of revenue (payments) minus the cost of goods sold ($25 for each item). Summarizing, the state of each agent cell i consists of the values of Product-Adopted$_i$, Expect$_{Ai}$, Expect$_{Bi}$, Investment$_{Ai}$, Investment$_{Bi}$, and Capital$_i$, all of which vary during a simulation. Each agent also has fixed weights w1$_i$ – w5$_i$ indicating the relative importance of the factors in its utility function. To provide reasonable estimates for these weight values, we collected data from 141 subjects on cell phone feature adoptions that allowed us to estimate their subjective weights using conjoint analysis. The details of this empirical study are given in the Appendix. The weights and thresholds used for each of the 600 agent cells in our simulations were randomly sampled, with replacement, from these 141 actual consumers.

Our model is stochastic for several reasons. First, agents are randomly placed, and their weights and thresholds are randomly assigned as described above. Second, when the product that maximizes the consumer's utility is determined, that product is adopted only with 85% probability to account for events outside the model (Arthur 1989). Third, expectations regarding the future size of the product networks (Expect$_{pi}$) are randomly assigned to each agent. These values are updated through the duration of the run to reflect the market's evolution. An agent's expectations for both technologies always sum to 100. Upon creation, each agent is set to change its expectation every 1 to 12 time ticks (i.e., 1 to 12 months). This number is randomly generated and fixed for that agent throughout the run. When it is time for an agent to change its expectation, it will update it based on which technology that the majority of its neighbors bought during the last time tick. The expectations are modified by a random number between 0 and 1. The expectation for the technology with the larger market share goes up by the fraction, while the expectation for its competitor goes down by the fraction. The expectations have a minimum of 0 and a maximum of 100.

2.2 Technology Sponsors

The two technology sponsors/firms for products A and B are able to take action simultaneously every three time periods (i.e., once per quarter) in order to react to the market conditions they observe. Specifically, pricing strategies are based primarily on two factors, market penetration and market share. *Market penetration* is the fraction of the population that has purchased any technology. The *market share* of some product p is the fraction of those individuals who have purchased any technology that have purchased product p. Each technology sponsor agent has two internal variables: MktShareTrigger which describes the market share this sponsor agent wants to achieve once at least 80% of the market has been penetrated; and PricingFactor which describes the percentage by which the sponsor will raise or lower the price of the technology. Based on these two factors, each technology sponsor follows a *penetration pricing strategy* as outlined in Figure 2. The sponsor agents' primary goal is to stay in business. Sponsor agents accumulate capital as long as they sell at a price above their costs. Sponsor agents may sell below their cost, but can only operate for 36 time periods (i.e., 36 months, three years, or 12 quarters) with negative cumulative capital. After 36 time periods with negative cumulative capital, technology sponsor agents will exit the market (i.e., go out of business.) Thus, the first step in executing the pricing rule is to check cumulative capital and, if it is negative, the sponsor agent will raise its price by PricingFactor. If cumulative income is positive, the sponsor

```
If cumulative profit is negative
   then raise price by PricingFactor        (stay in business)
   else                                     (profit is positive)
      if penetration < 80%
      then lower price by PricingFacto       (gain share)
      else                                  (penetration is > 80%)
         case(current share):
            > MktShareTarget
               then raise price by PricingFactor
                  (recoup investments)
            > 20% but < MktShareTarget
               then lower price by PricingFactor (gain share)
            < 20%
               then do nothing                (retain customers)
```

Fig. 2. Outline of penetration pricing strategy for inferior technology sponsor

firm proceeds with the pricing rule by observing the market share it and its competitor have achieved, as well as overall market penetration. When the market penetration is less than 80%, considered low market penetration, the technology sponsor will lower its price by the PricingFactor in order to attract more buyers. If the market penetration is high (over 80%) the technology sponsor will do one of three things based on its current market share and its MktShareTrigger. When the sponsor's market share is between 20% and its desired market share (MktShareTrigger), the technology sponsor will lower its price by the PricingFactor. When the sponsor's market share is above its desired market share, the technology sponsor will raise its price by the PricingFactor to recoup its earlier investment in obtaining this market share. When the sponsor's market share falls below 20%, the technology sponsor will do nothing in order to retain its loyal customers.

2.3 Experimental Methods

Simulations were run in a 5 (pricing strategies) x 2 (importance of network) x 2 (completeness of information) experimental design. We developed a menu of four pricing strategies to be enacted by the technology sponsors based on different MktShareTrigger and PricingFactor values and also tested a fifth scenario of "no action" where no pricing changes are made. We first examined competition between two sponsors of equal technologies who enact penetration pricing. We then examined two *unequal* technologies competing without undertaking penetration pricing strategies. Finally, we studied scenarios where the inferior technology enacts one of four pricing strategies and the superior technology enacts the least aggressive strategy. We examined the effectiveness of these strategies in terms of both market share and income gained.

For the four penetration pricing strategies, market share triggers for price cuts were either 20% or 65%, and penetration pricing factors were either 20% or 50%. In the "most aggressive" strategy, the technology sponsor will cut price by 50% when its market share drops below 65%. In the "least aggressive" strategy, the technology sponsor agent cuts price by 20% when its market share drops below 20%. In the "itchy but weak trigger finger" (IWTF) strategy, the technology sponsor is quick to cut prices (below 65% market share) but does so by a modest amount (20%). In the "desperation" strategy, the technology sponsor waits to cut prices until the market share has dropped below 20%, but at that point, cuts them significantly (50%). These four strategies, plus the strategy of "no action" or no price cutting, comprise the five strategies that were compared in simulations.

Network importance in a simulation was designated as either "low" or "high". With low network importance, agents used the weights derived from our experimental data on cell phone feature adoption as described earlier (network effects in this market are significant but relatively small). With high network importance, each agent's network importance weight $w3_i$ was doubled and then the agent's weights were normalized again to sum to 100 (details in the Appendix). Table 1 provides the average weights used across all utility functions for each condition. Each simulation was run 30 times for each experimental condition, so except where reported below, each table entry in the Results that follow is the mean over those 30 runs.

Table 1. Average Weights for Agents' Utility Functions (n=141)

Network Importance	Tech	Price	Network	Expect	Investment	Threshold
High	.16	.24	.39	.11	.10	15.54
Low	.19	.30	.26	.14	.12	15.54

3 Results

3.1 Incomplete Information

We first consider the effects of different pricing strategies and levels of network importance when consumer agents only have knowledge of adoption decisions of their local neighbors (i.e., individual consumers in their 11 x 11 neighborhood). Under such conditions an agent can only be influenced indirectly over time by more distant individuals. The first two rows of Table 2 summarize the results after five years (60 ticks) when two equal technologies (TechA = TechB = 70) compete in an environment where network externalities are important. Differences in left-to-right-adjacent table entries are statistically significant (p < .05) unless marked with an asterisk. As seen here, when network importance is high a more aggressive penetration pricing strategy by technology A leads to larger numbers of adopters and also to greater income. Even when met with meager price-cutting by the competition, this most-aggressive penetration pricing strategy proves successful both financially and from a market share standpoint. This can be contrasted with the outcomes when the network

Table 2. Equal Technologies, Incomplete Information, Firm A Implements Most Aggressive Strategy (n=30)

Network Importance	B's Strategy	Agents Adopting A	Agents Adopting B	A's Income	B's Income
High	No Action	477.2	9.6	176,690	87,245
High	Least Aggr.	422.9	61.5	171,085	61,056
Low	No Action	444.7	20.1	-21,307	121,205
Low	Least Aggr.	290.5	170.6	-16,880 *	-40,261 *

is less important in the consumer agents' utility function (Table 2, last two rows). In these cases, the aggressive network development strategy is somewhat successful from an adoption standpoint, but is always disastrous from a financial standpoint. Further, when the most aggressive strategy is countered by even the smaller price cutting strategy on the competitor's part, *both* firms lose money. Thus, while A's aggressive pricing strategy consistently gave it a larger market share, it only profited financially when network externalities were important.

What if, instead of two equal technologies, one is technically inferior to the other? We would expect a superior technology to dominate a market if it enacted a penetration pricing strategy as aggressive as its inferior competitor. The more interesting question is whether an inferior technology sponsor can win (gain greater market share or greater financial rewards) against a superior but more passive technology sponsor by enacting a penetration pricing strategy. Table 3 shows the results after five years when the firm with the inferior technology (Tech = 56) pursues different pricing strategies while the superior technology firm (Tech = 70) follows the least aggressive strategy. The base case of no action by either firm is also provided in the table for comparison. The superior technology gained a larger market share in all situations except when the inferior technology firm used a most aggressive pricing strategy in a high-importance network market; even then the inferior technology only obtained a marginal market share victory at the expense of a net negative income. The same pattern of results was found when simulations were allowed to continue for over 16 years (data not shown). Thus, our results provide no clear support for believing that inferior technologies can use penetration-pricing strategies effectively to capture the largest market share when consumers have incomplete information.

As also shown in Table 3, the financial rewards can vary substantially with network importance. When the inferior technology firm undertakes any penetration pricing strategy and the network importance is low, the inferior technology will, on average, receive as much or more income than the superior technology. In contrast, when the network is more important, the superior technology always dominates the inferior in terms of income achieved. The same pattern of results is found when we allow the simulations to continue to run for more than 16 years. This appears to be counter-intuitive until one examines the dynamics of the simulation.

Table 3. Unequal Technologies, Incomplete Information, Superior Technology Firm Uses Least Aggressive Strategy† (n=30)

Network Importance	Inf. Tech. Strategy	Agents Adopting Inf.Tech.	Agents Adopting Sup.Tech.	Inf. Tech. Income	Sup.Tech. Income
High	No Action†	47.4	216.6	210,290	904,935
High	Least Aggr.	74.1	405.0	18,021	207,037
High	Desperation	194.3	289.1	24,227	166,148
High	IWTF	92.4	391.1	-3,389	254,383
High	Most Aggr.	272.7[a]	207.6[a]	-3,040	180,200
Low	No Action†	45.8	121.0	207,500	518,640
Low	Least Aggr.	89.3	375.5	-8,402	-69,052
Low	Desperation	68.4	384.7	12,748	-85,833
Low	IWTF	77.5	389.3	-6,116	-79,691
Low	Most Aggr.	36.8	416.0	15,566	-81,373

† In the "No Action" case, no pricing action was undertaken by either sponsor.
[a] No significant difference.

When the network effects are minimal, we find that both the inferior and superior technologies are forced to cut prices severely to achieve their market share goals. In the least aggressive and the IWTF strategies, the average price for the inferior technology at tick 60 is below that of the superior technology, although both are below the cost of goods sold. However, the superior technology has a much larger number of

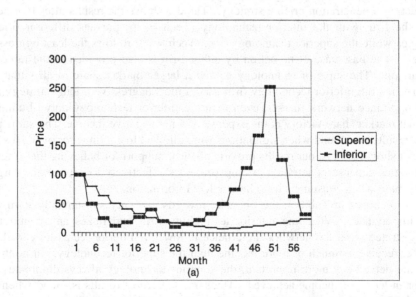

Fig. 3. Price (a), agents adopting (b) and income (c) with *low* network importance, incomplete information, and inferior (superior) technology enacting the most (least) aggressive strategy. (Figure 3 continues on the next page.)

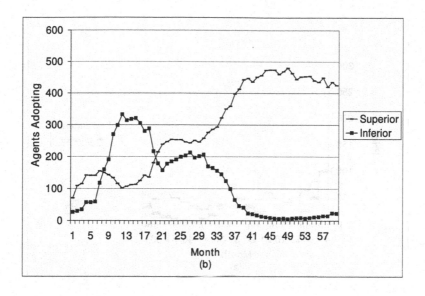

(b)

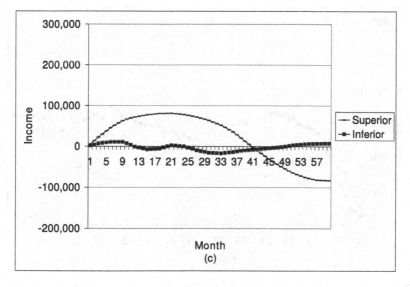

Month
(c)

Fig. 3. (*continued*)

adopters than the inferior, and each sale below cost drags the superior provider further into debt. Thus, for these two strategies, both firms have negative income but the superior technology's income is below the inferior's. For the desperation and the most aggressive strategies, the ultimate prices at tick 60 were comparable and typically just above cost of goods sold, but, again, because the superior technology has far more adopters than did the inferior, each dip into negative income pricing has a much larger overall effect on cumulative net income.

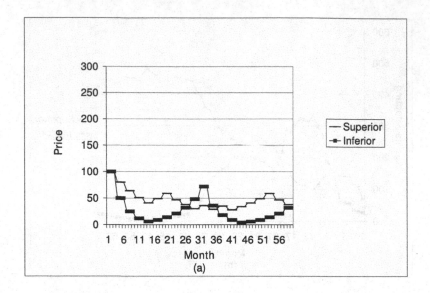

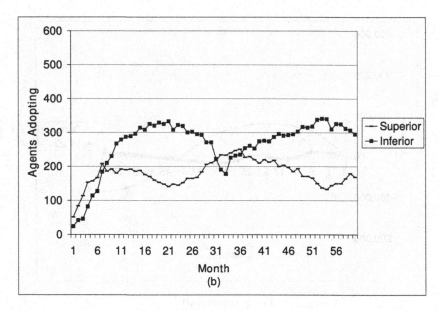

Fig. 4. Price (a), agents adopting (b) and income (c) with *high* network importance, incomplete information, and inferior (superior) technology enacting the most (least) aggressive strategy. (Figure 4 continues on the next page).

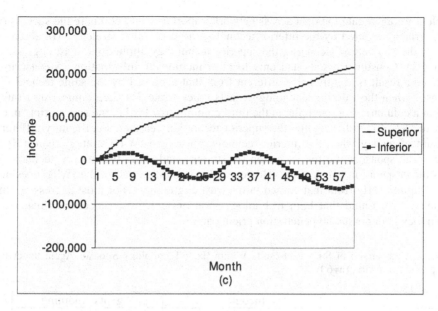

Fig. 4. (*continued*)

Figure 3 shows price, agents adopting, and cumulative income from one run of a low importance network simulation where the inferior technology performs the most aggressive strategy; Figure 4 shows the results from a similar run but with a high importance network. When the network importance is low, each price cut (Figure 3 (a)) adds additional adopters. But each additional adopter does not significantly raise the utility for other adopters; hence, the price cuts in a low network importance setting do not provide the extra boost to utility via increased network value. This can be seen in part (b) of Figures 3 and 4. In Figure 4 the number of agents adopting show a smoother growth, showing that additional agents are adopting even when price stays the same. Figure 3 part (b) shows more agents adding only in reaction to a price cut. Because the additional users do not add much value to one another in the low network importance simulation, technology sponsor agents have to find another way to provide enough value to get market penetration up to 80%. However, the only strategy that they have is price cutting. Because the superior technology sponsor practices the least aggressive strategy as its default, when it does begin to raise prices to recoup lost revenue, its price is raised modestly (by 20%) rather than at the more rapid rates (50%) of the most aggressive and desperation strategies, and it does not regain revenue quickly enough to stay in business. Thus, the importance of network externalities to consumers in a market had a major impact on income received (contrasted in parts (c) of Figures 3 and 4).

In the analysis described thus far, we adopted the perspective of asking "Who won?" in comparing the results of the inferior technology sponsor to the results achieved by the superior technology sponsor. However, the sponsor of an inferior technology may be less interested in "beating" the superior sponsor and more interested in knowing, regardless of the network importance, what strategy is likely to be most effective at gaining adopters and generating positive income for its firm. In

Table 4 we aggregate the data across network importance and compare the success of each strategy enacted by the inferior technology to each other strategy. (In all cases except the "no action" scenario, the superior technology applied the least aggressive strategy.) Consumer agents had only local or incomplete information. Superscripts indicate a result is significantly different from that achieved by the same technology sponsor when the inferior technology enacts other strategies (i.e., comparing results within a column). For example, in the third column, row IWTF, the superscript "n, d" next to "87,346" indicates that the superior technology earns a significantly different amount of money when the inferior technology uses the IWTF strategy than it does when both sponsors use the no action (n) strategy and when the inferior technology uses the desperation (d) strategy. However, the amount earned in the IWTF does not differ significantly from that earned in the least aggressive (l) or most aggressive (m) strategy. We found that both technology sponsors are most successful financially when they both pursue no penetration pricing at all.

Table 4. Comparison of Strategy Results Within Each Technology Sponsor Agent under Incomplete Information (n=60)

	Income		Agents Adopting	
	Inferior	Superior	Inferior	Superior
No action (n)	$208,895^{l,d,i,m}$	$711,788^{l,d,i,m}$	$46.58^{d,m}$	$168.78^{l,d,i,m}$
Least Aggr (l)	$4,810^{n}$	$68,993^{n}$	$81.70^{d,m}$	$390.27^{n,d,m}$
Desperation (d)	$18,488^{n}$	$40,158^{n,i}$	$131.35^{n,l,i}$	$336.93^{n,l,i}$
IWTF (i)	$-4,753^{n}$	$87,346^{n,d}$	$84.97^{d,m}$	$390.17^{n,l,i}$
Most Aggr (m)	$6,263^{n}$	$49,414^{n}$	$154.77^{n,l,i}$	$311.78^{n,l,i}$

On average, the superior technology achieves greater income in the no action case (mean $711,788) than with any pricing strategy as does the inferior (mean $208,895). Further, the superior technology sees significantly fewer agents adopting in the no action scenario compared to all other strategies, an average of 168.8. The inferior technology sees the fewest adopters under the no action strategy with an average of 46.6, but this is not significantly lower than in the IWTF (mean 85.0) or least aggressive strategies (81.7). The inferior technology obtains the greatest number of agents adopting by using the most aggressive strategy (154.8) or desperation strategy (131.4). The superior technology gains the most agents when the inferior technology employs the least aggressive (mean 390.3) or IWTF (mean 390.2) strategy. By the end of year five, the superior technology sponsor never goes out of business. The inferior technology sponsor agent does go out of business twice when using the desperation strategy and three times when using the most aggressive strategy; both occur when the network is less important. When we allow simulations to continue for over 16 years (data not shown), we see a much greater frequency of sponsors going out of business. The least aggressive and the IWTF strategies are most prone to putting the inferior technology sponsor out of business. Both of these strategies involve price cuts of only 20%. In contrast, the superior technology appears to be most likely to go out of business when the inferior technology sponsor adopts the desperation or most aggressive penetration pricing strategy. Both of these strategies involve price cuts of 50%.

3.2 Complete Information

In the results discussed above the agents have limited knowledge of other agents' actions because they can only see those nearby agents in their 11 x 11 neighborhood. Farrell and Saloner (Farrell and Saloner 1985) call such a constrained environment one of "incomplete information." How might the strategies influence adoption in a world of complete information, where each agent can see the actions of all other agents? Such a scenario could correspond to, for example, situations where market share information is published in mass media. We ran a second set of simulations where the neighborhood for each agent encompassed the entire *StandardScape* world and thus, each agent has complete information about product adoptions made by other agents.

When the two technologies are equal, our results (Table 5) are similar to what we find in the incomplete information simulations: penetration pricing is a financial success only when the network is important to the agents. When the network is less important, penetration pricing is not rewarded financially, although if the competition does not engage in any price cutting, it can still be successful from a market share perspective.

Table 5. Equal Technologies, Complete Information, Firm A Implements Most Aggressive Strategy (n=30)

Network Importance	B's Strategy	Agents Adopting A	Agents Adopting B	A's Income	B's Income
High	No Action	470.4	17.8	242,238	117,545
High	Least Aggr	397.6	94.0	221,688	70,150
Low	No Action	433.1	27.7	-11,984	155,970
Low	Least Aggr	238.5[a]	233.2[a]	-8,607[a]	-19,237[a]

[a] No significant difference.

We then simulated two unequal technologies in a complete information world (Table 6), repeating all of the simulations described above. When the network importance is high, we see that the inferior technology stands a better chance of gaining market share over the superior technology with complete information than with incomplete information. For each penetration pricing strategy the inferior technology gained as much market share (least aggressive strategy) or more market share (desperation, most aggressive, or IWTF strategies) than the superior technology. Also when the network importance is high, the inferior technology achieves positive income for all strategies but IWTF, but earns far less than its superior competitor. As in the incomplete information results, the inferior technology does as well as or better financially as the superior technology when network importance is low. Thus, penetration pricing appears to be more effective at increasing market share when consumers (agents) have information about what a larger number of other agents are adopting and when network externalities are important in their utility function. However, this larger neighborhood or more complete information does not necessarily translate into greater financial success for the inferior technology sponsor.

Table 6. Unequal Technologies, Complete Information, Superior Technology Firm Uses Least Aggressive Strategy† (n=30)

Network Importance	Inf. Tech. Strategy	Agents Adopting Inf. Tech.	Agents Adopting Sup. Tech.	Inf. Tech. Income	Sup.Tech. Income
High	No Action†	29.4	320.6	154,633	1,365,855
High	Least Aggr.	215.3*	272.0*	7,484	381,535
High	Desperation	280.7	202.4	84,223	236,726
High	IWTF	310.2	182.5	-14,394	395,622
High	Most Aggr.	334.7	149.5	68,370	208,440
Low	No Action†	54.7	167.1	255,170	725,845
Low	Least Aggr.	95.5	385.3	-5,582	-29,282
Low	Desperation	120.2	348.9	24,222	-58,687
Low	IWTF	129.2	352.3	-17,200 *	-17,502 *
Low	Most Aggr.	151.3	316.8	22,756	-56,427

† In the "No Action" case, no pricing action was undertaken by either sponsor.

We also compared the strategies to one another in the complete information scenarios in the same way that we did in the incomplete information simulations. We found that both sponsor firms make the most money when both take no penetration pricing actions (inferior mean $204,901 and superior mean $1,045,850). However, when we compare the outcomes of the penetration pricing strategies undertaken by the inferior technology, we found that for the inferior technology, enacting the most aggressive (mean $45,563) or desperation (mean $54,223) strategy was most financially rewarding and the least aggressive (mean $951) and IWTF (mean -$15,797) were the least successful financially. In terms of attracting adopters, the most aggressive (mean 243), IWTF (mean 219.7) and desperation (mean 200.5) all performed better than no action (mean 42.1), but not significantly different from one another. Further, the only time the inferior technology went of out business in the five year simulations was when it employed the most aggressive strategy. The superior technology's fortunes were the exact opposite of that of the inferior's technology sponsor.

In summary, the key difference we found in the complete information experiments is that in situations where agents are maximally influenced by other agents' choices—complete information and high network importance—the inferior technology is able to dominate the superior technology in the number of agents who adopt a particular technology. In the no action case and three penetration pricing strategies the inferior technology is able to gain more adopters than the superior technology. The inferior technology gains as many as the superior technology when using the least aggressive strategy. Thus, we see that when agents have complete information and care more about network choices, an inferior technology can gain a dominant market share.

4 Discussion

In this work we created an artificial society in which hundreds of agents representing consumers repeatedly made decisions about which of two competing technologies to adopt. The utility function that guided agent decisions was grounded in factors generally recognized to be important in people's decisions concerning product purchases, such as intrinsic value of the technology, product price, network influences, expectations, and switching costs (Besen and Farrell 1994; Frels, Shervani and Srivastava 2003; Farrell and Saloner 1986; Burnham, Frels and Mahajan 2003). Further, the weights assigned to each factor were derived from an empirical study that we conducted of the importance that real consumers attach to them, providing a loose coupling to the real world and making the individual agent policies heterogeneous. The results produced by the model were sometimes intuitive, but at other times surprising. While our model is of course limited by its size and simplicity relative to reality, we believe it adds to growing evidence that artificial society models may not only shed light on complex social, economic, and business questions, but may ultimately help guide policy makers in decision making.

Our simulations with the model produced a number of specific and informative results about technology adoption from a business management perspective. First, the importance of the network to consumers was found to be crucial in determining the success of penetration pricing strategies. A dramatic difference in results occurred when the network was more heavily weighted in the consumer's utility function versus when it was not. When networks were not important, network development strategies such as penetration pricing were not consistent with the consumer's utility function. Hence, cutting prices to attract adopters did not have the positive feedback effect it did in a more highly network-oriented market. Instead, the firm with the largest market share (typically the superior technology in our simulations) ended up losing more money per adopter through its price cutting without creating additional network utility for the consumer. Thus, misinterpreting the relevance of network externalities to a firm's potential market might allow a firm to experience successful market share development, but it might also lead to disastrous financial results. Also, when technologies were symmetric and a firm has "lost" the market share battle, little was to be gained by even mild price-cutting.

Second, when technologies were asymmetric, an inferior technology sponsor enacting a penetration pricing strategy might occasionally achieve market share success but that success was not necessarily tied to financial success. We explored an inferior technology's ability to dominate from a market share perspective as well as investigated its ability to profit from its gains in share. The link between market share and profitability has generally been considered to exist and to be positive (Buzzell and Gale 1987; Szymanski, Bharadwaj and Varadarajan 1993; Hellofs and Jacobson 1999). However, the wisdom of pursuing market share for its own sake as well as market share's positive relationship to profitability has been questioned by others (Slywotzky and Morrison 2001; Nagle and Holden 2002). The staying power of Apple Computer (even before the introduction of the iPod) has shown that market share alone is not the only determinant of success. Our results provide support for this disconnect.

Third, we found that the amount of information agents had about other agents' actions could significantly influence the effectiveness of penetration pricing strategies. An inferior technology operating in a market where network externalities were important and where agents had complete information could gain market share larger than or equal to a superior technology by using any of the pricing strategies we tested.

However, it still did not outperform the superior technology financially. When information was incomplete, even if the networks were important to consumers, the inferior technology might not be able to create enough value through growing its network to compensate for its inferior technology.

Finally, we observed that no penetration pricing at all brought the greatest financial reward to the sponsor agents, although it achieved very low market penetration. It is possible that this just reflects the manner in which we implemented the pricing rule; it is probably more reasonable to compare outcomes of different strategies to one another given that they are all implemented with the same biases. Within the strategies, we found that strategies with larger price reductions led to the largest number of adopters for the inferior technology. These same strategies, enacted by the inferior technology sponsor, were also most damaging to the superior technology's financial and market share position. Despite the market share results achieved with the desperation and most aggressive strategies, such price cuts did not necessarily lead to better or worse financial success for the inferior technology. However, in terms of punishing a competitor with a superior technology our results suggest the value of true penetration pricing rather than modest price cuts in a network market.

Future research should include additional elements of network effects such as network quality, network compatibility in addition to other networks such as the complements and producer network (Frels, Shervani and Srivastava 2003). Future studies might also consider additional network development strategies such as licensing to a low-cost provider, pre-announcement of product upgrades as well as product upgrades themselves.

Acknowledgment

This work was supported in part by the Marketing Science Institute Grant #4-1235, the Robert H. Smith School of Business Summer Research Grant program, and NSF award IIS-0325098.

References

Arthur, W.B.: Competing Technologies, Increasing Returns, and Lock-in by Historical Events. The Economic Journal 99, 116–131 (1989)

Axelrod, R.: The Dissemination of Culture. Journal of Conflict Resolution 41, 203–226 (1997)

Back, T., et al.: Modeling Urban Growth by Cellular Automata. In: Parallel Problem Solving from Nature IV, pp. 636-645. Springer, Heidelberg (1996)

Besen, S., Farrell, J.: Choosing How to Compete: Strategies and Tactics in Standardization. Journal of Economic Perspectives 8(2), 117–131 (1994)

Burnham, T., Frels, J., Mahajan, V.: Consumer Switching Costs. Academy of Marketing Science Journal, 109–126 (2003)

Buzzell, R., Gale, B.: The PIMS Principles. The Free Press, New York, N.Y (1987)

Chou, H., Reggia, J.: Emergence of Self-replicating Structures in a Cellular Automata Space. Physica D 110, 252–276 (1997)

Epstein, J., Axtell, R.: Growing Artificial Societies. MIT Press, Cambridge (1996)

Farrell, J., Saloner, G.: Standardization, Compatibility, and Innovation. Rand Journal of Economics 16(1), 70–83 (1985)

Farrell, J., Saloner, G.: Installed Base and Compatibility: Innovation, Product Preannouncements, and Predation. The American Economic Review 76(5), 940–955 (1986)

Financial Times, Shoppers Get Bargain with Subsidized PlayStation3, November 17, p. 15 (2006)

Fishbein, M., Ajzen, I.: Belief, Attitude, Intention, and Behavior: An Introduction to Theory and Research. Addison-Wesley Pub. Co, Reading (1975)

Frels, J., Shervani, T., Srivastava, R.: The Integrated Networks Model: Explaining Resource Allocations in Network Markets. Journal of Marketing 67(1), 29–45 (2003)

Gaylord, R., Nishidate, K.: Modeling Nature, pp. 25–35. Springer, Heidelberg (1996)

Goldenberg, J., Libai, B., Muller, E.: Riding the Saddle: How Cross-Market Communications can Create a Major Slump in Sales. Journal of Marketing 66, 1–16 (2002)

Hauert, C., Doebeli, M.: Spatial Structure Often Inhibits the Evolution of Cooperation in the Snowdrift Game. Nature 428, 643–646 (2004)

Hellofs, L., Jacobson, R.: Market Share and Perceptions of Quality: When Can Firms Grow Their Way to Higher Versus Lower Quality? Journal of Marketing 63, 16–25 (1999)

Hill, C.: Establishing a Standard: Competitive Strategy and Technological Standards in Winner-Take-All Industries. Academy of Management Executive 11, 7–25 (1997)

Lapizco-Encinas, G., Reggia, J.: Diagnostic Problem Solving Using Swarm Intelligence. In: Proc. IEEE Swarm Intelligence Symposium, pp. 365–372 (2005)

Libai, B., Muller, E., Peres, R.: The Role of Seeding in Multi-Market Entry. International Journal of Research in Marketing 22, 375–393 (2005)

Loia, V. (ed.): Soft Computing Agents. IOS Press, Amsterdam (2003)

Manazec, J.: Evaluating Perceptions-Based Marketing Strategies: An Agent-based Model and Simulation Experiment. Journal of Modeling in Management 1(1), 52–74 (2006)

Nagle, T., Holden, R.: The Strategy and Tactics of Pricing. Prentice Hall, Upper Saddle River, N.J (2002)

Orme, B. (ed.): CBC User Manual, Version 2. Sequim. Sawtooth Software, Wash (2001)

Reggia, J., Schulz, R., Wilkinson, G., Uriagereka, J.: Conditions Enabling the Evolution of Inter-Agent Signaling in an Artificial World. Artificial Life 7, 3–32 (2001)

Schelling, T.: Models of Segregation. American Economic Review 59, 488–493 (1969)

Slywotzky, A., Morrison, D.: The Profit Zone. Three Rivers Press, New York, N.Y (2001)

Szymanski, D., Bharadwaj, S., Varadarajan, P.R.: An Analysis of the Market Share-Profitability Relationship. Journal of Marketing 57(3), 1–18 (1993)

Tesfatsion, L.: Agent-Based Computational Economics. Artificial Life 8, 55–82 (2002)

Wagner, K., Reggia, J.: The Emergence of an Internally-Grounded Multi-referent Communication System. Interaction Studies 7, 103–129 (2006)

Wagner, K., Reggia, J., Uriagereka, J., Wilkinson, G.: Progress in the Simulation of Emergent Communication and Language. Adaptive Behavior 11, 37–69 (2003)

Appendix: Empirical Basis for Simulated Agents

As explained in the text, the weight or importance associated with each attribute in the consumer agents' utility functions in our simulations is based on empirical data that we gathered in a choice-based conjoint study involving 141 consumers. Our conjoint study presented subjects with multiple cell phone plans that offered a new feature: "push to talk." Cell phones and "push to talk" were chosen for several reasons. First, it was a technology-based product with which our subjects, undergraduate students, would have familiarity. Second, given the timing of the research, the "push to talk" feature was recently being marketed to consumers and specifically teens. Hence, we believed they would have some familiarity with the function, but the function would not yet be widely diffused. However, to ensure that each subject had at least a basic understanding of the feature, information on the feature was provided to all subjects. Third, the "push to talk" feature adds a network externality element to cell phones that does not exist otherwise. Users must be on the same "network" or use the same service provider in order to make use of this function. This was emphasized in the task as outlined below.

A choice conjoint design was used because it mirrored more closely the task of our agents in *StandardScape* and because it has been shown to more closely model actual choice contexts (Orme 2001). Providing the "none" option more clearly reflects realistic choices in the world and the choices our simulated agents have in *StandardScape*. Further, use of the "none" option allowed us to capture information on each subject's relative reservation utility or threshold.

The profiles presented varied on five attributes, reflecting the five elements simulated in the utility equation. The first attribute, technology, was defined as the reliability and quality of the service. This had three levels represented by 2.5, 3.5 and 4.5 stars. The second attribute, price of the service, was set at $29.99, $39.99 or $49.99. The third attribute, the network, was defined as the number of people in the subject's circle of friends who use the same service and thus, with whom one would be able to use the "push to talk" feature. The three levels of this attribute were "nobody you know uses this service," "about half of your friends use this service" and "pretty much all your friends use this service." The fourth attribute was the future success expected from this cell phone provider. The three levels of this attribute were represented as expectations by experts as to the long-term survivability of the service provider. Specifically, the low level was represented as "experts say the future prospects of this service provider are shaky, at best," mid-level as "experts are undecided about the future prospects of this service provider," and the high level as "experts believe this service provider will continue to provide the service in the future." Finally, switching costs were described as "the trouble it would be for you to switch to this service provider." This attribute had only two levels and was operationalized as "keep your same phone number and transfer all contact information" and "requires new cell phone number and you cannot transfer your saved numbers."

The experiment was computer-administered using Sawtooth Software CBC System (Orme 2001). Subjects used a computer in a behavioral lab and were presented with three screens of instructions. These instructed the subjects to imagine that they were choosing a new cell phone service provider that offered the "Push to Talk" feature. The attributes were explained. The subjects were given two practice choice sets each with two options and the "None" option. They were then informed that the actual experiment was beginning and provided with a brief reminder of the attributes' meanings. Subjects were presented with twenty choice tasks, each composed of three full-profile product descriptions along with the opportunity to select "I would choose none of the Services shown." The profiles were generated randomly for each subject with approximately random orthogonal designs, with each subject seeing a unique set of questions. Subjects indicated their preference for one of the three profiles or the "none" option by clicking on the box containing the profile with a mouse. Subjects could return to previous screens by clicking "Previous." Following the twenty choice tasks, subjects answered questions regarding their ownership and use of a cell phone as well as basic demographic questions.

The 141 subjects were drawn from a large mid-Atlantic undergraduate marketing course and were offered extra credit for participating. The choice data was transformed into individual importance weights for each attribute by using a hierarchical Bayes package. By including the "None" option, we were also able to obtain a unique "threshold" for each subject to use in the simulation. The weights and thresholds used for each of the 600 agents in the simulation were randomly sampled, with replacement, from these 141 actual consumers.

In order to draw conclusions from simulations using our empirically-based agents, we needed to calibrate the simulations with actual adoption witnessed in this group of subjects. To do this, the conjoint experiment was immediately followed by a series of questions including ones on whether or not the subject owned a cell phone. The results indicated that 82.6% of the subjects own a cell phone (9.4% of all subjects own a cell phone with the "push to talk" feature; 8% of the subjects do not own a cell phone; 2.2% responded that they were considering purchasing one.) Pilot simulations achieved market penetration rates of only 3 to 8%, far from the number of subjects who reported owning cell phones in our study. Thus, we calibrated the threshold of the agents and, to further focus the study on the network effects, we increased the relative weight of the network. The network weight was doubled and then all five weights were readjusted to sum to 100% again. Using a base price of 50, we adjusted the threshold of the agents downward by 10% of the original threshold until we could achieve market penetration rates of approximately 80 to 85%, mirroring the number of subjects who stated they owned cell phones. This occurred at 40% of the original threshold. The simulations reported here were then done using the modified threshold value. Thus, threshold values are fixed in our simulation, but we use both the original network weights ("minimal network") and the increased network weights ("augmented network") as a two-level factor in our experimental design.

An Evolutionary Programming Based Knowledge Ensemble Model for Business Risk Identification

Lean Yu[1,2], Kin Keung Lai[2], Shouyang Wang[1]

[1] Institute of Systems Science, Academy of Mathematics and Systems Science,
Chinese Academy of Sciences, Beijing 100080, P.R. China
[2] Department of Management Sciences, City University of Hong Kong, Tat Chee Avenue,
Kowloon, Hong Kong

Abstract. Business risk identification is one of the most important components in business risk management. In this study, a knowledge ensemble methodology is proposed to design an intelligent business risk identification system, which is composed of two procedures. First of all, some data mining and knowledge discovery algorithms are used to explore the implied knowledge about business risk hidden in the business data. Then the implied knowledge generated from different mining algorithms is aggregated into an ensemble output using an evolutionary programming (EP) technique. For verification, the knowledge ensemble methodology is applied to a real-world business risk dataset. The experimental results reveal that the proposed intelligent knowledge ensemble methodology provides a promising solution to business risk identification.

Keywords: Business risk identification, knowledge ensemble, data mining, soft computing, evolutionary programming, multivariate discriminant analysis, logit regression analysis, artificial neural network, support vector machines.

1 Introduction

Business risk management is a scientific field which many academic and professional people have been working for, at least, the last three decades. Almost all financial organizations, such as banks, credit institutions, clients, etc., need this kind of information for some firms in which they have an interest of any kind. However, business risk management is not an easy thing because business risk management is a very complex and challenging task from the viewpoint of system engineering. It contains many processes, such as risk identification and prediction, modeling, control and management. In this complex system analysis, risk identification is no doubt an important and crucial step (Lai et al., 2006a), which directly influences the later processes of business risk management. In this chapter, we focus only on the business risk identification and analysis.

For business risk identification and analysis, some approaches were presented during the past decades. Originally, the first approach to identify financial risk started with the use of empirical methods (e.g., the "three A" method, the "five C" method, and the "credit-men" method) proposed by large banks in USA (Lai et al., 2006). Then, the financial ratios methodology was developed for financial risk identification and prediction. These ratios have been long considered as objective indicators of firm insolvency risk (Lai et al., 2006a; Beaver, 1966; Courtis, 1978; Altman, 1993). The approach of the financial ratios (also called univariate statistical approach), gave rise to the methods for business risk prediction based on the multivariate statistical analysis. In 1968 already, Altman (1968) proposed to use the discriminant analysis (a discriminant function with five financial ratios has been assessed) for predicting the business failure risk. Subsequently, the use of this method has continued to spread out to the point that today we can speak of discriminant models of predicting business bankruptcy risk. At the same time, however, the generalization of the discriminant analysis resulted in some critical papers; interested readers can see Eisenbeis (1977), Ohlson (1980), and Dimitras et al (1996) for more details.

Since the work of Altman (1968), several studies proposing other methods have been used to overcome some disadvantages of the discriminant analysis so as to provide higher prediction accuracy. Among these studies, we can cite the study of Ohlson (1980) and Zavgren (1985) using logit analysis and the study of Zmijewski (1984) using probit analysis. In addition, Frydman et al. (1985) first employed the recursive partitioning algorithm while Gupta et al. (1990) used mathematical programming methods for the business failure prediction problem. Other methods used were survival analysis by Luoma and Latitinen (1991), expert systems by Messier and Hansen (1988), neural networks by Altman et al. (1994), Lee et al. (2005), and Lai et al. (2006b, c), genetic algorithm by Varetto (1998), rough set theory by Dimitras et al. (1999), support vector machine by Shin et al. (2005) and Lai et al. (2006d), multifactor model by Vermeulen et al. (1998) and some integrated methodology of rough set theory and neural network by Ahn et al (2000). Moreover, several methods were developed based on multi-criteria decision making (MCDM) methodology (Zopounidis, 1987; Siskos et al. (1994); Dimitras et al. (1995); Olmeda and Fernandez (1997)). They identify firms into categories according to their financial insolvency risk. But the use of MCDA methods in the financial risk prediction circumvents many of the problems that exist when using discriminant analysis (Eisenbeis, 1977; Dimitras et al., 1996).

Most of the methods mentioned above have already been investigated in the course of comparative studies related in several review articles, see Scott (1981), Zavgren (1983), Altman (1984), Jones (1987), Keasey and Watson (1991), and Dimitras et al. (1996) for more details. Of the review papers, Dimitras et al. (1996) gave a comprehensive review of methods used for the prediction of business failure risk and of new trends in this area. Due to the advancement of computer and information technology, these business risk identification techniques offer significant aid in the financial risk management to the managers of financial institutions.

Recent studies found that any unitary data mining technique may have their own disadvantages. For example, statistics and optimization techniques often require some assumptions about data. Neural networks are easy to trap into local minima, and the exertion of GA may take a lot of time. Even for new SVM technique, it may also suffer from overfitting problem in some situations (Tay and Cao, 2001). In order to overcome these drawbacks and further improve the classification accuracy, some hybrid or ensemble classifiers integrating two or more single classification methods have been proposed to overcome these drawbacks of individual models. Typical examples include Olmeda and Fernandez (1997), Piramuthu (1999), and Lai et al. (2006c). Empirical results have shown that these hybrid or ensemble classifiers have a higher classification accuracy and predictability than individual classifiers. Currently the research about hybrid or ensemble classifiers is flourishing in business risk analysis.

Motivated by the hybrid or ensemble classifiers, an intelligent knowledge ensemble approach is proposed to identify and analyze business risk. Actually, this proposed methodology is composed of two components. First of all, some data mining and knowledge discovery algorithms including traditional linear models and emerging intelligent techniques are used to explore the implied knowledge about business risk hidden in the business data. Then these implied knowledge generated from different data mining algorithms are aggregated into an ensemble output using an evolutionary programming (EP) technique. Typically, this chapter uses four data mining methods: multivariate discriminant analysis (MDA), logit regression analysis (LogR), artificial neural networks (ANN), and support vector machines (SVM), for business risk identification task. Because the results generated by data mining algorithms are assumed to be useful knowledge that can aid decision-making, the proposed approach is called as knowledge ensemble methodology.

The main motivation of this chapter is to design a high-performance business risk identification system using knowledge ensemble strategy and meantime compare its performance with other existing single approaches. The rest of the chapter is organized as follows. Section 2 introduces the formulation process of the proposed EP-based knowledge ensemble methodology. Section 3 gives the research data and experiment design. The experiment results are reported in Section 4. Section 5 concludes the chapter.

2   The EP-Based Knowledge Ensemble Methodology

In this section, an intelligent knowledge ensemble methodology is proposed for business risk identification. As earlier noted, the knowledge ensemble methodology proposed by this chapter consists of two main procedures. One is data mining and knowledge discovery procedure that explore the implied knowledge hidden in the business dataset. Another is knowledge ensemble procedure that is used to combine or aggregate the all implied knowledge into an ensemble output. The basic idea of the proposed knowledge ensemble originated from using all the valuable information hidden in all individual data mining and knowledge discovery algorithms, where each can contribute to the improvement of generalization.

2.1   Brief Introduction of Individual Data Mining Models

As previously mentioned, four typical data mining and knowledge discovery models, multivariate discriminant analysis (MDA), logit regression analysis (LogR), artificial neural networks (ANN), and support vector machines (SVM), are selected for business risk identification. Of course, other data mining and knowledge discovery algorithms are also selected. In this chapter, the four typical models are selected for illustration only. For reader's convenience, a brief introduction of each model is reviewed.

Multiple discriminant analysis. (MDA) tries to derive a linear combination of two or more independent variables that best discriminate among a priori defined groups, which in our case are failed and non-failed of the business enterprise. This is achieved by the statistical decision rule of maximizing the between-group variance relative to the within-group variance. This relationship is expressed as the ratio of the between-group to the within-group variance. The MDA derives the linear combinations from an equation that takes the following form

$$y = f(x) = w_1x_1 + w_2x_2 + \cdots w_nx_n \tag{1}$$

where y is a discriminant score, w_i ($i = 1, 2, \ldots, n$) are discriminant weights, and x_i ($i = 1, 2, \ldots, n$) are independent variables. Thus, each firm receives a single composite discriminant score which is then compared to a cut-off value, and with this information, we can determine to which group the firm belongs.

MDA does very well provided that the variables in every group follow a multivariate normal distribution and the covariance matrix for every group is equal. However, empirical experiments have shown that especially failed firms violate the normality condition. In addition, the equal group variance condition is often violated. Moreover, multi-collinearity among independent variables may cause a serious problem, especially when the stepwise procedures are employed (Hair et al., 1998).

Logistic regression analysis. (LogR) has also been used to investigate the relationship between binary or ordinal response probability and explanatory variables. The method fits linear logistic regression model for binary or ordinal response data by the method of maximum likelihood. Among the first users of Logit analysis in the context of financial distress was Ohlson (1980). Like MDA, this technique weights the independent variables and assigns a score in a form of failure probability to each company in a sample. The advantage of this method is that it does not assume multivariate normality and equal covariance matrices as MDA does. Logit regression analysis incorporates nonlinear effects, and uses the logistical cumulative function in identifying a business risk, i.e.

$$y(\text{Probability of failure}) = \frac{1}{1+e^{-Z}} = \frac{1}{1+e^{-(w_0+w_1x_1+\cdots+w_nx_n)}} \tag{2}$$

Logit regression analysis uses the stepwise method to select final variables. The procedure starts by estimating parameters for variables forced into the model, i.e. intercept and the first possible explanatory variables. Next, the procedure computes the adjusted chi-squared statistic for all the variables not in the model and examines the largest of these statistics. If it is significant at the specified level, 0.01 in our chapter, the variable is entered into the model. Each selection step is followed by one or more elimination step, i.e. the variables already selected into the model do not necessarily stay. The stepwise selection process terminates if no further variable can be added to the model, or if the variable just entered into the model is the only variable removed in the subsequent elimination. For more details, please refer to Ohlson (1980).

Artificial neural networks. (ANNs) are a new kind of intelligent learning algorithm and are widely used in some application domains. In this chapter, a standard three-layer feed-forward neural network (FNN) (White, 1990) based on error back-propagation algorithm is selected. Usually, a neural network can usually be trained by the in-sample dataset and applied to out-of-sample dataset to verification. The model parameters (connection weights and node biases) will be adjusted iteratively by a process of minimizing the error function. Basically, the final output of the FNN model can be represented by

$$y = f(x) = a_0 + \sum_{j=1}^{q} w_j \varphi(a_j + \sum_{i=1}^{p} w_{ij} x_i) \tag{3}$$

where x_i ($i = 1, 2, \ldots, p$) represents the input patterns, y is the output, a_j ($j = 0, 1, 2, \ldots, q$) is a bias on the jth unit, and w_{ij} ($i = 1, 2, \ldots, p; j = 1, 2, \ldots, q$) is the connection weight between layers of the model, $\varphi(\bullet)$ is the transfer function of the hidden layer, p is the number of input nodes and q is the number of hidden nodes.

In the identification classification problem, the neural network classifier can be represented by

$$F(x) = sign(f(x)) = sign\left(a_0 + \sum_{j=1}^{q} w_j \varphi(a_j + \sum_{i=1}^{p} w_{ij} x_i) \right) \tag{4}$$

The main reason of selecting ANN as a classifier is that an ANN is often viewed as a "universal approximator" (Hornik et al., 1989). Hornik et al. (1989) and White (1990) found that a three-layer back propagation neural network (BPNN) with an identity transfer function in the output unit and logistic functions in the middle-layer units can approximate any continuous function arbitrarily well given a sufficient amount of middle-layer units. That is, neural networks have the ability to provide flexible mapping between inputs and outputs.

Support vector machine. (SVM) was originally introduced by Vapnik (1995, 1998) in the late 1990s. Many traditional statistical theories had implemented the empirical risk minimization (ERM) principle; while SVM implements the structure risk mini-mization (SRM) principle. The former seeks to minimize the mis-classification error or deviation from correct solution of the training data but the latter searches to mini-mize an upper bound of generalization error. SVM mainly has two classes of applica-tions: classification and regression. In this chapter, application of classification is discussed.

The basic idea of SVM is to use linear model to implement nonlinear class bounda-ries through some nonlinear mapping the input vector into the high-dimensional feature space. A linear model constructed in the new space can represent a nonlinear decision boundary in the original space. In the new space, an optimal separating hy-perplane is constructed. Thus SVM is known as the algorithm that finds a special kind of linear model, the maximum margin hyperplane. The maximum margin hyperplane gives the maximum separation between the decision classes. The training examples that are closest to the maximum margin hyperplane are called support vectors. All other training examples are irrelevant for defining the binary class boundaries. Usu-ally, the separating hyperplane is constructed in this high dimension feature space. The SVM classifier takes the form as

$$y = f(x) = sgn(w \cdot \psi(x) + b) \tag{5}$$

where $\psi(x) = \varphi_1(x), \varphi_2(x), \ldots, \varphi_N(x)$ is a nonlinear function employed to map the original input space R^n to N-dimensional feature space, w is the weight vector and b is a bias, which are obtained by solving the following optimization problem:

$$\begin{cases} \text{Min} & \phi(w,\xi) = (1/2)\|w\|^2 + C\sum_{i=1}^{m}\xi_i \\ \text{s.t.} & y_i[\varphi(x_i)\cdot w + b] \geq 1 - \xi_i \\ & \xi_i \geq 0, i = 1,2,...,m \end{cases} \tag{6}$$

where $\|w\|$ is a distance parameter, C is a margin parameter and ξ_i is positive slack variable, which is necessary to allow misclassification. Through computation of Eq. (6), the optimal separating hyperplane is obtained in the following form:

$$y = \text{sgn}\left(\sum_{SV}\alpha_i y_i \varphi(x_i)\varphi(x_j) + b\right) \tag{7}$$

where SV represents the support vectors. If there exist a kernel function such that $K(x_i,x_j) = (\varphi(x_i),\varphi(x_j))$, it is usually unnecessary to explicitly know what $\varphi(x)$ is, and we only need to work with a kernel function in the training algorithm, i.e., the optimal classifier can be represented by

$$y = \text{sgn}\left(\sum_{SV}\alpha_i y_i K(x_i,x_j) + b\right) \tag{8}$$

Any function satisfying Mercy condition (Vapnik, 1995, 1998) can be used as the kernel function. Common examples of the kernel function are the polynomial kernel $K(x_i,x_j) = (x_i x_j^T + 1)^d$ and the Gaussian radial basis function $K(x_i,x_j) = \exp\left(-(x_i - x_j)^2/2\sigma^2\right)$. The construction and selection of kernel function is important to SVM, but in practice the kernel function is often given directly.

2.2 Knowledge Ensemble Based on Individual Mining Results

When different data mining and knowledge discovery algorithms are applied to a certain business dataset, different mining results are generated. The performances of these results are different due to the fact that each mining algorithm has its own shortcomings. To improve the identification performance, combining them into an aggregated result may generate a good performance. In terms of this idea, we utilize the ensemble strategy (Yu et al., 2005) to fuse their different results produced by individual data mining and knowledge discovery algorithms.

Actually, a simple way to take into account different results is to take the vote of the majority of the population of classifiers. In the existing literature, majority voting is the most widely used ensemble strategy for classification problems due to its easy implementation. Ensemble members' voting determines the final decision. Usually, it takes over half the ensemble to agree a result for it to be accepted as the final output of the ensemble regardless of the diversity and accuracy of each model's generalization. However, majority voting has several important shortcomings. First of all, it ignores the fact some classifiers that lie in a minority sometimes do produce the correct results. Second, if too many inefficient and uncorrelated classifiers are considered, the vote of the majority would lead to worse prediction than the ones obtained by using a single classifier. Third, it does not consider for their different expected performance when they are employed in particular circumstances, such as plausibility of outliers. At the stage of integration, it ignores the existence of diversity that is the motivation for ensembles. Finally, this method can not be used when the classes are continuous (Olmeda and Fernandez, 1997; Yang and Browne, 2004). For these reasons, an additive method that permits a continuous aggregation of predictions should be preferred. In this chapter, we propose an evolutionary programming based approach to realize the classification/prediction accuracy maximization.

Suppose that we create p classifiers and let c_{ij} be the classification results that classifier $j, j =1, 2, ..., p$ makes of sample $i, i = 1, 2, ..., N$. Without loss of generality, we assume there are only two classes (failed and non-failed firms) in the data samples, i.e., $c_{ij} \in \{0,1\}$ for all i, j. Let $C_i^w = Sign(\sum_{j=1}^{p} w_j c_{ij} - \theta)$ be the ensemble prediction of the data sample i, where w_j is the weight assigned to classifier j, θ is a confidence threshold and $sign(.)$ is a sign function. For corporate failure prediction problem, an analyst can adjust the confidence threshold θ to change the final classification results. Only when the ensemble output is larger than the cutoff, the firm can be classified as good or healthful firm. Let $A_i(w)$ be the associated accuracy of classification:

$$A_i(w) = \begin{cases} a_1 & \text{if } C_i^w = 0 \text{ and } C_i^s = 0, \\ a_2 & \text{if } C_i^w = 1 \text{ and } C_i^s = 1, \\ 0 & \text{otherwise.} \end{cases} \tag{9}$$

where C_i^w is the classification result of the ensemble classifier, C_i^s is the actual observed class of data sample itself, a_1 and a_2 are the Type I and Type II accuracy, respectively, whose definitions can be referred to Lai et al. (2006a, b, c).

The current problem is how to formulate an optimal combination of classifiers for ensemble prediction. A natural idea is to find the optimal combination of weights $w^* =(w_1^*, w_2^*, \cdots, w_p^*)$ by maximizing total classification accuracy including Type I and II accuracy. Usually, the classification accuracy can be estimated through k-fold cross-validation (CV) technique. With the principle of total classification accuracy maximization, the above problem can be summarized as an optimization problem:

$$(P)\begin{cases} \max_{w} A(w) = \sum_{i=1}^{M} A_i(w) \\ \text{s.t. } C_i^w = sign\left(\sum_{j=1}^{p} w_j c_{ij} - \theta\right), i = 1,2,\cdots M \\ A_i(w) = \begin{cases} a_1 & \text{if } C_i^w = 0 \text{ and } C_i^s = 0, \\ a_2 & \text{if } C_i^w = 1 \text{ and } C_i^s = 1, \\ 0 & \text{otherwise.} \end{cases} \end{cases} \tag{10}$$

where M is the size of cross-validation set and other symbols are similar to the above notations.

Since the constraint C_i^w is a nonlinear threshold function and the $A_i(w)$ is a step function, the optimization methods assuming differentiability of the objective function may have some problems. Therefore the above problem cannot be solved with classical optimization methods. For this reason, an evolutionary programming (EP) algorithm (Fogel, 1991) is proposed to solve the optimization problem indicated in (10) because EP is a useful method of optimization when other techniques such as gradient descent or direct, analytical discovery are not possible. For the above problem, the EP algorithm is described as follows:

(1) Create an initial set of L solution vectors $w_r = (w_{r1}, w_{r2}, \cdots, w_{rp}), r = 1, 2, \cdots, L$ for above optimization problems by randomly sampling the interval $[x, y]$, $x, y \in R$. Each population or individual w_r can be seen as a trial solution.

(2) Evaluate the objective function of each of the vectors $A(w_r)$. Here $A(w_r)$ is called as the fitness of w_r.

(3) Add a multivariate Gaussian vector $\Delta_r = N(0, G(A(w_r)))$ to the vector w_r to obtain $w_r' = w_r + \Delta_r$, where G is an appropriate monotone function. Re-evaluate $A(w_r')$. Here $G(A(w_r))$ is called as mutation rate and w_r' is called as an offspring of individual w_r.

(4) Define $\overline{w}_i = w_i, \overline{w}_{i+L} = w_i', i = 1, 2, \cdots, L, \overline{C} = \overline{w}_i, i = 1, 2, \cdots, 2L$. For every $\overline{w}_j$, $j = 1, 2, \cdots, 2L$, choose q vectors $\overline{w}_*$ from $\overline{C}$ at random. If $A(\overline{w}_j) > A(\overline{w}_*)$, assign $\overline{w}_j$ as a "winner".

(5) Choose the L individuals with more number of "winners" $w_i^*, i = 1, 2, \cdots, L$. If the stop criteria are not fulfilled, let $w_r = w_i^*, i = 1, 2, \cdots, L$, generation = generation +1 and go to step 2.

Using this EP algorithm, an optimal combination, w^*, of classifiers that maximizes the total classification accuracy is formulated. To verify the effectiveness of the proposed knowledge ensemble methodology, a real-world business risk dataset is used.

3   Research Data and Experiment Design

The research data used here is about UK corporate from the Financial Analysis Made Easy (FAME) CD-ROM database which can be found in the Appendix of Beynon and Peel (2001). It contains 30 failed and 30 non-failed firms. 12 variables are used as the firms' characteristics description:

(01) Sales
(02) ROCE: profit before tax/capital employed (%)
(03) FFTL: funds flow (earnings before interest, tax & depreciation)/total liabilities
(04) GEAR: (current liabilities + long-term debt)/total assets
(05) CLTA: current liabilities/total assets
(06) CACL: current assets/current liabilities

(07) QACL: (current assets − stock)/current liabilities

(08) WCTA: (current assets − current liabilities)/total assets

(09) LAG: number of days between account year end and the date the annual report and accounts were failed at company registry

(10) AGE: number of years the company has been operating since incorporation date.

(11) CHAUD: coded 1 if changed auditor in previous three years, 0 otherwise

(12) BIG6: coded 1 if company auditor is a Big6 auditor, 0 otherwise.

The above dataset is used to identify the two classes of business insolvency risk problem: failed and non-failed. They are categorized as "0" or "1" in the research data. "0" means failed firm and "1" represent non-failed firm. In this empirical test, 40 firms are randomly drawn as the training sample. Due to the scarcity of inputs, we make the number of good firms equal to the number of bad firms in both the training and testing samples, so as to avoid the embarrassing situations that just two or three good (or bad, equally likely) inputs in the testing sample. Thus the training sample includes 20 data of each class. This way of composing the sample of firms was also used by several researchers in the past, e.g., Altman (1968), Zavgren (1985) and Dimitras et al. (1999), among others. Its aim is to minimize the effect of such factors as industry or size that in some cases can be very important. Except from the above learning sample, the testing sample was collected using a similar approach. The testing sample consists of 10 failed and 10 non-failed firms. The testing data is used to test results with the data that is not utilized to develop the model.

In BPNN, this chapter varies the number of nodes in the hidden layer and stopping criteria for training. In particular, 6, 12, 18, 24, 32 hidden nodes are used for each stopping criterion because BPNN does not have a general rule for determining the optimal number of hidden nodes (Kim, 2003). For the stopping criteria of BPNN, this chapter allows 100, 500, 1000, 2000 learning epochs per one training example since there is little general knowledge for selecting the number of epochs. The learning rate is set to 0.15 and the momentum term is to 0.30. The hidden nodes use the hyperbolic tangent transfer function and the output node uses the same transfer function. This chapter allows 12 input nodes because 12 input variables are employed.

Similarly, the polynomial kernel and the Gaussian radial basis function are used as the kernel function of SVM. Tay and Cao (2001) showed that the margin parameter C and kernel parameter play an important role in the performance of SVM. Improper selection of these two parameters can cause the overfitting or the underfitting problems. This chapter varies the parameters to select optimal values for the best identification performance. Particularly, an appropriate range for σ^2 was set between 1 and 100, while the margin parameter C was set between 10 and 100, ccording to Tay and Cao (2001).

In addition, the identification performance is evaluated using the following criterion:

$$\text{Hit ratio} = \frac{1}{N}\sum_{i=1}^{N} R_i \tag{11}$$

where $R_i = 1$ if $IO_i = AO_i$; $R_i = 0$ otherwise. IO_i is the identification output from the model, and AO_i is the actual output, N is the number of the testing examples. To reflect model robustness, each class of experiment is repeated 20 times based on different samples and the final hit ratio accuracy is the average of the results of the 20 individual tests.

4 Experiment Results

Each of the identification models described in the previous section is estimated and validated by in-sample data and an empirical evaluation which is based on the in-sample and out-of-sample data. At this stage, the relative performance of the models is measured by hit ratio, as expressed by Eq. (11).

4.1 The Results of Individual Models

In terms of the previous data description and experiment design, the prediction performance of MDA and logit regression analysis model is summarized in Table 1. Note that the values in bracket are the standard deviation of 20 experiments.

Table 1. The identification results of MDA and logit regression models

Model	Identification Performance (%)	
	Training data	Testing data
MDA	86.45 [5.12]	69.50 [8.55]
LogR	89.81 [4.13]	75.82 [6.14]

As shown in Table 1, the logit regression analysis slightly outperforms MDA for both the training data and the testing data. The main reason leading to this effect may be that the logit regression can capture some nonlinear patterns hidden in the data.

For ANN models, Table 2 summarizes the results of three-layer BPNN according to experiment design.

As shown in Table 2, we can see that (1) the identification performance of training data tends to be higher as the learning epoch increases. The main reason is illustrated by Hornik et al. (1989), i.e., To make a three-layer BPNN approximate any continuous function arbitrarily well, a sufficient amount of middle-layer units must be given. (2) The performance of training data is consistently better than that of the testing data. The main reason is that the training performance is based on in-sample estimation and the testing performance is based on out-of-sample generalization. (3) The best prediction accuracy for the testing data was found when the epoch was 1000 and the number of hidden nodes was 24. The identification accuracy of the testing data turned out to be 77.67%, and that of the training data was 93.42%.

For SVM, there are two parameter, the kernel parameters σ and margin parameters C that need to be tuned. First, this chapter uses two kernel functions including the Gaussian radial basis function and the polynomial function. The polynomial function, however, takes a longer time in the training of SVM and provides worse results than the Gaussian radial basis function in preliminary test. Thus, this chapter uses the Gaussian radial function as the kernel function of SVM.

Table 2. The identification results of BPNN models with different designs

Learning epochs	Hidden nodes	Identification Performance (%)	
		Training data	Testing data
100	6	83.54 [7.43]	68.98 [8.12]
	12	84.06 [7.12]	67.35 [7.98]
	18	89.01 [6.45]	71.02 [8.21]
	24	86.32 [7.65]	70.14 [8.72]
	32	80.45 [8.33]	64.56 [8.65]
500	6	82.29 [7.21]	69.43 [8.27]
	12	85.64 [6.98]	70.16 [7.32]
	18	86.09 [7.32]	71.23 [8.01]
	24	88.27 [6.89]	74.80 [7.22]
	32	84.56 [7.54]	72.12 [8.11]
1000	6	87.54 [7.76]	70.12 [8.09]
	12	89.01 [7.13]	72.45 [7.32]
	18	90.53 [6.65]	75.45 [6.86]
	24	**93.42 [5.89]**	**77.67 [6.03]**
	32	91.32 [6.56]	74.91 [7.12]
2000	6	85.43 [7.76]	72.23 [8.09]
	12	87.56 [7.22]	70.34 [7.76]
	18	90.38 [7.67]	72.12 [8.45]
	24	89.59 [8.09]	71.50 [8.87]
	32	86.78 [7.81]	69.21 [8.02]

The chapter compares the identification performance with respect to various kernel and margin parameters. Table 3 presents the identification performance of SVM with various parameters according to the experiment design.

From Table 3, the best identification performance of the testing data is recorded when kernel parameter σ^2 is 10 and margin C is 75. Detailed speaking, for SVM model, too small a value for C caused under-fit the training data while too large a value of C caused over-fit the training data. It can be observed that the identification performance on the training data increases with C in this chapter. The identification performance on the testing data increases when C increase from 10 to 75 but decreases when C is 100. Similarly, a small value of σ^2 would over-fit the training data while a large value of σ^2 would under-fit the training data. We can find the identification performance on the training data and the testing data increases when σ^2 increase from 1 to 10 but decreases when σ^2 increase from 10 to 100. These results partly support the conclusions of Tay and Cao (2001) and Kim (2003).

Table 3. The identification performance of SVM with different parameters

C	σ^2	Identification Performance (%)	
		Training data	Testing data
10	1	83.52 [5.67]	67.09 [6.13]
	5	87.45 [6.12]	72.07 [6.66]
	10	89.87 [6.54]	74.21 [6.89]

Table 3. (*continued*)

	25	88.56 [6.33]	73.32 [6.76]
	50	84.58 [5.98]	69.11 [6.34]
	75	87.43 [6.57]	71.78 [7.87]
	100	82.43 [7.01]	67.56 [7.65]
25	1	84.58 [6.56]	72.09 [6.34]
	5	81.58 [7.01]	66.21 [7.33]
	10	89.59 [6.52]	70.06 [7.12]
	25	90.08 [6.11]	75.43 [6.65]
	50	92.87 [7.04]	77.08 [7.56]
	75	89.65 [6.25]	76.23 [6.71]
	100	87.43 [5.92]	74.29 [6.23]
50	1	85.59 [6.65]	72.21 [6.89]
	5	88.02 [6.67]	73.10 [6.92]
	10	89.98 [6.83]	74.32 [7.03]
	25	84.78 [6.95]	73.67 [7.32]
	50	89.65 [6.32]	76.35 [6.65]
	75	87.08 [6.47]	72.24 [6.67]
	100	85.12 [6.81]	71.98 [7.06]
75	1	87.59 [6.12]	74.21 [6.45]
	5	89.95 [6.56]	75.43 [6.78]
	10	**95.41 [6.09]**	**80.27 [6.17]**
	25	92.32 [6.48]	78.09 [6.75]
	50	93.43 [7.65]	77.61 [7.89]
	75	89.56 [6.13]	77.23 [6.69]
	100	88.41 [6.71]	74.92 [7.00]
100	1	88.12 [6.32]	75.47 [6.87]
	5	90.04 [6.87]	78.01 [7.21]
	10	93.35 [6.37]	79.17 [6.76]
	25	94.21 [6.98]	80.01 [6.29]
	50	91.57 [7.33]	78.23 [7.56]
	75	89.48 [6.56]	75.98 [7.12]
	100	88.54 [6.77]	74.56 [6.98]

4.2 Identification Performance of the Knowledge Ensemble

To formulate the knowledge ensemble models, two main strategies, majority voting and evolutionary programming ensemble strategy, are used. With four individual models, we can construct eleven ensemble models. Table 4 summarizes the results of different ensemble models. Note that the results reported in Table 4 are based on the testing data.

As can be seen from Table 4, it is not hard to find that (1) generally speaking, the EP- based ensemble strategy is better than the majority voting based ensemble strategy, revealing that the proposed EP-based ensemble methodology can give a promising results for business risk identification. (2) The identification accuracy increases as the number of ensemble member increase. The main reason is that different models may contain different useful information. However, the number of ensemble member

Table 4. Identification performance of different knowledge ensemble models

Ensemble model	Ensemble member	Ensemble Strategy	
		Majority voting	EP
E1	MDA+LogR	74.23 [6.12]	76.01 [5.67]
E2	MDA+ANN	75.54 [6.74]	77.78 [6.19]
E3	MDA+SVM	77.91 [5.98]	80.45 [6.03]
E4	LogR+ANN	78.76 [5.65]	79.87 [5.48]
E5	LogR+SVM	79.89 [6.33]	81.25 [5.92]
E6	ANN+SVM	81.87 [5.76]	84.87 [5.16]
E7	MDA+LogR+ANN	78.98 [5.87]	80.34 [6.42]
E8	MDA+LogR+SVM	80.67 [5.54]	83.56 [5.77]
E9	MDA+ANN+SVM	82.89 [6.12]	85.05 [6.22]
E10	LogR+ANN+SVM	85.01 [5.79]	**88.09 [5.56]**
E11	MDA+LogR+ANN+SVM	**85.35 [5.51]**	86.08 [6.78]

does not satisfy the principle of "the more, the better". For example, ensemble model E6 performs better than the ensemble models E7 and E8 for majority voting strategy while the ensemble model E11 is also worse than the ensemble model E10 for EP-based ensemble strategy.

4.3 Identification Performance Comparisons

Table 5 compares the best identification performance of four individual models (MDA, LogR, BPNN, and SVM) and two best ensemble models (E11 for majority voting based ensemble and E10 for EP-based ensemble) in terms of the training data and testing data. Similar to the previous results, the values in bracket are the standard deviation of 20 tests.

From Table 5, several important conclusions can be drawn. First of all, the ensemble models perform better than the individual models. Second, of the four individual models, the SVM model is the best, followed by BPNN, LogR and MDA, implying that the intelligent models outperform the traditional statistical models. Third, of the two knowledge ensemble models, the performance of the EP-based ensemble models is better than that of the majority-voting-based ensemble models, implying that the intelligent knowledge ensemble model can generate good identification performance.

In addition, we conducted McNemar test to examine whether the intelligent knowledge ensemble model significantly outperformed the other several models listed in this chapter. As a nonparametric test for two related samples, it is particularly useful

Table 5. Identification performance comparison with different models

Data	Individual Models				Ensemble Models	
	MDA	LogR	BPNN	SVM	Majority Voting	EP
Training data	86.45 [5.12]	89.81 [4.13]	93.42 [5.89]	95.41 [6.09]	95.71 [5.23]	98.89 [5.34]
Testing Data	69.50 [8.55]	75.82 [6.14]	77.67 [6.03]	80.27 [6.17]	85.35 [5.51]	88.09 [5.56]

Table 6. McNemar values (p values) for the performance pairwise comparison

Models	Majority	SVM	BPNN	LogR	MDA
EP Ensemble	1.696 (0.128)	3.084 (0.092)	4.213 (0.059)	5.788 (0.035)	7.035 (0.009)
Majority		1.562 (0.143)	2.182 (0.948)	5.127 (0.049)	6.241 (0.034)
SVM			1.342 (0.189)	1.098 (0.235)	3.316 (0.065)
BPNN				0.972 (0.154)	1.892 (0.102)
LogR					0.616 (0.412)

for before-after measurement of the same subjects (Kim, 2003; Cooper and Emory, 1995). Table 6 shows the results of the McNemar test to statistically compare the identification performance for the testing data among six models.

As revealed in Table 6, the EP-based knowledge ensemble model outperforms four individual models at 10% significance level. Particularly, the EP-based knowledge ensemble model is better than SVM and BPNN at 10% level, logit regression analysis model at 5% level, and MDA at 1% significance level, respectively. However, the EP-based knowledge ensemble model does not significantly outperform majority voting based knowledge ensemble model. Similarly, the majority voting based knowledge ensemble model outperforms the BPNN and two statistical models (LogR and MDA) at 10% and 5% significance level. Furthermore, the SVM is better than the MDA at 5% significance level. However, the SVM does not outperform the BPNN and LogR, which is consistent with Kim (2003). In addition, Table 9.6 also shows that the identification performance among BPNN, Logit regression analysis and MDA do not significantly differ each other.

5 Conclusions

In this chapter, we propose an intelligent knowledge ensemble model to identify business insolvency risk. First of all, some individual data mining and knowledge discovery algorithms are used to explore the implied knowledge hidden in the business dataset. Then an evolutionary programming based knowledge ensemble model is formulated in terms of the results produced by the individual models. As demonstrated in our empirical analysis, the proposed intelligent knowledge ensemble methodology is superior to the individual models and other ensemble methods in identifying the healthy condition of the business firms. This is a clear message for financial institutions and investors, which can lead to a right decision making. Therefore, this chapter also concluded that the proposed intelligent knowledge ensemble model provides a promising alternative solution to business insolvency risk identification.

Acknowledgements

This work described here is partially supported by the grants from the National Natural Science Foundation of China (NSFC No. 70221001, 70601029), the Chinese Academy of Sciences (CAS No. 3547600), the Academy of Mathematics and Systems Sciences (AMSS No. 3543500) of CAS, and the Strategic Research Grant of City University of Hong Kong (SRG No. 7001806).

References

Ahn, B.S., Cho, S.S., Kim, C.Y.: The integrated methodology of rough set theory and artificial neural network for business failure prediction. Expert Systems with Application 18, 65–74 (2000)

Altman, E.I.: Financial ratios, discriminant analysis and the prediction of corporate bankruptcy. The Journal of Finance 23, 589–609 (1968)

Altman, E.I.: The success of business failure prediction models: An international survey. Journal of Banking and Finance 8, 171–198 (1984)

Altman, E.I.: Corporate Financial Distress and Bankruptcy. John Wiley, New York (1993)

Altman, E.I., Marco, G., Varetto, F.: Corporate distress diagnosis: Comparison using discriminant analysis and neural networks (the Italian experience). Journal of Banking and Finance 18, 505–529 (1994)

Beaver, W.H.: Financial ratios as predictors of failure. Journal of Accounting Research 4, 71–111 (1966)

Beynon, M.J., Peel, M.J.: Variable precision rough set theory and data discretisation: An application to corporate failure prediction. Omega 29, 561–576 (2001)

Cooper, D.R., Emory, C.W.: Business Research Methods. Irwin, Chicago (1995)

Courtis, J.K.: Modeling a financial ratios categoric framework. Journal of Business Finance and Accounting 5, 371–386 (1978)

Dimitras, A.I., Slowinski, R., Susmaga, R., Zopounidis, C.: Business failure prediction using rough sets. European Journal of Operational Research 114, 263–280 (1999)

Dimitras, A.I., Zanakis, S.H., Zopounidis, C.: A survey of business failures with an emphasis on prediction methods and industrial applications. European Journal of Operational Research 90, 487–513 (1996)

Dimitras, A.I., Zopounidis, C., Hurson, C.H.: A multicriteria decision aid method for the assessment of business failure risk. Foundations of Computing and Decision Sciences 20, 99–112 (1995)

Eisenbeis, R.A.: Pitfalls in the application of discriminant analysis in business and economics. The Journal of Finance 32, 875–900 (1977)

Fogel, D.B.: System Identification through Simulated Evolution: A Machine Learning Approach to Modeling. Ginn Press, Needham, MA (1991)

Frydman, H., Altman, E.I., Kao, D.L.: Introducing recursive partitioning for financial classification: The case of financial distress. The Journal of Finance 40, 269–291 (1985)

Gupta, Y.P., Rao, R.P., Bagghi, P.K.: Linear goal programming as an alternative to multivariate discriminant analysis: A note. Journal of Business Finance and Accounting 17, 593–598 (1990)

Hair, J.F., Anderson, R.E., Tatham, R.E., Black, W.C.: Multivariate Data Analysis with Readings. Prentice Hall, Englewood Cliffs, NJ (1998)

Hornik, K., Stinchcombe, M., White, H.: Multilayer feedforward networks are universal approximators. Neural Networks 2, 359–366 (1989)

Jones, F.L.: Current techniques in bankruptcy prediction. Journal of Accounting Literature 6, 131–164 (1987)

Keasey, K., Watson, R.: Financial distress prediction models: a review of their usefulness. British Journal of Management 2, 89–102 (1991)

Kim, K.J.: Financial time series forecasting using support vector machines. Neurocomputing 55, 307–319 (2003)

Lai, K.K., Yu, L., Huang, W., Wang, S.Y.: A novel support vector machine metamodel for business risk identification. In: Yang, Q., Webb, G. (eds.) PRICAI 2006. LNCS (LNAI), vol. 4099, pp. 980–984. Springer, Heidelberg (2006a)

Lai, K.K., Yu, L., Wang, S.Y., Zhou, L.G.: Neural network metalearning for credit scoring. In: Huang, D.-S., Li, K., Irwin, G.W. (eds.) ICIC 2006. LNCS, vol. 4113, pp. 403–408. Springer, Heidelberg (2006b)

Lai, K.K., Yu, L., Wang, S.Y., Zhou, L.G.: Credit risk analysis using a reliability-based neural network ensemble model. In: Kollias, S., Stafylopatis, A., Duch, W., Oja, E. (eds.) ICANN 2006. LNCS, vol. 4132, pp. 682–690. Springer, Heidelberg (2006c)

Lai, K.K., Yu, L., Zhou, L.G., Wang, S.Y.: Credit risk evaluation with least square support vector machine. In: Wang, G.-Y., Peters, J.F., Skowron, A., Yao, Y. (eds.) RSKT 2006. LNCS (LNAI), vol. 4062, pp. 490–495. Springer, Heidelberg (2006d)

Lee, K., Booth, D., Alam, P.: A comparison of supervised and unsupervised neural networks in predicting bankruptcy of Korean firms. Expert Systems with Applications 29, 1–16 (2005)

Luoma, M., Laitinen, E.K.: Survival analysis as a tool for company failure prediction. Omega 19, 673–678 (1991)

Messier, W.F., Hansen, J.V.: Including rules for expert system development: An example using default and bankruptcy data. Management Science 34, 1403–1415 (1988)

Ohlson, J.A.: Financial ratios and the probabilistic prediction of bankruptcy. Journal of Accounting Research 3, 109–131 (1980)

Olmeda, I., Fernandez, E.: Hybrid classifiers for financial multicriteria decision making: The case of bankruptcy prediction. Computational Economics 10, 317–335 (1997)

Piramuthu, S.: Financial credit-risk evaluation with neural and neurofuzzy systems. European Journal of Operational Research 112, 310–321 (1999)

Scott, J.: The probability of bankruptcy: A comparison of empirical predictions and theoretical models. Journal of Banking and Finance 5, 317–344 (1981)

Shin, K.S., Lee, T.S., Kim, H.J.: An application of support vector machines in bankruptcy prediction model. Expert Systems with Applications 28, 127–135 (2005)

Siskos, Y., Zopounidis, C., Pouliezos, A.: An integrated DSS for financing firms by an industrial development bank in Greece. Decision Support Systems 12, 151–168 (1994)

Tay, F.E.H., Cao, L.J.: Application of support vector machines in financial time series forecasting. Omega 29, 309–317 (2001)

Vapnik, V.N.: The Nature of Statistical Learning Theory. Springer, New York (1995)

Vapnik, V.N.: Statistical Learning Theory. Wiley, New York (1998)

Varetto, F.: Genetic algorithms applications in the analysis of insolvency risk. Journal of Banking and Finance 2, 1421–1439 (1998)

Vermeulen, E.M., Spronk, J., van der Wijst, N.: The application of the multi-factor model in the analysis of corporate failure. In: Zopounidis, C. (ed.) Operational Tools in the Management of Financial Risks, pp. 59–73. Kluwer Academic Publisher, Dordrecht (1998)

White, H.: Connectionist nonparametric regression: Multilayer feedforward networks can learn arbitrary mappings. Neural Networks 3, 535–549 (1990)

Yang, S., Browne, A.: Neural network ensembles: Combining multiple models for enhanced performance using a multistage approach. Expert Systems 21, 279–288 (2004)

Yu, L., Wang, S.Y., Lai, K.K.: A novel nonlinear ensemble forecasting model incorporating GLAR and ANN for foreign exchange rates. Computers and Operations Research 32, 2523–2541 (2005)

Zavgren, C.V.: The prediction of corporate failure: The state of the art. Journal of Financial Literature 2, 1–37 (1983)

Zavgren, C.V.: Assessing the vulnerability to failure of American industrial firms: A logistic analysis. Journal of Business Finance and Accounting 12, 19–45 (1985)

Zmijewski, M.E.: Methodological issues related to the estimation of financial distress prediction models. Studies on Current Econometric Issues in Accounting Research, 59–82 (1984)

Zopounidis, C.: A multicriteria decision making methodology for the evaluation of the risk of failure and an application. Foundations of Control Engineering 12, 45–67 (1987)

The Application of Fuzzy Decision Trees in Company Audit Fee Evaluation: A Sensitivity Analysis

Malcolm J. Beynon

Cardiff Business School, Cardiff University, Colum Drive, Cardiff, CF10 3EU, Wales, UK
BeynonMJ@Cardiff.ac.uk

Abstract. This chapter investigates the appropriateness of the application of fuzzy decision trees on the evaluation of company audit fees, with attention to the sensitivity of the results. With the rudiments of fuzzy decision trees in a fuzzy environment, it implies a linguistic emphasis on the concomitant analysis, allowing readability in the fuzzy decision rules constructed. Two processes for the construction of membership functions (MFs) used to define the linguistic terms characterising the linguistic variables considered allowing the impact of considering alternative MFs. The tutorial fuzzy decision tree analysis clearly allows the construction processes to be exposed.

1 Introduction

This chapter considers fuzzy decision trees, one area of data analysis that has benefited from its development in a fuzzy environment. The general methodology of fuzzy set theory (FST), introduced in Zadeh (1965), includes the opportunities to develop techniques that incorporate vagueness and ambiguity in their operation. It is therefore closely associated with the notion of soft computing through the prevalence of imprecision and uncertainty, as well as its close resemblance with human like decision-making (Pal and Ghosh, 2004).

Decision trees, within crisp and fuzzy environments, are concerned with the classification of objects described by a number of attributes. The general tree structure starts with a root node, from which paths emanate down to leaf nodes, which offer levels of classification of objects to defined decisions/outcomes. An early fuzzy decision tree reference is attributed to Chang and Pavlidis (1977), with developments continually introduced in the research and applied domains (see for example, Abdel-Galil *et al.*, 2005; Wang *et al.*, 2007).

An important feature of fuzzy decision trees is the concomitant sets of fuzzy '*if* .. *then* ..' decision rules constructed, found from progressing down the individual paths in the tree to the leaf nodes. It has been suggested that the comprehensibility of fuzzy decision rules viewed by human users is an important criterion in designing a fuzzy rule-based system (Ishibuchi *et al.*, 2001). The fuzziness associated with these rules brings with it the pertinent qualities of readability and interpretability to any analysis, offering an efficient insight into the considered problem, a further facet of soft computing (Bodenhofer *et al.*, 2007).

The suggested fuzzy approach employed here was presented in Yuan and Shaw (1995) and Wang *et al.* (2000), and attempts to include the cognitive uncertainties evident in the attributes describing the attributes. Moreover, cognitive uncertainty,

B. Prasad (Ed.): Soft Computing Applications in Business, STUDFUZZ 230, pp. 73–91, 2008.
springerlink.com   © Springer-Verlag Berlin Heidelberg 2008

unlike its statistical counterpart, is the uncertainty that deals with phenomena arising from human thinking, reasoning, cognition and perception processes, or cognitive information in general (see Yuan and Shaw, 1995). Rudimentary to the ability to consider fuzzy decision trees is the *a priori* construction of fuzzy membership functions (MFs), which enable the linguistic representation of the numerical attributes describing the objects (Ishibuchi *et al.*, 2001; Kecman, 2001).

The example business application considered here concerns the discernment of the levels of audit fees outlaid by companies, based on certain financial and non-financial attributes (Yardley *et al.*, 1992; Chan *et al.*, 1993; Beynon *et al.*, 2004). The use of fuzzy decision trees here is pertinent since the linguistic oriented fuzzy decision rules constructed open up the ability to understand the relationship between a company's attributes and their predicted level of audit fees.

A feature in this chapter is an attempt to elucidate the effect on a fuzzy decision tree analysis when there are changes to the MFs associated with the attributes describing the objects. Moreover, the prescribed sensitivity analysis included here looks at the variation in the constructed fuzzy decision trees (and subsequent fuzzy decision rules), using relatively naïve and more informative approaches to the MF construction process *a priori* to the fuzzy decision tree analysis (Beynon *et al.*, 2004).

The contribution of this chapter to the reader is the clear understanding of the advantages of the utilization of fuzzy decision trees in a business related application. A small hypothetical example, also included, allows the reader the opportunity to clearly follow the included analytical rudiments of a fuzzy decision tree analysis. The larger, previously described, application based analysis demonstrates the interpretability allowed through the use of this approach along with the concomitant technical sensitivity.

2 Background on Fuzzy Decision Trees

The domain of a fuzzy decision tree analysis is a universe of objects, described and classified by condition and decision attributes, respectively. As such this kind of analysis approach is included in the general theme of object classification techniques; see Breiman (2001) for a general discussion.

Central to the utilization of fuzzy set theory (Zadeh, 1965) in decision tree based analyses is the finite set of membership functions (MFs), linguistically defining a numerical attribute's domain, both condition and decision attributes in this case. This allows an attribute to be subsequently denoted a linguistic variable (Herrera *et al.*, 2000). The number of descriptors (MFs) in a linguistic term set (also called fuzzy label in this chapter), which define a linguistic variable, determines the granularity of the characterization of an attribute's domain.

The definition, in terms of structure, of MFs is a hurdle that requires solving. Indeed, it is a research issue in itself (see for example, Grauel and Ludwig 1999; Hu, 2006), with here piecewise linear MFs adopted (Dombi and Gera, 2005). A visual representation of a piecewise linear MF $\mu(\cdot)$, the type utilised here, with its *defining values* $[\alpha_{j,1}, \alpha_{j,2}, \alpha_{j,3}, \alpha_{j,4}, \alpha_{j,5}]$, is presented in Figure 1.

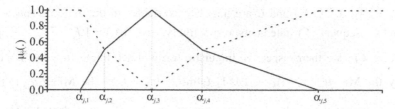

Fig. 1. General definition of a linear piecewise MF (including use of defining values)

In Figure 1, how a linear piecewise MF is structured through the concomitant defining values is clearly shown. The support associated with the MF is where $\mu(x) > 0$, shown to be over the domain $\alpha_{j,1} \le x \le \alpha_{j,5}$. Further, its majority support, where $\mu(x) > 0.5$ is given by the interval $[\alpha_{j,2}, \alpha_{j,4}]$. The dotted lines also shown in Figure 1 illustrate its neighbouring MFs, which collectively would form a linguistic variable describing the variable x.

A small example series of MFs is next presented which demonstrates the fuzzy modelling of a numerical attribute into a linguistic variable, see Figure 2.

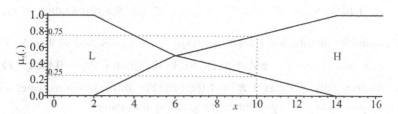

Fig. 2. Example of linguistic variable defined by two MFs (labelled L and H)

The two MFs presented in Figure 2 are described by the respective sets of defining values, $[-\infty, -\infty, 2, 6, 14]$ for $\mu_L(x)$ and $[2, 6, 14, \infty, \infty]$ for $\mu_H(x)$. It follows, any x value can be described by its levels of membership to the linguistic terms, L (low) and H (High), evaluated using $\mu_L(x)$ and $\mu_H(x)$. In Figure 2 also, the example linguistic variable expression of an x value is given, for $x = 10$, $\mu_L(x) = 0.25$ and $\mu_H(x) = 0.75$, indicating its majority support with the linguistic term H. It is linguistic variables like those described in Figure 1, but described with three MFs (linguistic terms) that are used in the fuzzy decision tree analyses later described.

A detailed description on the current work relating to fuzzy decision trees is presented in Olaru and Wehenkel (2003). Recent examples of their successful application include in the areas of; optimizing economic dispatch (Roa-Sepulveda *et al.*, 2003), medical decision support (Chiang *et al.*, 2005) and text-to-speech synthesis (Odéjobí et al., 2007).

Following the approach first outlined in Yuan and Shaw (1995) and later references, the underlying knowledge within a fuzzy decision tree, related to a decision outcome C_j can be represented as a set of fuzzy '*if*.. *then* ..' decision rules, each of the form:

$$\text{If } (A_1 \text{ is } T_{i_1}^1) \text{ and } (A_2 \text{ is } T_{i_2}^2) \ldots \text{ and } (A_k \text{ is } T_{i_k}^k) \text{ then } C \text{ is } C_j,$$

where $A = \{A_1, A_2, .., A_k\}$ and C are linguistic variables in the multiple antecedents (A_i's) and consequent (C) statements, respectively, and $T(A_k) = \{T_1^k, T_2^k, .. T_{S_i}^k\}$ and $\{C_1, C_2, ..., C_L\}$ are their respective linguistic terms. Each linguistic term T_j^k is defined by the MF $\mu_{T_j^k}(x): A_k \to [0, 1]$ (similar for a C_j). The MFs, $\mu_{T_j^k}(x)$ and $\mu_{C_j}(y)$, represent the grade of membership of an object's antecedent A_j being T_j^k and consequent C being C_j, respectively.

A fuzzy set A in a universe of discourse U is characterized by a membership function μ_A which takes values in the interval $[0, 1]$, as described in Figures 1 and 2 (including their description as linguistic terms). For all $u \in U$, the intersection $A \cap B$ of two fuzzy sets is given by $\mu_{A \cap B} = \min[\mu_A(u), \mu_B(u)]$. A membership function $\mu(x)$ from the set describing a fuzzy linguistic term Y defined on X, can be viewed as a possibility distribution of Y on X, that is $\pi(x) = \mu(x)$, for all $x \in X$ (also normalized so $\max_{x \in X} \pi(x) = 1$).

The possibility measure $E_\alpha(Y)$ of ambiguity is defined by $E_\alpha(Y) = g(\pi) = \sum_{i=1}^{n} (\pi_i^* - \pi_{i+1}^*) \ln[i]$, where $\pi^* = \{\pi_1^*, \pi_2^*, ..., \pi_n^*\}$ is the permutation of the normalized possibility distribution $\pi = \{\pi(x_1), \pi(x_2), ..., \pi(x_n)\}$, sorted so that $\pi_i^* \geq \pi_{i+1}^*$ for $i = 1, .., n$, and $\pi_{n+1}^* = 0$ (see Zadeh, 1978). In the limit, if $\pi_2^* = 0$ then $E_\alpha(Y) = 0$ indicates no ambiguity, whereas if $\pi_n^* = 1$ then $E_\alpha(Y) = \ln[n]$, which indicates all values are fully possible for Y, representing the greatest ambiguity.

The ambiguity of attribute A, over the objects $u_1, ..., u_m$, is given as, $E_\alpha(A) = \frac{1}{m} \sum_{i=1}^{m} E_\alpha(A(u_i))$, where $E_\alpha(A(u_i)) = g(\mu_{T_s}(u_i) / \max_{1 \leq j \leq s}(\mu_{T_s}(u_i)))$, with $T_1, ..., T_s$ the linguistic terms of an attribute (antecedent) with m objects. When there is overlapping between linguistic terms (MFs) of an attribute, then ambiguity exists. The fuzzy subsethood $S(A, B)$ measures the degree to which A is a subset of B, and is given by, $S(A, B) = \sum_{u \in U} \min(\mu_A(u), \mu_B(u)) / \sum_{u \in U} \mu_A(u)$.

Given fuzzy evidence E, the possibility of classifying an object to a consequent C_i can be defined as, $\pi(C_i|E) = S(E, C_i) / \max_j S(E, C_j)$, where $S(E, C_i)$ represents the degree of truth for the associated classification rule ('if E then C_i'). With a single piece of evidence (a fuzzy number from an attribute describing an object), then the classification ambiguity based on this fuzzy evidence is defined as, $G(E) = g(\pi(C|E))$, which is measured based on the possibility distribution $\pi(C|E) = (\pi(C_1|E), ..., \pi(C_L|E))$.

The classification ambiguity with fuzzy partitioning $P = \{E_1, ..., E_k\}$ on the fuzzy evidence F, denoted as $G(P|F)$, is the weighted average of classification ambiguity with each subset of partition, $G(P|F) = \sum_{i=1}^{k} w(E_i|F)G(E_i \cap F)$, where $G(E_i \cap F)$ is the classification ambiguity with fuzzy evidence $E_i \cap F$, and where $w(E_i|F)$ is the

weight which represents the relative size of subset $E_i \cap F$ in F, $w(E_i| F) =$
$$\sum_{u \in U} \min(\mu_{E_i}(u), \mu_F(u)) \Big/ \sum_{j=1}^{k} \left(\sum_{u \in U} \min(\mu_{E_j}(u), \mu_F(u)) \right).$$

To summarize the underlying process for fuzzy decision tree construction, the attribute associated with the root node of the tree is that which has the lowest classification ambiguity $(G(E))$. Attributes are assigned to nodes down the tree based on which has the lowest level of classification ambiguity $(G(P| F))$. A node becomes a leaf node if the level of subsethood $(S(E, C_i))$ associated with the evidence down the path, is higher than some truth value β assigned across the whole of the fuzzy decision tree. Alternatively a leaf node is created when no augmentation of an attribute improves the classification ambiguity associated with that down the tree. The classification from a leaf node is to the decision outcome with the largest associated subsethood value.

The truth level threshold β controls the growth of the tree; lower β may lead to a smaller tree (with lower classification accuracy), higher β may lead to a larger tree (with higher classification accuracy). The construction process can also constrain the effect of the level of overlapping between linguistic terms that may lead to high classification ambiguity. Moreover, evidence is strong if its membership exceeds a certain significant level, based on the notion of α-cuts (Yuan and Shaw, 1995).

To demonstrate the prescribed construction process of a fuzzy decision tree, a small example data set is considered, consisting of three objects, u_1, u_2 and u_3, making up the universe of discourse, described by the two attributes, T1 and T2 (each with three fuzzy labels L, M and H) and one decision attribute C (with two fuzzy labels L and H), see Table 1.

Table 1. Small example fuzzy data set

Object	T1 = [T1$_L$, T1$_M$, T1$_H$]	T2 = [T2$_L$, T2$_M$, T2$_H$]	C = [C$_L$, C$_H$]
u_1	[0.1, 0.3, **0.6**]	[0.2, **0.6**, 0.2]	[0.3, **0.7**]
u_2	[0.3, **0.4**, 0.3]	[**0.6**, 0.4, 0.0]	[0.3, **0.7**]
u_3	[**0.5**, 0.2, 0.3]	[0.0, 0.4, **0.6**]	[**0.8**, 0.2]

Using the information presented in Table 1, for the construction of a fuzzy decision tree, the classification ambiguity of each condition attribute with respect to the decision attributes is first considered, namely the evaluation of the $G(E)$ values. Prior to undertaking this, a threshold value of $\beta = 0.9$ was used throughout this construction process, required for a node to be defined a leaf node.

The evaluation of a $G(E)$ value is shown for the first attribute T1 $(= g(\pi(C| T1)))$, where it is broken down to the fuzzy labels L, M and H, so for L; $\pi(C| T1_L) =$ $S(T1_L, C_i)/ \max_j S(T1_L, C_j)$, considering C_L and C_H with the information in Table 1:

$$S(T1_L, C_L) = \sum_{u \in U} \min(\mu_{T1_L}(u), \mu_{C_L}(u)) \Big/ \sum_{u \in U} \mu_{T1_L}(u)$$
$$= (0.1 + 0.3 + 0.5)/0.9 = 0.9/0.9 = 1.000,$$

whereas $S(T1_L, C_H) = 0.667$. Hence $\pi = \{0.667, 1.000\}$, giving $\pi^* = \{1.000, 0.667\}$, with $\pi_3^* = 0$, then:

$$G(T1_L) = g(\pi(C| T1_L)) = \sum_{i=1}^{2}(\pi_i^* - \pi_{i+1}^*)\ln[i]$$
$$= (1.000 - 0.667)\ln[1] + (0.667 - 0.000)\ln[2] = 0.462,$$

with $G(T1_M) = 0.616$ and $G(T1_H) = 0.567$, then $G(T1) = (0.462 + 0.616 + 0.567)/3 = 0.548$. Compared with $G(T2) = 0.452$, it follows the T2 attribute, with less associated classification ambiguity, forms the root node for the desired fuzzy decision tree. The subsethood values in this case are; for T2: $S(T2_L, C_L) = 0.625$ and $S(T2_L, C_H) = \mathbf{1.000}$; $S(T2_M, C_L) = 0.714$ and $S(T2_M, C_H) = \mathbf{0.857}$; $S(T2_H, C_L) = \mathbf{1.000}$ and $S(T2_H, C_H) = 0.500$.

For each path from the root node, the larger subsethood value (given in bold) is the possible resultant node's classification. For the paths, $T2_L$ and $T2_H$ these are to C_H and C_L respectively, with largest subsethood values above the desired truth value of 0.9 (= β previously defined). In the case of $T2_M$, its largest subsethood value is 0.857 (for $S(T2_M, C_H)$), below the $\beta = 0.9$ value, hence is determined not able to be a leaf node, so further possible augmentation needed to be considered.

With only two attributes in the example data set, the possible augmentation from $T2_M$ is through T1, since $G(T2_M) = 0.578$, the classification ambiguity with partition evaluated for T1 ($G(T2_M \text{ and } T1| C)$) has to be less than this value, where;

$$G(T2_M \text{ and } T1| C) = \sum_{i=1}^{k} w(T1_i | T2_M)G(T2_M \cap T1_i).$$

Starting with the weight values, in the case of $T2_M$ and $T1_L$, it follows $w(T1_L| T2_M) = 0.8/2.9 = 0.276$, similarly $w(T1_M| T2_M) = 0.310$ and $w(T1_H| T2_M) = 0.414$. Hence $G(T2_M \text{ and } T1| C) = 0.569$, which is lower than the concomitant $G(T2_M) = 0.578$ value so less ambiguity would be found if the T1 attribute was augmented to the T2 = M path. The subsequent subsethood values in this case for each new path are; $T1_L$; $S(T2_M \cap T1_L, C_L) = \mathbf{1.000}$ and $S(T2_M \cap T1_L, C_H) = 0.750$; $T1_M$: $S(T2_M \cap T1_M, C_L) = 0.889$ and $S(T2_M \cap T1_M, C_H) = \mathbf{1.000}$; $T1_H$: $S(T2_M \cap T1_H, C_L) = 0.750$ and $S(T2_M \cap T1_H, C_H) = \mathbf{0.917}$. With each path the largest subsethood value is above the truth level threshold of 0.9, so a completed tree has been achieved.

The resultant fuzzy decision tree in this case is presented in Figure 3.

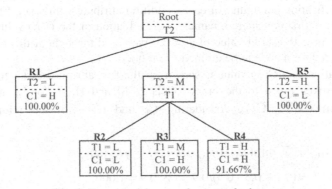

Fig. 3. Fuzzy decision tree for example data set

3 Audit Fees and Fuzzy Decision Tree Analyses

The main thrust of this chapter is the application of the previously described fuzzy decision tree approach to modelling the predicted classification of levels of audit fees on companies (see Beynon *et al.*, 2004). Technical details included here surround the employment of two different approaches to the construction of certain membership functions (MFs), used in the linguistic variables that themselves replace the attributes describing the companies in the audit fee problem.

Motivation for examining audit fees is that there is a well developed theoretical and empirical literature relating to their determinants (Chan *et al.*, 1993). Hence the data analysed, which includes financial ratios, is typical of that used in accounting and business applications (see Beynon *et al.*, 2004). Here 1,000 companies are considered, with each company described by a number of independent financial and non-financial condition attributes, see Table 2.

Table 2. Description of condition and decision attributes in audit fee problem

Attribute	Label	Definition	Correlation with ADT
c_1	TRN	Turnover (£1,000s)	0.349
c_2	PBT	Profit (Loss) before Taxation (£1,000s)	−0.071
c_3	NTA	Net Tangible Assets (Liab.) (£1,000s)	0.230
c_4	GRN	Gearing Ratio (%)	0.028
c_5	NOE	Number of Employees	0.324
c_6	LQD	Liquidity Ratio (%)	0.068
c_7	SHF	Shareholders Funds (£1,000s)	0.269
c_8	ROC	Return on Capital Employed (%)	−0.080
c_9	NAF	Non-Audit Fee (£1,000s)	0.573
c_{10}	AGE	Age	0.039
d_1	ADT	Audit fee (£1,000s)	-

In Table 2, ten condition attributes are reported, which are all continuous in nature, along with the decision attribute of audit fee (ADT). Accompanying their descriptions are their respective Pearson correlation coefficients values, which offer an early (recognised) indication of the strength of association of a condition attribute with ADT. A brief inspection identifies the level of non-audit fees (NAF) for a company is most associated with ADT (has highest correlation value), followed by TRN and NOE, then the others.

For the audit fee problem, three linguistic terms (MFs) are used to describe each attribute, condition and decision. It follows, there are three intervals used to identify an initial partitioning of the companies based on their levels of ADT, defined in linguistic terms, as; low (L), medium (M) and high (H). An equal-frequency partitioning of the companies is employed to initially intervalise them, with the midpoint between neighbouring, ordered, ADT values in different interval groups identified and used as the interval boundary points, see Table 3.

Table 3. Relevant information on the three groups of companies based on their ADT values

	Low (L)	Medium (M)	High (H)
Interval	L ≤ 11.014 (358)	11.014 < M ≤ 18.040 (329)	18.040 < H (313)
Mean	7.478	14.454	31.623

Along with the interval boundary values presented in Table 3, in parentheses are the actual number of companies included in each defined interval (supporting the 'near' equal-frequency approach adopted, affected by multiplicity). To illustrate the presented information, for a medium level of audit fees, its interval domain is between 11.014 and 18.040 (£1,000s), for which there are 329 companies associated with this interval (ADT = M). These interval boundary values, and the within group means also shown in the table are used as the defining values in the construction of the associated MFs for the decision attribute ADT.

From Figure 1, for a set of defining values $[\alpha_{j,1}, \alpha_{j,2}, \alpha_{j,3}, \alpha_{j,4}, \alpha_{j,5}]$, $\alpha_{j,3}$ is the modal value of the MF associated with the jth interval (single value where the MF equals one), here $\alpha_{M,3}$ = 14.454 (using the information in Table 3). The neighbouring defining values, $\alpha_{j,2}$ and $\alpha_{j,4}$, around $\alpha_{j,3}$, are simply the left and right boundary values of that interval, here $a_{M,2}$ = 11.014 and $a_{M,4}$ = 18.040 (when $j = 1$ or k then the respective $\alpha_{j,2}$ and $\alpha_{j,4}$ are special cases). The final outer defining values, $\alpha_{j,1}$ and $\alpha_{j,5}$, are the modal defining values of its neighbouring intervals, here $\alpha_{L,3}$ = $\alpha_{M,1}$ = 7.478 and $\alpha_{M,5}$ = $\alpha_{H,3}$ = 31.623. From these definitions of the defining values, the associated linear piecewise MFs, defining the linguistic terms describing ADT, can be constructed, see Figure 3.

The MFs graphically represented in Figure 4 (similar to those described in Figures 1 and 2), show the association of an ADT value of a company to the three linguistic terms, low (L), medium (M) and high (H). For example, when ADT = 10.000 (£1,000s), $S_{ADT}(10.000) = \mu_{ADT,L}(10.000) + \mu_{ADT,M}(10.000) + \mu_{ADT,H}(10.000) = 0.643 + 0.357 + 0.000 = 1.000$.

From Table 2, all ten of the considered condition attributes are continuous in nature, and so they require sets of three MFs to be constructed. Here, the same approach as for the decision attribute ADT is followed, to create these sets of MFs, see Figure 5.

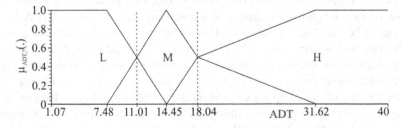

Fig. 4. Graphical representation of a set of MFs, defining the linguistic terms associated with the decision attribute ADT (using details in Table 3)

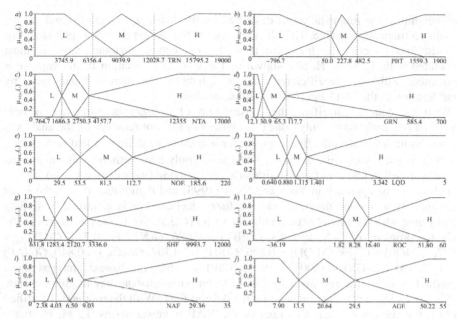

Fig. 5. Graphical representation of sets of MFs for the condition attributes

The sets of MFs presented in Figure 5 define the linguistic variables characterizing the companies. Moreover, they allow the condition attributes of any company to be described through linguistic terms. To demonstrate this, a single company is next considered, namely o_{169}. The fuzzification of this company's details is reported in Table 4.

In Table 4, the original crisp attribute values describing the company can be replaced by triplets of fuzzy values, in each case representing the grades of

Table 4. Attribute values and fuzzification with final labels presented for company o_{169}, when using naïve approach to construction of MFs

o_{169}	Crisp value	Fuzzy values [L, M, H]	Label
TRN	4307.684	[**0.892**, 0.108, 0.000]	L
PBT	84.391	[0.403, **0.597**, 0.000]	M
NTA	2147.343	[0.283, **0.717**, 0.000]	M
GRN	79.211	[0.000, **0.867**, 0.133]	M
NOE	71.000	[0.185, **0.815**, 0.000]	M
LQD	0.533	[**1.000**, 0.000, 0.000]	L
SHF	1679.883	[0.263, **0.737**, 0.000]	M
ROC	3.930	[0.337, **0.663**, 0.000]	M
NAF	4.233	[0.458, **0.542**, 0.000]	M
AGE	10.000	[**0.813**, 0.187, 0.000]	L
ADT	10.000	[**0.643**, 0.357, 0.000]	L

membership to the linguistic terms described by the MFs, $\mu_{,L}(x)$, $\mu_{,M}(x)$ and $\mu_{,H}(x)$, defining a linguistic variable. The MF value that is the largest in each triplet is shown in bold and governs the linguistic term the respective attribute value is most associated with, L, M or H, also shown in Table 4. The variation in the vagueness associated with the fuzzification of the attributes is demonstrated here by consideration of the LQD and NAF condition attributes.

With the data all fuzzified, the construction of a fuzzy decision tree is next described (the early stages only), one prior further constraint required by the analyst is the minimum level of truth (β) acceptable for a node to be defined a leaf node. Here, $\beta = 0.66$ is used, indicating at the technical level, only a subsethood value evaluated above 0.66 will define a leaf node (with an accompanying fuzzy decision rule).

Following the approach in Yuan and Shaw (1995) and demonstrated with the small example data set earlier, the root node of a fuzzy decision tree is identified by finding the attribute with the lowest class ambiguity value ($G(E)$). For the ten attributes in the audit fee problem; $G(\text{TRN}) = 0.683$, $G(\text{PBT}) = 0.886$, $G(\text{NTA}) = 0.714$, $G(\text{GRN}) = 0.887$, $G(\text{NOE}) = 0.713$, $G(\text{LQD}) = 0.896$, $G(\text{SHF}) = 0.707$, $G(\text{ROC}) = 0.840$, $G(\text{NAF}) = 0.659$ and $G(\text{AGE}) = 0.904$. With $G(\text{NAF}) = 0.645$ the lowest amongst those identified, it infers the non-audit fees (NAF) condition attribute has the lowest class ambiguity and so forms the root node for the desired FDT. With the root node identified (NAF), the subsethood $S(A, B)$ values of the NAF linguistic terms, L, M, H, to the classes of the decision attribute ADT (also L, M, H) are investigated, see Table 5.

Table 5. Subsethood values of NAF linguistic terms to ADT classes

NAF	ADT = L	ADT = M	ADT = H
L	**0.570**	0.419	0.158
M	0.383	**0.578**	0.361
H	0.205	0.423	**0.585**

In each row in Table 5, the largest value identifies the decision class each of the paths from the root node should take. In the case of the path NAF = L, the largest value is 0.570 (highlighted in bold with ADT = L), since this value is below the minimum truth level of 0.66 earlier defined, there is no leaf node defined here. This is the case for all the paths from the root node ADT, which means further checks for the lessening of ambiguity are made for each path, see Table 6.

In Table 6, the $G(\text{NAF} = \cdot)$ column of values identifies the level of classification ambiguity with each subset partition from the NAF root node. As shown before, these values are important since they benchmark the value below which a weighted average of classification ambiguity with each subset of partition $G(P| F)$ must be to augment a node to a particular path.

In the case of NAF = L, the $G(\text{NAF} = L) = 0.622$, the subsequent lowest $G(\cdot | \text{NAF} = L)$ value was associated with TRN, with $G(\text{LQD} | \text{NAF} = L) = 0.567$, hence the path NAF = L is augmented with TRN. In the cases of NAF = M and NAF = H, the lowest $G(\cdot | \text{NAF} = \cdot)$ values are, $G(\text{GRN} | \text{NAF} = M) = 0.700$ and $G(\text{NTA} | \text{NAF} = H) = 0.683$, respectively. Hence in the case of NAF = M, its further augmentation is with

Table 6. Classification ambiguity values for attributes, down the path NAF = ·

NAF		$G(\cdot \mid NAF = \cdot)$				
$G(NAF = \cdot)$		TRN	PBT	NTA	GRN	NOE
	L	**0.567**	0.722	0.647	0.718	0.607
L: 0.622	M	0.727	0.703	0.726	**0.700**	0.737
	H	0.706	0.704	**0.683**	0.744	0.693
M: 0.713		LQD	SHF	ROC	AGE	
	L	0.730	0.636	0.718	0.704	
H: 0.643	M	0.703	0.728	0.708	0.702	
	H	0.748	0.686	0.733	0.740	

GRN, but for NAF = H with $G(NTA \mid NAF = H) = 0.683 > 0.622 = G(NAF = H)$ there is no further possible augmentation possible that would reduce ambiguity so it is a leaf node with less than appropriate truth level. For the path, NAF = L, the subsethood values associated with TRN, to the possible classification to ADT = L, M or H, are reported in Table 7.

Table 7. Subsethood values of NAF = L and TRN paths to ADT groups

NAF = L then TRN	ADT = L	ADT = M	ADT = H
L	**0.776**	0.353	0.077
M	0.558	**0.593**	0.216
H	0.347	**0.647**	0.345

The results in Table 7 concern the progression from the NAF = L then TRN path, the only largest subsethood value greater than the 0.65 boundary of minimum truth is associated with TRN = L (value 0.776). It follows, the path NAF = L and TRN = L classifies to ADT = L with 0.776 level of truth. For the other two paths there would be further checks needed on the possibility of minimising ambiguity by further augmenting other attribute to the respective path.

This described process is repeated until all paths end with a leaf node, found from either achieving the required level of subsethood, or no further augmentation is possible. The resultant fuzzy decision tree formed through this approach is presented in Figure 6.

The fuzzy decision tree presented in Figure 6 indicates that a total of 19 leaf nodes were established (fuzzy '*if .. then ..*' decision rules, **R1** to **R19**). Further, the depth of the constructed fuzzy decision tree shows each company is classified using a maximum of five condition attributes (rules **R9**, **R10** and **R11**). To illustrate the practicality of the derived fuzzy decision rules, the rule labeled **R10** interprets to:

"If NAF = M, GRN = M, AGE = L, TRN = L and SHF = M then ADT = M with truth level 0.701."

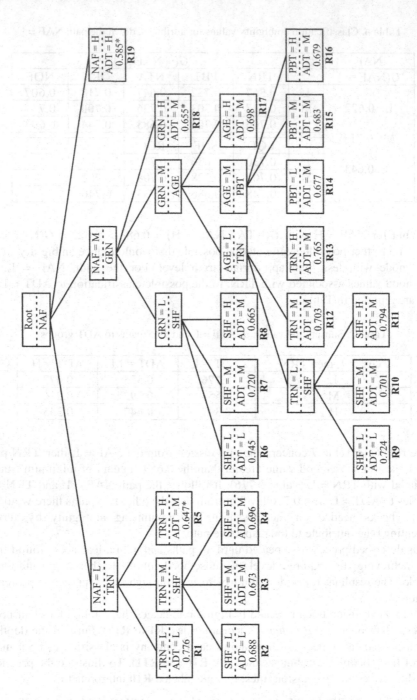

Fig. 6. Constructed fuzzy decision tree using naïve constructed MFs for condition attributes in the audit fee problem

Alternative interpretations to this rule, and other rules, can be given using the information in Figure 5. To look at the matching of the rules, it follows, the company o_{169} described in Table 4 is matched with this rule (based on the linguistic terms it has majority support to).

The results of this matching is that the company has a predicted classification to ADT = M, a medium level of audit fees. However, from Table 4, its fuzzified ADT value indicates majority support to having low audit fees ($\mu_{ADT,L}(10.000) = 0.643$ compared with $\mu_{ADT,M}(10.000) = 0.357$), hence the rule **R10** confers an incorrect predicted classification of ADT = M for the company. This miss-classification is only partial, since there is a level of membership to ADT = L. Further interpretation can be given to the tree structure and fuzzy decision rules, here further sensitivity based investigation is next described.

The second fuzzy decision tree analysis considered here utilizes a variation on the approach previously considered for the construction of MFs to describe the condition attributes associated with the companies in the audit fee problem. The approach taken here was first reported in Beynon et al. (2004), and takes into account the spread of the condition attribute values describing the companies. This approach starts with the construction of estimated distributions, in the form of probability density functions (*pdfs*), for each of the previously defined intervals of a condition attribute. The connection between probability distributions and MFs has been an active issue since the research of Zadeh (1978), see Singpurwalla and Booker (2004) for a relevant discussion.

While it is possible for the constructed *pdfs* themselves to be used as the desired MFs, the philosophy of Kovalerchuk and Vityaev (2000) is followed in specifying a compact representation of the fuzzy linguistic terms describing each attribute. Here, the MFs are constructed from information given in the *pdfs*, rather than using the *pdfs* themselves (hence the compactness of representation). Moreover, only the centre of area (C-o-A) value for each *pdf* is used to evaluate certain of the defining values for the required sets of MFs (and the interval boundary values found previously).

To construct the estimated distributions for each of the decision classes, the method of Parzen windows is employed (Parzen, 1962) to construct an associated *pdf*, based on the values in the domain of the interval. In its general form - assuming each attribute value x_i is represented by a zero mean, unit variance, univariate density function (see Thompson and Tapia, 1990), the estimated *pdf* is given by;

$$pdf(x) = \frac{1}{m}\sum_{i=1}^{m}\frac{1}{h}\frac{1}{\sqrt{2\pi}}\exp\left[-\frac{1}{2}\left(\frac{x-x_i}{h}\right)^2\right] \tag{1}$$

where m is the number of values in the interval and h is the window width. Duda and Hart (1973) suggest $h = h_j/\sqrt{m}$, where h_j here is the range of the individual values in the interval under consideration. Defining I_j to be the jth interval, then $h_j = \max(I_j) - \min(I_j)$, where $\min(I_j)$ and $\max(I_j)$ signify the smallest and largest values in the jth interval, respectively. It follows, the associated *pdf* ($pdf_j(x)$) for the jth interval is given by:

$$pdf_j(x) = \frac{1}{\sqrt{2m_j\pi}(\max(I_j) - \min(I_j))} \sum_{i=1}^{m_j} \exp\left[-\frac{m_j}{2}\left(\frac{x - x_i}{\max(I_j) - \min(I_j)}\right)^2\right] \quad (2)$$

where m_j is the number of values in I_j. Hence the $pdf_j(x)$ function is the mean of the univariate density functions centred at each of the values in the jth interval. To demonstrate the construction of such pdf functions, one of the condition attributes is considered, namely TRN. The interval boundary values that describe the initial partitioning of the TRN values for all the companies are the same as those used in the previous analysis (see Figure 4). For the TRN condition attribute, the respective estimated distributions are reported in Figure 7, together with the values that partition the intervals.

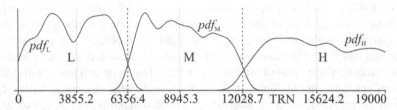

Fig. 7. Estimated distributions of values in the three intervals partitioning the TRN domain

The estimated distributions given in Figure 7 are described over the same domain as in Figure 5a. It follows, the interval partition values of 6356.4 and 12028.7 are the same as before, the other values found contribute to the construction of the concomitant MFs, next described.

To summarise, the construction of the MFs for the TRN condition attribute the functional forms for evaluating each of the defining values are now described. From Figure 7, the defining value $\alpha_{j,3}$ is the top of the MF (a single value where the MF equals one) of the jth interval and is the C-o-A of the respective pdf_j, that is the point at which the area under the pdf is partitioned into two equal halves, given by:

$$\int_{-\infty}^{\alpha_{j,3}} pdf_j(x)\,dx = \int_{\alpha_{j,3}}^{\infty} pdf_j(x)\,dx = \frac{1}{2} \quad (3)$$

The neighbouring defining values $\alpha_{j,2}$ and $\alpha_{j,4}$ around $\alpha_{j,3}$ are simply the left and right boundary values of that interval, respectively. The final outer defining values $\alpha_{j,1}$ and $\alpha_{j,5}$ are the middle (C-o-A) defining values of the neighbouring intervals, that is $\alpha_{j,1} = \alpha_{j-1,3}$ and $\alpha_{j,5} = \alpha_{j+1,3}$. Hence these values can be found using expression (3) given earlier (also shown in Figure 6).

This approach can be adopted for each of the ten condition attributes describing the companies, the resulting sets of MFs are reported in Figure 8.

The sets of MFs constructed in Figure 8 are comparable with those found following the naïve approach of using the simple means of attribute values in each interval, see Figure 5. On first inspection there is little difference in their structure, to aid their comparison the first series of MFs are also given in the graphs in Figure 8, denoted with dotted lines. The identified differences in the MFs are mostly with respect to

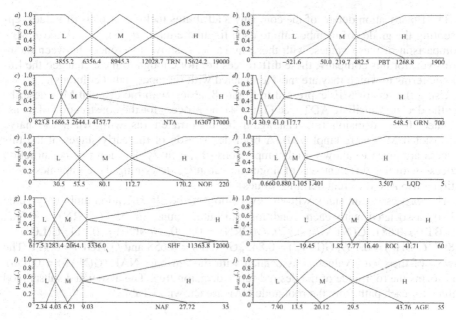

Fig. 8. Graphical representation of sets of MFs for condition attributes

those associated with the high MFs (labeled H), where there is a noticeable change in the respective defining values inherent. There are also differences in two of the MFs associated with the low (L) linguistic terms, namely for the condition attributes PBT (8*b*) and ROC (8*h*).

To further demonstrate the possible variation the results in Figure 8 could confer, the company o_{169} is again considered. The fuzzification of this company's condition attribute values is reported in Table 8 (the fuzzification of the decision attribute is the same as before, see Table 4).

Table 8. Attribute values and fuzzification with final labels presented for company o_{169}, when using informative approach to construction of MFs

o_{169}	Crisp value	Fuzzy values [L, M, H]	Label
TRN	4307.684	[**0.910**, 0.090, 0.000]	L
PBT	84.391	[0.399, **0.601**, 0.000]	M
NTA	2147.343	[0.259, **0.741**, 0.000]	M
GRN	79.211	[0.000, **0.839**, 0.161]	M
NOE	71.000	[0.171, **0.829**, 0.000]	M
LQD	0.533	[**1.000**, 0.000, 0.000]	L
SHF	1679.883	[0.246, **0.754**, 0.000]	M
ROC	3.930	[0.323, **0.677**, 0.000]	M
NAF	4.233	[0.452, **0.548**, 0.000]	M
AGE	10.000	[**0.812**, 0.188, 0.000]	L
ADT	10.000	[**0.643**, 0.357, 0.000]	L

The fuzzification details of the company's attributes includes triplets of values representing the grades of membership to each linguistic term, $\mu_{,L}(x)$, $\mu_{,M}(x)$ and $\mu_{,H}(x)$. Comparison of these values, with those in Table 4, show differences between them. However, for this company, these differences are not enough to have changed the linguistic terms to which they are most associated with (compare label columns).

Using the sets of MFs expressed in Figure 8, along with the original MFs describing the decision attribute ADT (see Figure 4) the process for the construction of an associated fuzzy decision tree can again be undertaken (as with when the naïve constructed MFs were employed). A brief description of the early stages of the this process are given to allow initial comparisons of the fuzzy decision tree construction process with this different sets of linguistic variables describing the companies' condition attributes (different defining values).

This process again starts with the evaluation of the classification ambiguity value ($G(\cdot)$), associated with each condition attribute, found to be; $G(\text{TRN}) = 0.682$, $G(\text{PBT}) = 0.881$, $G(\text{NTA}) = 0.720$, $G(\text{GRN}) = 0.890$, $G(\text{NOE}) = 0.711$, $G(\text{LQD}) = 0.897$, $G(\text{SHF}) = 0.709$, $G(\text{ROC}) = 0.843$, $G(\text{NAF}) = 0.658$ and $G(\text{AGE}) = 0.911$. The lowest of these $G(\cdot)$ values is associated with the attribute NAF ($G(\text{NAF}) = 0.658$), and forms the root node of the desired fuzzy decision tree. The associated subsethood values for each path from the root node NAF are shown in Table 9.

Table 9. Subsethood values of NAF groups to ADT groups

NAF	ADT = L	ADT = M	ADT = H
L	**0.572**	0.415	0.155
M	0.384	**0.575**	0.355
H	0.207	0.422	**0.580**

With the adoption of the minimum truth level required of 0.66 (as before), inspection of Table 9 shows no path from the NAF root node forms an immediate leaf node. Hence further inspection of each path, NAF = L, NAF = M and NAF = H, to see whether the augmentation of another condition attribute will improve (lessen) the ambiguity currently associated with that path. This process was further described when the first fuzzy decision tree was constructed (see the discussion surrounding Figure 6), here the final fuzzy decision tree is next given, see Figure 9.

In Figure 9, the fuzzy decision tree constructed has a depth of four non-leaf nodes (maximum of four conditions to any concomitant fuzzy decision rule), and a total of sixteen rules associated with it (**R1, R2, ..., R16**). When compared with the FDT constructed previously, there has been a reduction here in both of the structural measures, maximum number of conditions in a rule and total number of rules, when compared with the fuzzy decision tree shown in Figure 6.

There are a number of differences in the decision trees surrounding the condition attributes at certain nodes, here these differences are demonstrated by further consideration of the company o_{169}. Utilizing the fuzzy decision tree in Figure 9, with the fuzzification details of the company o_{169} presented in Table 8, it follows the rule **R9** defines its classification. This rule can be written out as;

"If NAF = M, GRN = M, LQD = L and TRN = L, then ADT = L with truth level 0.679."

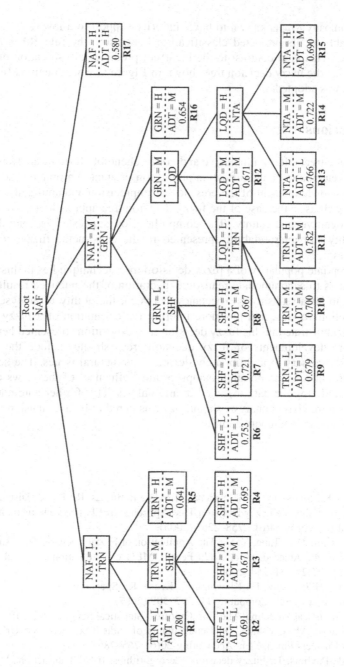

Fig. 9. Constructed fuzzy decision tree using informative constructed MFs for condition attributes in the audit fee problem

Since the company o_{169} is known to have majority support to a low (L) level of audit fees (see Table 8), its predicted classification here using the rule **R9** is ADT = L, hence correct. This is in contrast to the incorrect predicted classification of the company o_{169} using the fuzzy decision tree shown in Figure 6, when a naïve MF construction process was adopted.

4 Conclusions

Fuzzy decision trees offer an inclusive and comprehendible form of analysis for those interested in elucidating a classification/prediction orientated problem. The inclusiveness is an acknowledgement of the presence of imprecision and ambiguity inherent in the data, as well as in the case of the fuzzy decision tree approach employed here the notion of cognitive uncertainty. The comprehension suggested is centrally due to the readability and interpretability presented in the constructed fuzzy '*if .. then ..*' decision rules.

The appropriate popularity of a fuzzy decision trees technique lies in this suggested appropriateness to accommodate realistic expectation of the intended results. That is, there needs to be a realism of expectancy between, the ability to analyse data in a crisp and definitive way, and the more interpretive orientation of a fuzzy based approach, as demonstrated in this fuzzy decision tree expositions presented here.

The future developments of fuzzy decision trees should include the intelligent mechanisms necessary to appropriately define the structural issues. The accompanying aim is to mitigate the possibly inappropriate 'influence' of the views of experts, but by fully utilizing the data included in an analysis. This further automation of the fuzzy decision tree based analysis should be considered only an extension to its functionality and not an incumbent addition.

References

Abdel-Galil, T.K., Sharkawy, R.M., Salama, M.M.A., Bartnikas, R.: Partial Discharge Pattern Classification Using the Fuzzy Decision Tree Approach. IEEE Transactions on Instrumentation and Measurement 54(6), 2258–2263 (2005)

Beynon, M.J., Peel, M.J., Tang, Y.-C.: The Application of Fuzzy Decision Tree Analysis in an Exposition of the Antecedents of Audit Fees. OMEGA - International Journal of Management Science 32(2), 231–244 (2004)

Bodenhofer, U., Hüllermeier, E., Klawonn, F., Kruse, R.: Special issue on soft computing for information mining. Soft Computing 11, 397–399 (2007)

Breiman, L.: Statistical Modeling: The Two Cultures. Statistical Science 16(3), 199–231 (2001)

Chan, P., Ezzamel, M., Gwilliam, D.: Determinants of audit fees for quoted UK companies. Journal of Business Finance and Accounting 20(6), 765–786 (1993)

Chang, R.L.P., Pavlidis, T.: Fuzzy decision tree algorithms. IEEE Transactions Systems Man and Cybernetics SMC-7(1), 28–35 (1977)

Chiang, I.-J., Shieh, M.-J., Hsu, J.Y.-J., Wong, J.-M.: Building a Medical Decision Support System for Colon Polyp Screening by using Fuzzy Classification Trees. Applied Intelligence 22, 61–75 (2005)

Dombi, J., Gera, Z.: The approximation of piecewise linear membership functions and Łu-kasiewicz operators. Fuzzy Sets and Systems 154, 275–286 (2005)

Duda, R.O., Hart, P.E.: Pattern classification and Scene Analysis. Wiley, New York (1973)

Grauel, A., Ludwig, L.A.: Construction of differentiable membership functions. Fuzzy Sets and Systems 101, 219–225 (1999)

Herrera, F., Herrera-Viedma, E., Martinez, L.: A fusion approach for managing multi-granularity linguistic term sets in decision making. Fuzzy Sets and Systems 114(1), 43–58 (2000)

Hu, Y.-C.: Determining membership functions and minimum fuzzy support in finding fuzzy association rules for classification problems. Knowledge-Based Systems 19, 57–66 (2006)

Ishibuchi, H., Nakashima, T., Murata, T.: Three-objective genetics-based machine learning for linguistic rule extraction. Information Sciences 136, 109–133 (2001)

Kecman, V.: Learning and Soft Computing: Support Vector Machines, Neural Networks, and Fuzzy Logic. MIT Press, London, UK (2001)

Kovalerchuk, B., Vityaev, E.: Data Mining in Finance: Advances in Relational and Hybrid Methods. Kluwer Academic Publishers, Dordrecht (2000)

Odéjobí, O.A., Wong, S.H.S., Beaumont, A.J.: A fuzzy decision tree-based duration model for Standard Yorùbá text-to-speech synthesis. Computer Speech and Language 21(2), 325–349 (2006)

Olaru, C., Wehenkel, L.: A complete fuzzy decision tree technique. Fuzzy Sets and Systems 138(2), 221–254 (2003)

Pal, S.K., Ghosh, A.: Soft computing data mining. Information Sciences 176, 1101–1102 (2004)

Parzen, E.: On Estimation of a probability density function mode. Annals of Mathematical Statistics 33, 1065–1076 (1962)

Roa-Sepulveda, C.A., Herrera, M., Pavez-Lazo, B., Knight, U.G., Coonick, A.H.: Economic dispatch using fuzzy decision trees. Electric Power Systems Research 66, 115–122 (2003)

Singpurwalla, N.D., Booker, J.M.: Membership Functions and Probability Measures of Fuzzy Sets. Journal of the American Statistical Association 99(467), 867–877 (2004)

Thompson, J.R., Tapia, R.A.: Nonparametric Function Estimation, Modeling, and Simulation, Society for Industrial and Applied Mathematics, Philadelphia (1990)

Wang, X., Chen, B., Qian, G., Ye, F.: On the optimization of fuzzy decision trees. Fuzzy Sets and Systems 112(1), 117–125 (2000)

Wang, X., Nauck, D.D., Spott, M., Kruse, R.: Intelligent data analysis with fuzzy decision trees. Soft Computing 11, 439–457 (2007)

Yardley, J.A., Kauffman, N.L., Clairney, T.D., Albrecht, W.D.: Supplier behaviour in the US audit market. Journal of Accounting Literature 24(2), 405–411 (1992)

Yuan, Y., Shaw, M.J.: Induction of fuzzy decision trees. Fuzzy Sets and Systems 69(2), 125–139 (1995)

Zadeh, L.A.: Fuzzy Sets. Information and Control 8(3), 338–353 (1965)

Zadeh, L.A.: Fuzzy sets as a basis for a theory of possibility. Fuzzy Sets and Systems 1, 3–28 (1978)

An Exposition of NCaRBS: Analysis of US Banks and Moody's Bank Financial Strength Rating

Malcolm J. Beynon

Cardiff Business School, Cardiff University, Colum Drive, Cardiff, CF10 3EU, Wales, UK
BeynonMJ@Cardiff.ac.uk

Abstract. This chapter centres on a novel classification technique called NCaRBS (N-state Classification and Ranking Belief Simplex), and the analysis of Moody's Bank Financial Strength Rating (BFSR). The rudiments of NCaRBS are based around uncertain reasoning, through Dempster-Shafer theory. As such, the analysis is undertaken with the allowed presence of ignorance throughout the necessary operations. One feature of the analysis of US banks on their BFSR, is the impact of missing values in the financial characteristic values describing them. Using NCaRBS, unlike other traditional techniques, there is no need to externally manage their presence. Instead, they are viewed as contributing only ignorance. The comparative results shown on different versions of the US bank data set allows the impact to the presence of missing values to be clearly exposited. The use of the simplex plot method of visualizing data and analysis results furthers the elucidation possible with NCaRBS.

1 Introduction

NCaRBS (N-state Classification and Ranking Belief Simplex) is a development on the nascent CaRBS technique (Beynon, 2005a, 2005b) and is a novel approach for object classification, which operates in the presence of ignorance. The inspiration for the introduction of the original CaRBS technique was to enable the classification of objects described by a number of characteristics that are considered low level measurements (Safranek et al., 1990; Gerig et al., 2000).

The information from the characteristics describing the objects are represented by bodies of evidence (BOEs), made up of belief (mass) values, whose definition comes from the general methodology called the Dempster-Shafer theory of evidence (DST) (Dempster, 1968; Shafer, 1976). The positioning of DST within the general soft computing domain is with respect to its operation through uncertain reasoning (see for example Velagic et al., 2006).

Pal and Ghosh (2004) consider soft computing as a consortium of methodologies, which works to, amongst other things, exploit the tolerance for imprecision and uncertainty, possibly over a domain of data that is frequently ambiguous and incomplete. These facets of soft computing are clearly shown to be present when employing the NCaRBS technique, due to its functional association with DST (Beynon, 2005a).

This chapter applies the NCaRBS technique to the classification of an exhaustive group of U.S. banks that have been assigned the bank financial strength rating (BFSR), introduced by Moody's Investors Services in 1995. A number of example characteristics, including financial ratios, are utilised to describe each bank, and considered to be including the evidence that will classify the banks to a reduced number of BFSR classes.

B. Prasad (Ed.): Soft Computing Applications in Business, STUDFUZZ 230, pp. 93–111, 2008.
springerlink.com   © Springer-Verlag Berlin Heidelberg 2008

The employment of the NCaRBS technique, to BFSR classify the considered banks, is through the effective configuration of NCaRBS systems that optimally predict their classifications, this configuration process defined as a constrained optimisation problem. The solution to this configuration process is through the employment of trigonometric differential evolution (TDE) (Fan and Lampinen, 2003), itself a developing evolutionary algorithm technique to achieve optimisation. An objective function (OB) is considered, necessary when applying TDE, which aims to quantify the level of classification of the objects through the minimisation of only ambiguity, in the subsequent classification of the banks here, and not ignorance, a facet of the role played by DST in the NCaRBS technique and demonstrative of the inherent 'soft computing' attitude (Beynon, 2005b).

The NCaRBS technique is also shown to be totally inclusive of the quality of the data, with respect to its completeness in particular, through the allowed presence of ignorance, without the requirement to undertake inhibiting management of the missing values present amongst their descriptive characteristics (see Schafer and Graham, 2002). To demonstrate this, alternative versions of the original bank data set are considered, namely when a level of incompleteness in the data values is conferred, NCaRBS analyses are then undertaken when the missing values are retained and when they are externally managed. The simplex plot method of data representation is an integral part of the NCaRBS system, and here offers a visual presentation of the results elucidated, allowing a clear exposition of them accrued from the different versions of the bank data set considered.

The intention of this chapter is to offer a benchmark in the undertaking of object classification in the presence of ignorance, through the expositional analysis using NCaRBS, perceived in terms of the incompleteness of the analysed data. The results are pertinently comparable, with identified variations systematic of the possible impact of the presence of managing missing values in data. Through the utilization of the TDE evolutionary algorithm the underlying processes also adhere to the philosophy that soft computing solutions are effective if the strengths of the different techniques are combined into hybrid systems (see Bodenhofer et al., 2007).

2 Background on the NCaRBS Technique

The N-State Classification and Ranking Belief Simplex (NCaRBS), is a development on the original CaRBS (Beynon, 2005a, 2005b), whereby objects are described by characteristics which potentially offer evidence in the classification of the objects to one of a number of decision classes, unlike with the original CaRBS when only two decision classes could be considered. The foundations of the NCaRBS technique are based around the Dempster-Shafer theory of evidence (DST), previously considered a generalization on the well known Bayesian probability calculus (Shafer and Srivastava, 1990).

The applications of the original CaRBS technique have included; the exposition of the antecedents of credit ratings classification (Beynon, 2005b), the investigation of osteoarthritic knees in patient subjects (Jones et al., 2006) and temporal E-learning efficacy (Jones and Beynon, 2007). In these analyses there has been a balance presented between the numerical and visual results exposited, including also, comparisons with other techniques such as Neural Networks (see Beynon, 2005b; Jones et al., 2006).

More formally, the remit of NCaRBS is the attempted classification of n_O objects $(o_1, o_2, \ldots)$, to one of n_D decision classes $(d_1, d_2, \ldots)$, based on their description by n_C characteristics $(c_1, c_2, \ldots)$. Informally, the build up of evidence from the characteristics, starts with the construction of *constituent* BOEs (bodies of evidence), which model the evidence from a characteristic value $v_{i,j}$ (i^{th} object, j^{th} characteristic) to discern between an object's classification to a decision class (say $\{d_h\}$), its complement ($\{\neg d_h\}$) and a level of concomitant ignorance ($\{d_h, \neg d_h\}$).

The technical rudiments of NCaRBS in the construction of a constituent BOE, defined $m_{i,j,h}(\cdot)$, discerning between $\{d_h\}$ and $\{\neg d_h\}$, can be best described by reference to Figure 1.

In Figure 1; stage *a*) shows the transformation of a characteristic value $v_{i,j}$ (i^{th} object, j^{th} characteristic) into a confidence value $cf_{j,h}(v_{i,j})$, using a sigmoid function, with control variables $k_{j,h}$ and $\theta_{j,h}$.

Stage *b*) transforms a $cf_{j,h}(v_{i,j})$ value into a *constituent* BOE $m_{i,j,h}(\cdot)$, made up of the three mass values (and accompanying focal elements in brackets), $m_{i,j,h}(\{d_h\})$, $m_{i,j,h}(\{\neg d_h\})$ and $m_{i,j,h}(\{d_h, \neg d_h\})$, defined by (from Safranek *et al.*, 1990; Gerig *et al.*, 2000):

$$m_{i,j,h}(\{d_h\}) = \max\left(0, \frac{B_{j,h}}{1 - A_{j,h}} cf_{j,h}(v_{i,j}) - \frac{A_{j,h} B_{j,h}}{1 - A_{j,h}}\right),$$

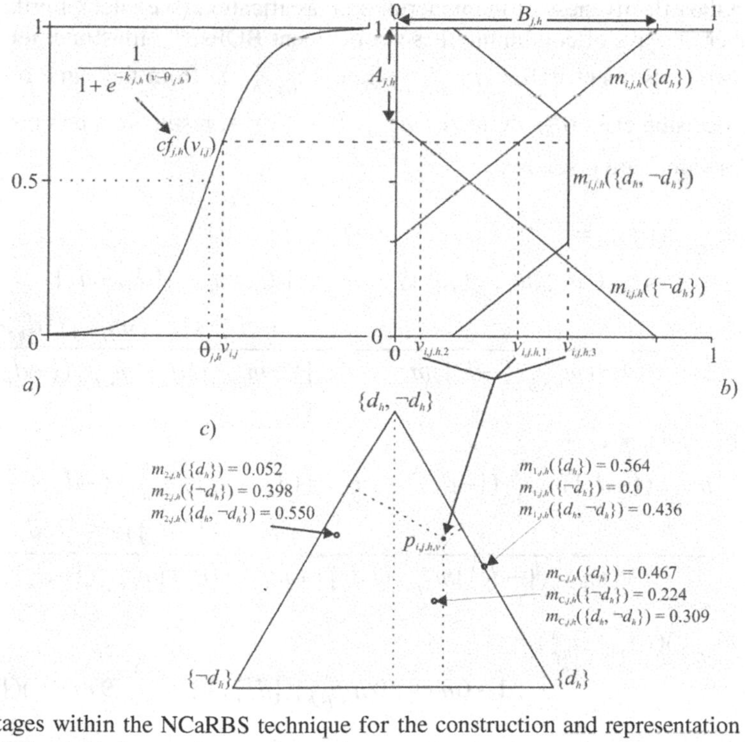

Fig. 1. Stages within the NCaRBS technique for the construction and representation of a constituent BOE $m_{i,j,h}(\cdot)$, from a characteristic value $v_{i,j}$

$$m_{i,j,h}(\{\neg d_h\}) = \max\left(0, \frac{-B_{j,h}}{1-A_{j,h}} cf_{j,h}(v_{i,j}) + B_{j,h}\right),$$

$$\text{and } m_{i,j,h}(\{d_h, \neg d_h\}) = 1 - m_{i,j,h}(\{d_h\}) - m_{i,j,h}(\{\neg d_h\}),$$

where, $A_{j,h}$ and $B_{j,h}$, are two further control variables. From DST, these mass values confer the evidential levels of exact belief towards the focal elements $\{d_h\}$, $\{\neg d_h\}$ and $\{d_h, \neg d_h\}$. This representation of belief needs strengthening for the reader not familiar with DST, semantically, a $m_{i,j,h}(\cdot)$ value represents a portion of our total belief allocated to one of the focal elements in accordance with a piece of evidence, here that which is inherent in a characteristic value. Further, a positive $m_{i,j,h}(\cdot)$ value does not automatically imply a finer re-allocation of it into any subsets of the included focal element in particular with respect to $\{d_h, \neg d_h\}$ here (see Srivastava and Liu, 2003).

Stage c) shows a BOE $m_{i,j,h}(\cdot)$; $m_{i,j,h}(\{d_h\}) = v_{i,j,h,1}$, $m_{i,j,h}(\{\neg d_h\}) = v_{i,j,h,2}$ and $m_{i,j,h}(\{d_h, \neg d_h\}) = v_{i,j,h,3}$, can be represented as a simplex coordinate $(p_{i,j,h,v})$ in a simplex plot (equilateral triangle). That is, a point $p_{i,j,h,v}$ exists within an equilateral triangle such that the least distance from $p_{i,j,h,v}$ to each of the sides of the equilateral triangle are in the same proportion (ratio) to the values, $v_{i,j,h,1}$, $v_{i,j,h,2}$ and $v_{i,j,h,3}$.

For an individual object, with the constituent BOEs constructed from their characteristic values, Dempster's rule of combination is used to combine them to allow a final classification. This combination is in part dictated by the objective function used to measure the effectiveness of the undertaken classification (see later). Further, there are a number of ways of combining these constituent BOEs. To illustrate, the combination of two constituent BOEs, $m_{i,j_1,h}(\cdot)$ and $m_{i,j_2,h}(\cdot)$, from the same object (o_i) and single decision class (d_h), defined $(m_{i,j_1,h} \oplus m_{i,j_2,h})(\cdot)$, results in a combined BOE whose mass values are given by;

$$(m_{i,j_1,h} \oplus m_{i,j_2,h})(\{d_h\}) =$$

$$\frac{m_{i,j_1,h}(\{d_h\})m_{i,j_2,h}(\{d_h\}) + m_{i,j_2,h}(\{d_h\})m_{i,j_1,h}(\{d_h, \neg d_h\}) + m_{i,j_1,h}(\{d_h\})m_{i,j_2,h}(\{d_h, \neg d_h\})}{1 - (m_{i,j_1,h}(\{\neg d_h\})m_{i,j_2,h}(\{d_h\}) + m_{i,j_1,h}(\{d_h\})m_{i,j_2,h}(\{\neg d_h\}))},$$

$$(m_{i,j_1,h} \oplus m_{i,j_2,h})(\{\neg d_h\}) =$$

$$\frac{m_{i,j_1,h}(\{\neg d_h\})m_{i,j_2,h}(\{\neg d_h\}) + m_{i,j_2,h}(\{d_h, \neg d_h\})m_{i,j_1,h}(\{\neg d_h\}) + m_{i,j_2,h}(\{\neg d_h\})m_{i,j_1,h}(\{d_h, \neg d_h\})}{1 - (m_{i,j_1,h}(\{\neg d_h\})m_{i,j_2,h}(\{d_h\}) + m_{i,j_1,h}(\{d_h\})m_{i,j_2,h}(\{\neg d_h\}))},$$

$$(m_{i,j_1,h} \oplus m_{i,j_2,h})(\{d_h, \neg d_h\}) =$$

$$1 - (m_{i,j_1,h} \oplus m_{i,j_2,h})(\{d_h\}) - (m_{i,j_1,h} \oplus m_{i,j_2,h})(\{\neg d_h\}).$$

The example combination of two BOEs, $m_{1,j,h}(\cdot)$ and $m_{2,j,h}(\cdot)$, is shown in Figure 1c, where the simplex coordinate based versions of these BOEs are shown with that also of the combined BOE $m_{C,j,h}(\cdot)$.

This combination process can then be performed iteratively to combine all the characteristic based evidence describing an object to a single decision class, producing a *class* BOE for that object. For an object o_i and decision class d_h, a *class* BOE is defined $m_{i,-,h}(\cdot)$, with the 'dashed line' subscript showing the index not kept constant during the combination process, in this case the different characteristics. It is these class BOEs that are used in the objective function later described. Alternatively, for a single object o_i and characteristic c_j, the discernment over the different decision classes produces a *characteristic* BOE, defined $m_{i,j,-}(\cdot)$.

The effectiveness of the NCaRBS technique is governed by the values assigned to the incumbent control variables $k_{j,h}$, $\theta_{j,h}$, $A_{j,h}$ and $B_{j,h}$, $j = 1, ..., n_C$ and $h = 1, ..., n_D$. This necessary configuration is considered as a constrained optimisation problem, solved here using trigonometric differential evolution (TDE) (Fan and Lampinen, 2003). In summary, TDE is an evolutionary algorithm that creates a population of solutions, and then iteratively generates improved solutions to a problem through the marginal changes in previous solutions augmented with the differences between the pairs of other solutions.

When the classification of a number of objects to some hypothesis and its complement is known, the effectiveness of a configured NCaRBS system can be measured by a defined objective function (OB), later described. Indeed it is the objective function that is used by TDE to identify the improving solution generated through its running.

For the objective function used here, a series of 'decision' equivalence classes are defined $E(d_h)$ $h = 1, ..., n_D$, which each contain those objects associated with a single decision class. Then an interim objective function, defined $OB(d_h)$, is evaluated for each decision class d_h, using the class BOEs of the objects in the respective equivalence class $E(d_h)$, given by;

$$OB(d_h) = \frac{1}{4}\left(\frac{1}{|E(d_h)|} \sum_{o_i \in E(d_h)} (1 - m_{i,-,h}(\{d_h\}) + m_{i,-,h}(\{\neg d_h\})) \right.$$

$$\left. + \frac{1}{|E(\neg d_h)|} \sum_{o_i \in E(\neg d_h)} (1 + m_{i,-,h}(\{d_h\}) - m_{i,-,h}(\{\neg d_h\})) \right).$$

It is noted, in the intermediate objective functions $OB(d_h)$, maximising a difference value such as $(m_{i,-,h}(\{d_h\}) - m_{i,-,h}(\{\neg d_h\}))$ only affects the associated ignorance ($m_{i,-,h}(\{d_h,\neg d_h\})$), rather than making it a direct issue. The single final OB is the sum of these interim objective functions over the different decision classes, so OB = $\frac{1}{n_D} \sum_{h=1}^{n_D} OB(d_h)$, in the limit, $0 \le OB \le 1$.

Once the minimum OB is found, through employing the TDE approach, the respective class BOEs are combined in a similar way to the combination of the constituent BOEs. Moreover, for each object o_i, an object BOE is evaluated, defined $m_{i,-,-}(\cdot)$ (for brevity this is often reduced to $m_i(\cdot)$), from which their classification to the n_D decision classes is evident.

To offer a level of final classification for each object to each decision class, the pignistic probability function - BetP($\cdot$) is utilised (Denœux and Zouhal, 2001). More formally, for the object o_i and decision class d_h it is given by $$\text{BetP}_i(d_h) = \sum_{s_j \subseteq \{d_1, d_2, ...\}, s_j \cap d_h \neq \varnothing} m_{i,-,-}(s_j)/|s_j|.$$ It follows, the largest of the BetP$_i(d_h)$ $h = 1$, ..., n_D, dictates the predicted classification of the object o_i.

Within NCaRBS, if a characteristic value is missing, the associated constituent BOE supports only ignorance, namely $m_{i,j,h}(\{d_h, \neg d_h\}) = 1$ (with $m_{i,j,h}(\{d_h\}) = 0$ and $m_{i,j,h}(\{\neg d_h\}) = 0$). That is, a missing value is considered an ignorant value and so offers no evidence in the subsequent classification of that object. This means that the missing values can be retained in the analysis rather than having to be imputed or managed in some way, which would change the data set considered (see later).

3 The Bank Rating Problem

The main analysis of this chapter is the use of NCaRBS in a business application. Moreover, the rating of banks in the U.S. is considered, through their investigation with the Bank Financial Strength Rating (BFSR). Introduced in 1995 by Moody's Investors Services, it represents Moody's opinion of a bank's intrinsic safety and soundness - a measure of the likelihood that a bank will require assistance from third parties (Moody's, 2004; Poon et al., 1999). An exhaustive group of 507 large U.S. banks is established (found through the use of Bankscope (2006)).

The actual BFSR classification of the individual banks, can be rated from a high of A to a low of E - 13 ordinal values (+ and − modifiers below and above A and E respectively, Moody's (2004)). Here, the U.S. bank BFSR data set ratings are partitioned into three classes to allow analysis using NCaRBS that avails itself formally to the graphical representation of results (more class partitions could be analysed using NCaRBS), see Table 1.

Table 1. Description of the intervalisation of banks into three BFSR classes; Low, Medium and High

Interval	Low - L (below C−)	Medium - M (between C− and C+)	High - H (above C+)
Number of banks	173	210	124

In Table 1, the distribution of the banks across the three classes of BFSR are slightly imbalanced, with the medium BFSR class having a noticeable number more than the high class in particular (as well as low class). This imbalanced set of BFSR classes defined is not a critical problem with the NCaRBS technique, since the objective function employed, described previously, weights the measurement of errors appropriately across the different equivalence classes of banks (L, M or H). There is one important note, namely that it is clear the three classes are not nominal, which is what the NCaRBS technique should work on, rather they are ordinal.

Table 2. Description of characteristics of banks

Index	Label	Description	Mean	Standard deviation
c_1	ETA	Equity / Total Assets (%)	8.38	4.70
c_2	RAA	Return on Average Assets (ROAA) (%)	1.37	1.07
c_3	RAE	Return on Average Equity (ROAE) (%)	16.36	8.54
c_4	LCS	Liquid Assets / Cust & ST Funding (%)	13.84	13.68
c_5	NTA	Net Loans / Total Assets (%)	60.04	16.42
c_6	LNR	Loan Loss Prov / Net Int Rev (%)	11.15	13.60
c_7	IGL	Impaired Loans / Gross Loans (%)	2.93	3.95
c_8	TCR	Total Capital Ratio (%)	13.44	4.60

The large U.S. banks considered are each described by a number of characteristics, including those associated with the well known areas of; profitability, leverage, liquidity and risk (see for example Poon *et al.*, 1999). Moreover, here, eight financial ratios are used to describe them, see Table 2.

In Table 2, the eight financial ratios, constituting the characteristics, are explicitly described, alongside their mean and standard deviation values (of the whole population of banks).

3.1 NCaRBS Analysis of Original BFSR Data Set

This subsection of the chapter elucidates the first NCaRBS analysis of the U.S. bank data set in its original complete form. The desired goal is the optimum classification of the banks to the defined, low (L), medium (M) and high (H) BFSR classes introduced previously, based on their characteristic values. This classification is quantified through the measured objective function (OB). With the intention to configure a NCaRBS system.

From the definition of OB this necessitates the individual minimization of the three intermediate objective functions, OB(L), OB(M) and OB(H), which individually consider the discernibility of banks with a BFSR class, L, M or H, against the others, termed their respective complements. The TDE approach to optimization was run to minimize the intermediate objective functions using the relevant parameters; amplification control $F = 0.99$, crossover constant $CR = 0.85$, trigonometric mutation probability $M_t = 0.05$ and number of parameter vectors $NP = 10 \times$ number of control variables $= 240$ (Fan and Lampinen, 2003).

The results of the TDE run was the convergence of the overall objective function value to OB $= (OB(L) + OB(M) + OB(H))/3 = (0.32 + 0.39 + 0.35)/3 = 0.35$. The control variables evaluated for the subsequently configured NCaRBS system are reported in Table 3.

In Table 3, the sets of control variables are presented that minimize the respective intermediate objective functions OB(d_h), with respect to the optimal ability to discern, low (L), medium (M) and high (H), BFSR classed banks. These control variables are specifically used to construct the constituent BOEs, from the banks' characteristic values, from which the different classifications of are then derived (through their combination etc.).

Table 3. Control variable values associated with the intermediate objective functions, OB(L), OB(M) and OB(H)

$d_h,$ $\neg d_h$		ETA	RAA	RAE	LCS	NTA	LNR	IGL	TCR
L, ¬L	$k_{j,h}$	-2.96	3.00	-1.06	3.00	2.30	3.00	3.00	3.00
	$\theta_{j,h}$	0.21	-0.39	1.76	-0.30	0.39	0.25	-0.20	0.07
	$A_{j,h}$	1.00	0.87	0.92	0.28	0.99	0.61	0.26	0.39
M, ¬M	$k_{j,h}$	-3.00	-3.00	3.00	3.00	-3.00	3.00	3.00	3.00
	$\theta_{j,h}$	-0.52	0.89	-0.26	-0.34	0.52	0.11	0.11	0.06
	$A_{j,h}$	0.28	0.94	0.30	0.27	0.26	0.35	0.22	0.69
H, ¬H	$k_{j,h}$	3.00	3.00	-3.00	3.00	3.00	3.00	3.00	3.00
	$\theta_{j,h}$	-0.50	-0.36	1.36	-0.92	-0.13	0.63	-0.22	0.13
	$A_{j,h}$	0.31	0.95	0.99	0.83	0.99	0.84	0.24	0.29

The evaluation of constituent BOEs is briefly illustrated by considering the classification evidence surrounding the bank b_{453}, known to be associated with the high BFSR class. Further, the construction of a constituent BOE is shown, associated with the characteristic LCS and its evidence to discern between the BFSR classes {L} and {M, H}. Starting with the evaluation of the confidence value $cf_{LCS,L}(\cdot)$ (see Figure 1a), with the LCS characteristic value 1.92 for the bank b_{453}, which when standardized gives $v_{453,LCS} = -0.87$, then;

$$cf_{LCS,L}(-0.87) = \frac{1}{1+e^{-3.00(-0.87+0.30)}} = \frac{1}{1+5.61} = 0.15,$$

using the control variables reported in Table 3. This confidence value is used in the expressions making up the mass values in the associated constituent BOE, in this case, $m_{453,LCS,L}(\cdot)$, namely; $m_{453,LCS,L}(\{L\})$, $m_{453,LCS,L}(\{M, H\})$ and $m_{453,LCS,L}(\{L, M, H\})$, found to be;

$$m_{453,LCS,L}(\{L\}) = \max\left(0, \frac{0.6}{1-0.28}0.15 - \frac{0.28\times0.6}{1-0.28}\right) = 0.00,$$

$$m_{453,LCS,L}(\{M, H\}) = \max\left(0, \frac{-0.6}{1-0.28}0.15 + 0.6\right) = 0.47,$$

$$m_{453,LCS,L}(\{L, M, H\}) = 1 - 0.00 - 0.47 = 0.53.$$

Inspection of this constructed constituent BOE, shows positive mass values associated with the focal elements {M, H} and {L, M, H}, the latter being the concomitant ignorance in the evidence from this characteristic. Since the bank b_{453} is known to have high BFSR class, then with $m_{453,LCS,I}(\{M, H\}) = 0.47 > 0.00 = m_{453,LCS,L}(\{L\})$, there is correct supporting evidence from this constituent BOE.

For the bank b_{453}, this constituent BOE is representative of the constituent BOEs $m_{453,j,h}(\cdot)$ found over the different characteristics and classes of BFSR being discerned, see Table 4.

Table 4. Details of the characteristics of the bank b_{453}, and the constituent BOEs for its classification to the BFSR classes, L, M and H, against their respective complements

BOE	ETA	RAA	RAE	LCS	NTA	LNR	IGL	TCR
b_{453}	4.99	1.11	19.8	1.92	76.2	7.15	0.11	10.80
Standardized b_{453}	−0.72	−0.25	0.41	−0.87	0.98	−0.29	−0.71	−0.57
$m_{453,j,L}(\{L\})$	0.00	0.00	0.00	0.00	0.00	0.00	0.00	0.00
$m_{453,j,L}(\{M, H\})$	0.00	0.00	0.00	0.47	0.00	0.45	0.46	0.48
$m_{453,j,L}(\{L, M, H\})$	1.00	1.00	1.00	0.53	1.00	0.65	0.54	0.52
$m_{453,j,M}(\{M\})$	0.06	0.00	0.00	0.46	0.44	0.38	0.54	0.35
$m_{453,j,M}(\{L, H\})$	0.30	0.29	0.50	0.00	0.00	0.00	0.00	0.00
$m_{453,j,M}(\{L, M, H\})$	0.64	0.71	0.50	0.54	0.56	0.61	0.46	0.65
$m_{453,j,H}(\{H\})$	0.31	0.00	0.00	0.00	0.00	0.39	0.45	0.51
$m_{453,j,H}(\{L, M\})$	0.03	0.00	0.00	0.00	0.00	0.00	0.00	0.00
$m_{453,j,H}(\{L, M, H\})$	0.66	1.00	1.00	1.00	1.00	0.61	0.55	0.49

In Table 4, the triplets of mass values making up each of the constituent BOEs, are shown, constructed for the classification of the bank b_{453}. Inspection of these triplets of mass values shows many of the characteristics' evidences infer the contribution of total ignorance in discerning a bank with respect to a BFSR class against its complement (cases where $m_{453,j,h}(\{L, M, H\}) = 1.00$, the rest 0.00). When discerning between BFSR class L and its complement ($\neg L = M, H$), the characteristics consistently offer correct supporting evidence to the classification of the bank, with $m_{453,j,L}(\{M, H\}) > 0$ and $m_{453,j,L}(\{L\}) = 0$ with the characteristics, LCS (shown previously), LNR, IGL and TCR.

In the case of when discerning the BFSR class M from its complement ($\neg M$), the constituent BOEs shown show a mixture of supporting evidence towards {M} and {L, H}, which could infer more ambiguity when their evidences are combined. Interestingly, there are no characteristics offering total ignorance amongst these constituent BOEs. When discerning the BFSR class H from its complement ($\neg H = L, M$), there is a majority of evidence to {H}, which is correct supporting evidence for this bank with its known association to the H BFSR class.

The constituent BOEs in Table 4 can be combined, using Dempster's combination rule, to form the respective class BOEs, defined $m_{453,-,h}(\cdot)$ $h = L, M$ and H, showing the evidence from the characteristics to separately discern each BFSR class from its respective complement, given below;

$$m_{453,-,L}(\{L\}) = 0.00, \; m_{453,-,L}(\{M, H\}) = 0.90, \; m_{453,-,L}(\{L, M, H\}) = 0.10,$$

$$m_{453,-,M}(\{M\}) = 0.82, \; m_{453,-,M}(\{L, H\}) = 0.14, \; m_{453,-,M}(\{L, M, H\}) = 0.04,$$

$$m_{453,-,H}(\{H\}) = 0.88, \; m_{453,-,H}(\{L, M\}) = 0.01, \; m_{453,-,H}(\{L, M, H\}) = 0.11.$$

Each of the class BOEs shown is also made up of a triplet of mass values, offering levels of exact belief on the classification of the bank to a BFSR class, its complement and concomitant ignorance. It is noticeable that in the three class BOEs presented, the levels of ignorance present are less than what was present in their respective constituent BOEs (as previously illustrated in the example contained in Figure 1c).

In terms of the evidence these class BOEs offer for the classification of the bank b_{453}, since it is known to have high BFSR class, correct supporting evidence would come from the mass values of, $m_{453,-,L}(\{M, H\})$, $m_{453,-,M}(\{L, H\})$ and $m_{453,-,H}(\{H\})$ in the respective class BOEs. Inspection shows the class BOEs, $m_{453,-,L}(\cdot)$ and $m_{453,-,H}(\cdot)$, do majority support its correct classification, but $m_{453,-,M}(\cdot)$ suggests more evidence to the bank being classified to the medium BFSR class.

The three class BOEs can themselves be combined to give a final bank BOE, defined $m_{453,-,-}(\cdot)$, for the bank b_{453}, shortened to $m_{453}(\cdot)$, made up of the following series of focal elements and concomitant mass values;

$$m_{453}(\{L\}) = 0.00, m_{453}(\{M\}) = 0.35, m_{453}(\{H\}) = 0.63,$$
$$m_{453}(\{L, M\}) = 0.00, m_{453}(\{M, H\}) = 0.02, m_{453}(\{L, H\}) = 0.00, m_{453}(\{L, M, H\}) = 0.00.$$

The presented bank BOE is made up of seven mass values partitioning the exact belief between seven distinct focal elements, a consequence of the combination process.

To enable a final classification to be assigned to the bank b_{453}, this bank BOE is further operated upon to partition the beliefs to the individual BFSR classes of, L, M and H, using the pignistic probability function (BetP.($\cdot$)), described previously. The respective BetP$_{453}(\cdot)$ values of this bank BOE can be found, given to be;

$$\text{BetP}_{453}(L) = 0.00, \text{BetP}_{453}(M) = 0.36, \text{BetP}_{453}(H) = 0.64.$$

Inspection of the three BetP$_{453}(\cdot)$ values shows BetP$_{453}(H) = 0.64$ to be the largest, hence the predicted classification of the bank b_{453} is to the high BFSR class.

These series of numerical results describe the build up of the classification evidence for the bank b_{453}; from constituent BOEs to class BOEs and then a bank BOE. Graphical representations of all this evidence can be presented, using the standard domain of the simplex plot (describe in Figure 1c), see Figure 2.

In Figures 2a, 2b and 2c, the individual simplex plots show the evidence from the characteristics to the classification of the bank b_{453} to a BFSR class and its complement, along with the respective class BOE (labelled $m_{453,-,h}$). To describe their presentation, in Figure 2a, the attempt to discern the bank b_{453} from being low BFSR class ({Low}) or its complement ({Medium, High}) shows the characteristics, LCS, LNR, IGL and TCR, offer correct supporting evidence, since the bank b_{453} is known to have high BFSR class. These findings from the visual results are understandably the same as those from the numerical results previously discussed.

The final classification results for the bank b_{453} to the separate BFSR classes, L, M or H, is shown in Figure 2d, along with the contribution from the individual characteristics (in constructed characteristic BOEs briefly defined earlier).

All the points shown in Figure 2d are the BetP($\cdot$) triplet based representations of the respective characteristic and bank BOEs (as given earlier for the case of the bank BOE - labeled m_{453}). Greater correct supporting evidence for this bank is towards the High vertex, hence here the characteristics ETA and TCR offer the strongest supporting evidence. With the position of the m_{453} bank BOE nearest the High vertex, the

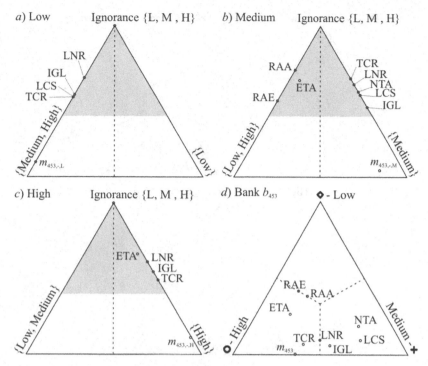

Fig. 2. Classification details of a single bank b_{453}, when discerning them from a BFSR class and its complement (2a, 2b and 2c) as well as its final classification (2d), based on a NCaRBS analysis of the original BFSR data set

visual information infers predicted classification to the high BFSR class, the correct classification in this case.

This analysis approach can be repeated for any bank, culminating in a series of bank BOEs each of which can then be represented as a triplet of BetP$_i(\cdot)$ values that offer the levels of individual classification of a bank to the three BFSR classes of L, M and H. Further, the BetP$_i(\cdot)$ triplets of the bank BOEs, can each be represented as simplex coordinates in a simplex plot, see Figure 3.

In the simplex plots presented in Figure 3, the final classification of all the banks are shown, separated based on whether a bank is known to have a BFSr class of; Low (◆) in 3a, Medium (✚) in 3b, or High (⬤) in 3c. The dashed lines in the presented simplex plots partition where there is majority classification association to one BFSR class (described at the respective vertex). The simplex coordinate representations of the triplet of BetP$_i(\cdot)$ values associated with the banks denoted by the respective symbols, ◆, ✚ and ⬤.

A breakdown of the classification results identifies, 130 out of 173 low (75.15%), 115 out of 210 medium (54.76%) and 60 out of 124 high (48.39%) BFSR classed banks were correctly classified, giving an overall total correct classification of 305 out of 507 banks (60.16%).

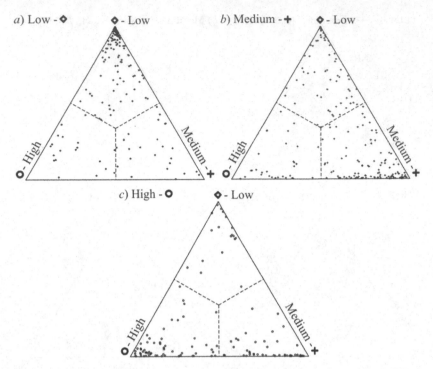

Fig. 3. Final classification of banks, based on a NCaRBS analysis of the original BFSR data set

3.2 NCaRBS Analysis of Incomplete Version of BFSR Data Set

In this section an incomplete version of the BFSR data set is analysed, to demonstrate the ability of the NCaRBS technique to analyse data with missing values, without the need for their management. This first analysis here is then benchmarked with a similar NCaRBS analysis when the missing values are managed in some way, namely through the mean imputation of them (see later). These analyses are briefly exposited, with emphasis on the graphical representation of results (more technical details were presented previously on the original BFSR data set).

A noticeable level of incompleteness is imposed on the original 'complete' BFSR data set, to allow the fullest demonstration of the variation in the effect of having to or not having to manage the missing values present. Moreover, here, 40.00% of the original data set was randomly removed, with 507 banks and eight characteristics describing them, this means the available data items went down from 4056 to 2434 (with 1622 missing values now present), see Table 5.

Table 5. Description of characteristics with missing values in an incomplete version of the BFSR data set

Characteristic	ETA	RAA	RAE	LCS	NTA	LNR	IGL	TCR
Number missing	215	217	207	176	209	201	195	202
Mean	8.45	1.35	16.45	13.62	60.57	11.34	2.88	13.25
Standard deviation	5.07	1.06	9.00	13.10	16.17	13.26	3.78	4.19

In Table 5, the breakdown of the number of missing values present shows a level of variation of the missing values associated with each characteristic. The initial impact of this imposed incompleteness is shown in the descriptive measures, mean and standard deviation, reported, where there is variation of these values with respect to those shown in Table 2, associated with the original BFSR data set.

As mentioned earlier, the NCaRBS system is able to analyse an incomplete data set without the need for the management of the missing values now present, by assigning them as ignorant values. It follows, a similar analysis can be undertaken as previously presented, which commences with the employment of TDE to optimise the intermediate objective functions, OB(L), OB(M) and OB(H). When the TDE was run, the overall objective function was found to converge to OB = (OB(L) + OB(M) + OB(H))/3 = (0.35 + 0.41 + 0.38)/3 = 0.38.

Post this optimisation process, using the evaluated control variables, the intermediate and final classification of a bank, namely b_{453} can be considered, the same bank as whose results were described previously. Before these results are presented, the characteristic values associated with this bank are reported in Table 6.

Table 6. Details of the characteristics of the bank b_{453}, and the constituent BOEs for its classification to the BFSR class L, against its complement

BOE	ETA	RAA	RAE	LCS	NTA	LNR	IGL	TCR
b_{453}	-	1.11	-	-	76.2	-	-	-
Standardized b_{453}	-	−0.25	-	-	0.98	-	-	-
$m_{453,j,L}(\{L\})$	0.00	0.00	0.00	0.00	0.00	0.00	0.00	0.00
$m_{453,j,L}(\{M, H\})$	0.00	0.00	0.00	0.00	0.43	0.00	0.00	0.00
$m_{453,j,L}(\{L, M, H\})$	1.00	1.00	1.00	1.00	0.57	1.00	1.00	1.00

In Table 6, the dashes presented show where the characteristic values are now missing, for this bank b_{453} there are six out of the eight characteristic values missing, with only RAA and NTA still present. Even with a high level of missing values associated with the bank, those present still are enough for it to have been included in the NCaRBS analysis.

Also shown in Table 6 are the constituent BOEs associated with the bank's discernment from the low BFSR class against its complement. For the missing characteristic values, their constituent BOEs are consistently showing evidence of total ignorance. In the case of the characteristic RAA, it also contributes only ignorance, but not because of it being missing but from the optimisation process, only the characteristic NTA offers non-ignorant evidence. A visual representation of the classification evidence associated with the bank b_{453} is given in Figure 4.

In Figure 4, the four simplex plots again show the intermediate classifications of the bank b_{453} to each BFSR class against its respective complement (4a, 4b and 4c), as well as its final classification (4d). What is immediately noticeable in the classifications is the dearth of non-ignorant evidence shown from the characteristics, with only RAA and NTA contributing, a consequence of the other characteristics all being missing in this incomplete data set. The ignorance contributions of the missing characteristics are shown with their simplex coordinates at the {L, M, H} vertices in Figures 4a,

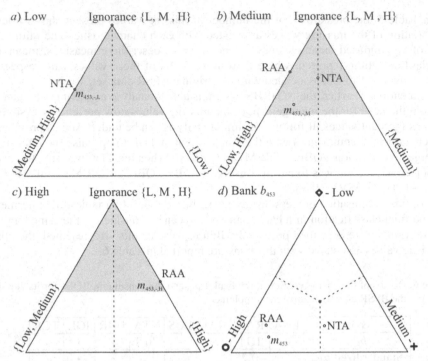

Fig. 4. Classification details of a single bank b_{453}, when discerning them from a BFSR class and its complement (4*a*, 4*b* and 4*c*) as well as its final classification (4*d*), based on a NCaRBS analysis of the incomplete BFSR data set

4*b* and 4*c*, and at the centroid of the simplex plot in Figure 4*d*. The final classification of the bank, defined by its bank BOE m_{453}, shown in Figure 4*d* is nearest to the High vertex, again showing a correct predicted classification.

The BetP$_i(\cdot)$ triplets of all the bank BOEs of the banks in the incomplete bank data set are next considered, represented as simplex coordinates in a simplex plot, see Figure 5.

In Figure 5, the classifications of all the 507 banks are separately shown depending on their known BFSR class. Comparison with the results in the Figure 3, which elucidated classification results on the original BFSR data set, are the presence of an increased number of banks whose triplets of BetP$_i(\cdot)$ values mean their positions are more towards the centre (centroid) of the simplex plot domain. This is a consequence of the probable increased ignorance associated with many of the banks due to the number of their characteristics now being missing.

A breakdown of the classification results identifies, 116 out of 173 low (67.05%), 111 out of 210 medium (52.86%) and 62 out of 124 high (50.00%) BFSR classed banks were correctly classified, giving an overall total correct classification of 289 out of 507 banks (57.00%).

The second NCaRBS analysis on the 'incomplete' BFSR data set, allows a level of external management to take place on the missing values present, prior to its employment. The relevance of this analysis is that it is the norm, rather than the exception, to have to manage the presence of missing values. It follows, any comparisons, in particular

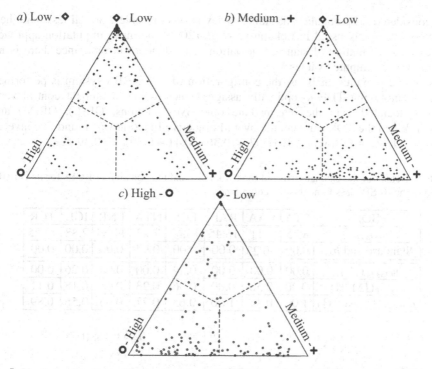

Fig. 5. Final classification of banks, based on a NCaRBS analysis of the incomplete BFSR data set

variations, in the results next presented with those from the previous two NCaRBS analyses will highlight the potential impact of such management. Techniques to manage their presence form the central debate within research on missing values (Schafer and Graham, 2002), along with the reasons for their presence (Rubin, 1976).

Here, since the incompleteness was contrived for the purposes of the chapter, one of the most well known management approaches, called imputation, is employed. Huisman (2000) clearly suggest why imputation is a popular approach, despite its dangers, because it allows the researcher to use standard complete-data methods of analysis on the filled-in data. Moreover, mean imputation is employed, whereby, missing values are replaced with surrogates, namely the mean of the characteristic values from those which are present (Lokupitiya *et al.*, 2006). The effect on the now completed data set is exposited in Table 7.

In Table 7, the descriptive measures, mean and standard deviation, on the data set, post mean imputation, are shown. Inspection shows them to be again different to those presented in Tables 2 and 5, associated with the other versions of the BFSR data

Table 7. Description of characteristics with missing values in an incomplete BFSR data set

Characteristic	ETA	RAA	RAE	LCS	NTA	LNR	IGL	TCR
Mean	8.45	1.35	16.45	13.62	60.57	11.34	2.88	13.25
Standard deviation	3.85	0.80	6.92	10.58	12.40	10.30	2.97	3.25

sets considered. In particular, the standard deviations in Table 7 are all less than their counterparts, as discussed in Lokupitiya *et al.* (2006) the mean imputation approach causes underestimation of standard deviations and standard errors, since there is no variation in the imputed values.

As in the previous analyses, the configuration of a NCaRBS system is performed through employing TDE to optimally assign values to the incumbent control variables, associated with the intermediate objective functions, OB(L), OB(M) and OB(H). When the TDE was run, the overall objective function was found to converge to OB = (OB(L) + OB(M) + OB(H))/3 = (0.36 + 0.41 + 0.39)/3 = 0.39.

Table 8. Details of the characteristics of the bank b_{453}, and the constituent BOEs for its classification to the BFSR class L, against its complement

BOE	ETA	RAA	RAE	LCS	NTA	LNR	IGL	TCR
b_{453}	8.45	1.11	16.45	13.62	76.2	11.34	2.88	13.25
Standardized b_{453}	0.00	−0.23	0.00	0.00	0.97	0.00	0.00	0.00
$m_{453,j,L}(\{L\})$	0.00	0.00	0.00	0.29	0.00	0.00	0.26	0.00
$m_{453,j,L}(\{M, H\})$	0.00	0.56	0.00	0.06	0.23	0.05	0.18	0.11
$m_{453,j,L}(\{L, M, H\})$	1.00	0.44	1.00	0.65	0.77	0.95	0.56	0.89

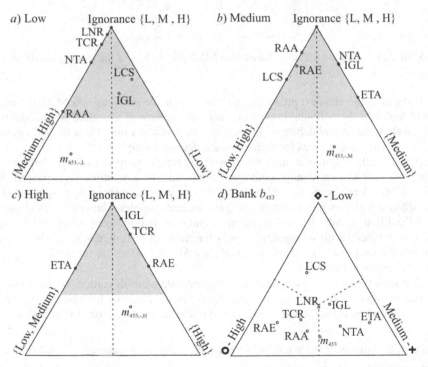

Fig. 6. Classification details of a single bank b_{453}, when discerning them from one BFSR class and the others (*a*, *b* and *c*) as well as its final classification (*d*), based on a NCaRBS analysis of the completed BFSR data set

To exposit the subsequent configured NCaRBS system, the intermediate and final classifications of the bank b_{453}, considered previously can again be reported. Before this, details on the bank b_{453} are first presented, see Table 8.

In Table 8, there is shown to be no missing characteristic values describing the bank b_{453}, but what has actually happened is that the six characteristics previously defined missing have now taken the respective mean values presented in Table 7. The associated constituent BOEs found when discerning the low BFSR class against its complement, show six out of the eight of them contribute evidence, even though many of them were missing. This is an example of the clear effect of managing missing values through imputation, indeed here they are shown to include offering false evidence.

A visual representation of the classification evidence associated with the bank b_{453} is given in Figure 6.

The findings in Figure 6 initially do not look too dissimilar to those reported previously (see Figures 2 and 4). However, in Figures 6a, 6b and 6c, there are more contributions of evidence from the different characteristics than there was previously. For example, when discerning the bank from the low BFSR class against its complement, in Figure 6a, six characteristics are shown to contribute evidence, rather than total ignorance, compared to only four when the original data set was analysed (see Figure 2a). The reasons for this are the same as in the discussion surrounding the numerical results presented in Table 8.

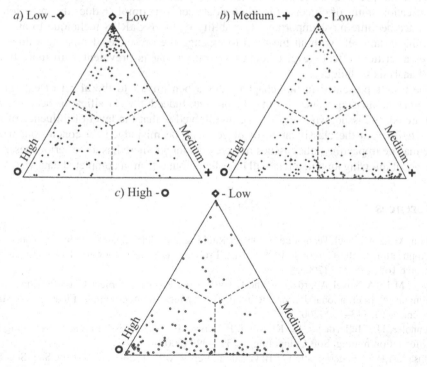

Fig. 7. Final classification of banks, based on a NCaRBS analysis of the completed BFSR data set

The BetP$_i$($\cdot$) triplets of all the bank BOEs of the banks in the completed BFSR data set are next considered, represented as simplex coordinates in a simplex plot, see Figure 7.

A breakdown of the classification results in Figure 7 identifies, 106 out of 173 low (61.27%), 87 out of 210 medium (41.43%) and 70 out of 124 high (56.45%) BFSR classed banks were correctly classified, giving an overall total correct classification of 263 out of 507 banks (51.87%). The overall level of accuracy shown here (51.87%) is noticeably less than the accuracy found on the incomplete data set (57.00%), indicating in this case the further detrimental effect of employing imputation to manage the presence of missing values.

4   Conclusions

Object classification is a central issue for many real life problems, including in particular in business. This chapter has exposited one classification technique, commonly called NCaRBS. The NCaRBS technique described operates closely with respect to the notion of uncertain reasoning, a fundamental trait of soft computing, and here due to the operational rudiments of NCaRBS being based on the methodology of Dempster-Shafer theory. The demonstration of this inherency of uncertainty is the description of evidence contribution and final classification of objects (in general), includes a level of ignorance.

While the presence of ignorance, at the very least, adds a new dimension to the classification status of objects, it is when a data set is incomplete does the inclusion of ignorance become most important. The ability of the NCaRBS technique to analyse incomplete data sets without the need to manage the presence of missing values removes a 'critical effect' level of data preparation that is prevalent with more traditional analysis techniques.

The results presented in this chapter offer a benchmark to the affect of externally managing the missing values that may be present. Indeed, the classification accuracy results are not the issue, what needs to be highlighted is that the level of influence of the characteristics in the classification of objects can be misjudged. In conclusion, when there may be impacting policy issues from results found, there has to be serious reservations about inferring anything that will have been found from a completed data set.

References

Beynon, M.J.: A Novel Technique of Object Ranking and Classification under Ignorance: An Application to the Corporate Failure Risk Problem. European Journal of Operational Research 167, 493–517 (2005a)

Beynon, M.J.: A Novel Approach to the Credit Rating Problem: Object Classification Under Ignorance. International Journal of Intelligent Systems in Accounting, Finance and Management 13, 113–130 (2005b)

Bodenhofer, U., Hüllermeier, E., Klawonn, F., Kruse, R.: Special issue on soft computing for information mining. Soft Computing 11, 397–399 (2007)

Dempster, A.P.: A generalization of Bayesian inference (with discussion). J. Roy. Stat. Soc. Series B 30, 205–247 (1968)

Denœux, T., Zouhal, L.M.: Handling possibilistic labels in pattern classification using evidential reasoning. Fuzzy Sets and Systems 122, 409–424 (2001)

Fan, H.-Y., Lampinen, J.: A Trigonometric Mutation Operation to Differential Evolution. Journal of Global Optimization 27, 105–129 (2003)

Gerig, G., Welti, D., Guttman, C.R.G., Colchester, A.C.F., Szekely, G.: Exploring the discrimination power of the time domain for segmentation and characterisation of active lesions in serial MR data. Medical Image Analysis 4, 31–42 (2000)

Huisman, M.: Imputation of Missing Item Responses: Some Simple Techniques. Quality & Quantity 34, 331–351 (2000)

Jones, P., Beynon, M.J.: Temporal support in the identification of e-learning efficacy: An example of object classification in the presence of ignorance. Expert Systems 24(1), 1–16 (2007)

Jones, A.L., Beynon, M.J., Holt, C.A., Roy, S.: A novel approach to the exposition of the temporal development of post-op osteoarthritic knee subjects. Journal of Biomechanics 39(13), 2512–2520 (2006)

Lokupitiya, R.S., Lokupitiya, E., Paustian, K.: Comparison of missing value imputation methods for crop yield data. Environmetrics 17, 339–349 (2006)

Moody's: Rating Definitions - Bank Financial Strength Ratings (2004), http://www.moodys.com

Pal, S.K., Ghosh, A.: Soft computing data mining. Information Sciences 176, 1101–1102 (2004)

Poon, W.P.H., Firth, M., Fung, M.: A multivariate analysis of the determinants of Moody's bank financial strength ratings. Journal of International Financial Markets Institutions and Money 9, 267–283 (1999)

Rubin, D.B.: Inference and missing data. Biometrika 78, 609–619 (1976)

Safranek, R.J., Gottschlich, S., Kak, A.C.: Evidence Accumulation Using Binary Frames of Discernment for Verification Vision. IEEE Transactions on Robotics and Automation 6, 405–417 (1990)

Schafer, J.L., Graham, J.W.: Missing Data: Our View of the State of the Art. Psychological Methods 7(2), 147–177 (2002)

Shafer G.A.: Mathematical theory of Evidence. Princeton University Press, Princeton (1976); Shafer G., Srivastava R.: The Bayesian and belief-function formalisms: A general perspective for auditing. Auditing: A Journal of Practice and Theory (1990); Reprinted in Pearl J, Shafer G, (eds.); Readings in Uncertain Reasoning. pp. 482–521. Morgan Kaufmann, San Mateo, CA)

Srivastava, R.P., Liu, L.: Applications of Belief Functions in Business Decisions: A Review. Information Systems Frontiers 5(4), 359–378 (2003)

Velagic, J., Lacevic, B., Perunicic, B.: A 3-level autonomous mobile robot navigation system designed by using reasoning/search approaches. Robotics and Autonomous Systems 54, 989–1004 (2006)

A Clustering Analysis for Target Group Identification by Locality in Motor Insurance Industry

Xiaozhe Wang[1] and Eamonn Keogh[2]

[1] Department of Computer Science and Software Engineering, the University of Melbourne, VIC 3010, Australia
[2] Computer Science & Engineering Department, University of California, Riverside, CA 92521

Abstract. A deep understanding of different aspects of business performance and operations is necessary for a leading insurance company to maintain its position on the market and make further development. This chapter presents a clustering analysis for target group identification by locality, based on a case study in the motor insurance industry. Soft computing techniques have been applied to understand the business and customer patterns by clustering data sets sourced from policy transactions and policyholders' profiles. Self organizing map clustering and k-means clustering are used to perform the segmentation tasks in this study. Such clustering analysis can also be employed as a predictive tool for other applications in the insurance industry, which are discussed in this chapter.

1 Introduction

The insurance industry has become very competitive in recent years with legal regulatory changes and globalization requirements. With more and more banks and brokerages joining the game to provide insurance products, it has become harder for leading traditional insurance companies to control market conditions and maintain leading positions. In such a current tough insurance market, it is very hard for companies to retain a perfect balance between market growth and profitability, which are the key factors for a successful business. In particular, a motor insurance company is one type of property and casualty insurance business which focuses on insuring automobile assets. When an insurance company offers a policy to a customer, the basic premium is determined not only based on the value of the motor vehicle, but also on many other characteristics of the owner. As a general practice, a risk level and a population group will be applied for each policyholder and their car. Actuarial analysis is commonly employed for pricing decision making and claim cost analysis. A overview of actuarial analysis for policy pricing, risk and claim analysis can be found in (Webb, Harrison et al. 1992).

Statistical analysis, expert knowledge, and human insight are the fundamental components used in actuarial analysis to develop models and identify risk populations. As the insurance industry becomes more competitive with unprecedented changes, insurers must adopt new technologies to charge appropriate rate for policyholders according to their true level of risk. As such, soft computing techniques have emerged in the insurance applications early this decade. Neural networks, fuzzy systems, and genetic algorithms have been successfully applied to many insurance industry problems including

B. Prasad (Ed.): Soft Computing Applications in Business, STUDFUZZ 230, pp. 113–127, 2008.
springerlink.com

fraud detection (Williams and Huang 1997; Araya, Silva et al. 2004), underwriting (Collins, Ghosh et al. 1988; Nikolopoulos and Duvendack 1994), and customer segmentation (Weber 1996).

From the business intelligence and management perspective, there is also a strong desire for a good understanding of the customer and business patterns with deeper insight, which becomes the necessity and benefits to the business survival and development. Various commercial analytical packages have emerged onto the market tailored for the insurance industry which also proved the need for data mining in the insurance industry.

Insurance companies collect hundreds of data fields (variables) for each policy underwritten and millions of policy entries are recorded into databases on a daily basis. Given the globalization and modernization changes that insurance businesses going through, the data kept in the insurances companies become high dimensional. Valuable information and hidden patterns are buried in large data warehouses, thus soft computing techniques are considered as suitable tools for discovering knowledge and patterns before business strategies are put into action.

Many business goals, such as achieving a good combination of market growth and profitability and running successful marketing campaigns, all require an understanding of the customer behavior and patterns. This chapter focuses on highlighting the capability of a particular soft computing technique, namely clustering techniques, in pattern analysis for insurance business and customers. Two clustering techniques, K-means clustering and self-organizing map are demonstrated in this case study for an insurance company. The insurance business and customer data can be explored by clustering techniques without prior knowledge, in other words, under an unsupervised approach. Consequently, insurers can gain insights from the clusters discovered and natural patterns of the business and customers activities can be revealed.

Identification of target groups using clustering techniques in the insurance industry is discussed in the following section before the detailed description of insurance data analysis and clustering analyses are introduced in Section 3 and Section 4 respectively. In the rest of the chapter, a discussion on how the patterns and knowledge discovered from the clustering analysis can be beneficial and implemented in further insurance business applications is presented.

2   Identification of Target Groups in Insurance Business Context

The process of knowledge discovery by applying data techniques in business databases commonly involves four steps: 1. identify the problems to be solved or analyzed in the business; 2. collect data that needs to be analyzed; 3. establish business actions to be taken based on the analysis result; 4. the outcomes of the action can be measured (Berry and Linoff 1997). Steps 1 and 3 are carried out by business experts and operational teams, and steps 2 and 4 are the focus for knowledge miners or data mining researchers. In this section, the business objectives and problems are discussed before data analysis using soft computing techniques to be demonstrated in following Section 3.

In the insurance industry, target group identification generally means individual policyholders are distinguished into groups and subgroups with certain characteristics, according to specific objectives; these groups will become target groups for business actions. The identification target group appears as a main component in risk management, marketing campaigning, pricing adjustment and underwriting techniques. For instance, a group of customers who have high claim cost and frequency verses low pure premium payment have been identified as 'high risk' customers. In the motor insurance industry, for instance, there is a well-known group, 'male driver under age 25 who purchases third party policy product only', has been identified as 'high risk' group. Based on the characteristics of this group of customers, pricing factors and marketing strategies are determined accordingly.

In the study demonstrated in this chapter, the objectives of target group by locality are to:

- understand the business performances and other manageable facts by locations, scaled on the postcode level;
- identify different groups of customers based on the locality and their specific characteristics;
- provide analyses and recommendations that can help to improve the business for the groups identified;
- for specific business stratigies and targets, various models can be developmed and analyses results can be deployed immediate to make the changes that matter.

3 Data Collection, Preprocessing and Exploratory Analysis

In the 4-step process of knowledge discovery for business applications as discussed in previous Section 2, the data collection, analysis and measurements are the primary interests for data miners and researchers. Within the process of data analysis, there are also four steps to follow. In a simplified description, they are: data preparation, exploratory data analysis, knowledge discovery and business application modeling. In the common practice of data analysis on real-world data, data preparation is a very time consuming procedure especially when the data set comes from multiple sources in a large size with low quality. As such, the data preparation step plays a very important role to achieve high quality data analysis results. In the rest of the section, detailed data preparation steps are discussed and some exploratory analysis results are demonstrated before clustering techniques are applied as a knowledge discovery process in Section 4.

3.1 Data Collection

There are thousands of data fields or variables stored in the databases or data warehouses in the insurance company. Before performing any analysis, a discussion with business experts can benefit the data collection process because they can provide more specific directions of business problems. Therefore, the data collection process becomes more focused and many unnecessary data acquisition steps can also be avoided.

In the study discussed in this chapter, business experts suggested to consider four business perspectives based on which the target group patterns should be analyzed and insight should be gained. They are: financial management, market and sales, product development, and customer and services. More specific factors under each perspective are listed below:

Financial Management
- Volume of business opportunities
- Paid rate
- Termination rate
- Average claim cost
- Incident rate
- Loss ratio

Market & Sales
- Number of vehicles registered
- Unites number of earned premium
- Penetration

Product Development
- Pure premium
- Pure optional benefit
- Standard excess option
- Flexible excess option
- Vehicle's financial situation
- Payment option

Customer & Services
- Website channel option
- Age of policy holders
- Age of insured vehicles

Because the objective of the analysis is based on locality and for groups instead of individual customers, much market and demographic information is required and important. Unfortunately, such information is normally not available in the organization at all. In order to obtain the analysis items discussed above, data integration from various sources is required.

However, customer/product data are generally available within the organization already, which are records of past and current transactions or customer records. This data can be used to understand past behavior, analyze the current situation, or predict the future direction. However, this data only contains the information that the company owns, not covering anything for competitors or the overall industry. Census/survey data is a common source of outsourced data used by individual companies for analytic tasks. It is collected from a large population in order to fill specific information needs with descriptive information, attitudes, or other intangible information. Such information can facilitate new product development and concept and scenario testing. In this study, a survey database for locality, (for instance, the postcode) and a census database for motor vehicle registration are used.

Due to reasons of confidentiality, a simulated dataset based on real business transactions from a leading automobile insurance company has been used in this analysis for demonstration. The real business transactions were extracted from an in-house data warehouse. The company runs insurance businesses in different geographical regions, but the data collected for this study is only for 1 region within a time period of 1 calendar year. The time frame of 1 year for the analysis is determined by business experts based on the product and vehicle registration on an annual business cycle basis. Since the company offers many different types of insurance products, this study is limited to the most popular product within the company. Transactions are based on individual policy holders which have financial transactions within the studied time frame. Generally, the customers included in the dataset consist of existing policy holders, new policy holders, and terminated policyholders.

An overview of the data collected is:

From company data warehouse:
- more than 9 million observations from company data warehouse
- each observation has about 150 variables

From postcode survey database
- around 17000 observations are stored
- each observation has 10 variables

From census data
- more than 1 million observations are collected
- each observation has 6 variables

3.2 Data Preprocessing

Data Cleaning

Data cleaning is a must have step in data preprocessing. SQL and other database query tools can be used for auditing the data quantitatively. The auditing process also includes data source and quality examination, hypothesis testing for variable metrics and preliminary variable selection based on business objectives.

The sizes of the three data sets after basic data cleaning step are:

- business transaction dataset: 830855 observations * 31 variables;
- postcode survery dataset: 3307 observations * 2 variables;
- census registration dataset: 3750 observations * 4 variables.

Data Aggregation and Integration

Data aggregation is a enhancing or enriching step in date preprocessing. This process includes summarizing or aggregating values for variables on a higher level, calculating additional variables from existing variables, and transforming categorical variables when required.

In this study, the three data sets described in the previous section are summarized to a yearly level from daily transactions, and additional variables are calibrated. The integration of the three data sets produced one dataset which has around 700 observations (on the postcode level) and each observation has about 60 variables.

In addition to the steps discussed as above, further operations are also taken into action to deal with missing value and outliers.

3.3 Data Exploratory Analysis

Data exploratory analysis is normally based on fundamental statistic analysis, and produces initial descriptive statistics of the data being examined. The outcomes from this exploratory analysis can provide prior information on variables and their relationships before they are selected to feed into other soft computing techniques for further analysis.

In this study, frequency distribution analysis on single variables in terms of location, variability, and shape measuring were performed. Cross-tabulation analysis on two or more variables is also completed. Due to the limitation of the simulated data, the distributions and trends identified as below are only for demonstration purpose which contains no real business values:

- Penctration rate is in a range 0-29, with an average of 13;
- Average pure premium is $420 in the region over the year, a large group of postcodes have average pure premium in a range of $290 -$390, the other group falls in $500-$600;
- Average termination rate is 9%, with most of the postcodes appear in the range of 6-12%;
- Average age of insured vehicle is 7 year old;
- The average policy holder age is 47 in the region;
- Website channel has not been utilized yet in this financial year;
- Among the policyholders who selected flexible excess, the choices of option 1, 3, and 6 are the popular ones being undertaken.

Although the data exploratory analysis not only provide some fundamental statistics of the data, it is also very useful for understanding the variables' background before they are to be selected for further steps in pattern discovery process, for example, the clustering analysis.

4   Clustering Analysis

Compared with other soft computing techniques, clustering analysis is a type of data-driven approach with unsupervised learning. Rather than using predefined classes, groups and categories, clustering analysis can provide an understanding of data itself naturally. Clustering methods are designed to classify data into group structures which have maximum homogeneity within groups and maximum heterogeneity between groups. In recently analytical analysis on pricing, claims, and customer profiling for insurance companies, there is a trend towards employing clustering techniques on the most updated data instead of relying merely on business expert knowledge, or existing categories. Data-driven clustering analysis can benefit business applications with more insights and deeper understandings in a dynamic environment.

4.1   Clustering Methods

K-means Clustering

K-means (MacQueen 1967) is one of the simplest unsupervised learning clustering algorithms and is widely used for classification / clustering problems. The procedure follows a simple method in order to classify a given data set into a certain number of clusters (assume *K* clusters). The main idea is to define *K* centroids, one for each cluster. These centroids are placed in a cunning way because different locations can cause different result. *K* observations are randomly selected from amongst all observations in the data set. They are the centroids of initial clusters. The next step is to take each point belonging to a given data set and associate it with the nearest centroid in terms of Euclidean distance. When all observations are assigned to the K clusters, the first step is completed. Then, in the second step, K new centroids need to be recomputed within each cluster and as in the previous step, a new binding has to be done between the same data set points and the nearest new centroids. The loop based on second step is repeated until K centroids experience no more change in their locations. Finally, this algorithm aims at minimizing an objective function, in this case a squared error function.

Although it can be proved that the procedure will always terminate, the K-means algorithm does not necessarily find the optimal configuration, corresponding to the global objective function minimum. The algorithm is also significantly sensitive to the initial randomly selected cluster centers. The K-means algorithm can be run multiple times to reduce this effect. K-means is a simple algorithm that has been adapted to many problem domains.

Self Organizing Map Clustering

The Self Organizing Map (SOM) is both a projection method which maps high-dimensional data space into low-dimensional space, and a clustering method such that similar data samples tend to map to nearby neurons. Since its introduction in the early 1980s (Kohonen 1997), the SOM has widely been adopted as a statistical tool for multivariate analysis (Honkela 1997). Compared with other clustering methods, such as hierarchical clustering and K-means clustering, SOM has advantage in superior visualization in addition to its robustness in parameter selection, natural clustering results.

Like other neural network models, the learning algorithm for the SOM follows the basic steps of presenting input patterns, calculating neuron output, and updating weights. The difference between the SOM and the more well-known (supervised) neural network algorithms lies in the method used to calculate the neuron output (a similarity measure), and the concept of a neighborhood of weight updates (Smith 1999).

Unsupervised learning can be achieved by reinforcement learning and competitive learning. The learning method used in the SOM is an extension of competitive learning by involving the neighborhood function for weight updating. When the winning neuron receives the most learning, the weights of neurons in the neighborhood of the winning neuron are also updated with less learning. The neighborhood size is defined to be quite large at the initial stage and gets smaller as learning progresses until it reduces to zero. When the learning passes through a global to local search procedure, localized responses for the neurons in the neighborhood are also varied by receiving different amount of learning which are lesser for the neurons further away from the winning neuron.

The two clustering algorithms discussed above have been used in insurance applications. K-means clustering is used to identify high claiming policyholders in a motor vehicle insurance portfolio (Williams and Huang 1997). SOM is presented in a framework which performs change of representation in knowledge discovery problems using insurance policy data (Hsu, Auvil et al. 1999).

4.2 Clustering Analysis on Motor Insurance Data

Experimentations

For clustering analysis experimentation, we used SAS enterprise Miner which is a data mining package implementing many data mining techniques for clustering, classification, association rules, and decision tree analysis. In addition to the basic processes for applying various techniques to the database, it also provides several other functions such as insight and assessment which provide statistics on original or transformed datasets and comparison of outcomes using different techniques on the same dataset. The clustering analysis process in this study involves following steps:

- Choose input database source / files
- Select variables for transformation
- Statistical analysis
- Clustering
  - SOM clustering
  - K-means clusteirng
- Results assessment and comparsion

After the basic data exploratory analysis, as discussed in previous section, only a limited number of variables are selected as inputs for the clustering analysis. Many unnecessary variables carrying redundant information have been eliminated. For the variables to be used in the clustering process, their missing values and outliers have also been processed to achieve a better solution without undesirable side effects. Table 1 below shows the variables after the data variable selection step has been performed and their corresponding transformation applied. There are two types of data formats obtained from the variable transformation step: categorical and continuous. The only variable in categorical format is 'Risk postcode', which is a targeting variable instead of input variable those are the inputs for clustering techniques. After the clustering results obtained from the clustering techniques, this 'postcode' variable is used to label or map the original data records with their identified cluster group memberships. For other variables in continuous format, a logarithmic transformation has been applied to those which have highly skewed distributions to normalize the data, for example, a variable called 'number of business opportunities'. For variables in a ratio form, a linear 0-1 transformation has been applied.

Once the variables have been selected and transformed into appropriate formats and scales, they are ready to feed into the clustering algorithm. As opposed to classification problems, the number of clusters is unknown. The total number of clusters should not be too few because this could result in a lack of discriminating information for adequate segmentation. In the meantime, the total number of clusters cannot be too many which cause too less number of observations in each cluster. In the experimentation, the number of observations in the smallest cluster should not be less than 10% of the overall data population, to ensure there are enough cases fall in one cluster

Table 1. Variables used in the clustering analysis

Variable	Data type	Transformation
Risk postcode	Categorical	No
Number of vehicles registered	Continuous	Log transformation
Number of unit pure policies	Continuous	Log transformation
Number of business opportunities	Continuous	Log transformation
Termination rate	Continuous	[0,1]
Penetration rate	Continuous	[0,1]
Incidence rate	Continuous	[0,1]
Paid rate	Continuous	[0,1]
Loss ratio	Continuous	[0,1]
Excess option 1	Continuous	[0,1]
Excess option 2	Continuous	[0,1]
Flexible premium option 1 – 6	Continuous	[0,1]
Finance status 1 – 6	Continuous	[0,1]
Website channel (Yes) percentage	Continuous	[0,1]
Website channel (No) percentage	Continuous	[0,1]
Payment per annual	Continuous	[0,1]
Payment per monthly	Continuous	[0,1]
Average amount of optional benefit	Continuous	Log transformation
Average amount of pure premium	Continuous	Log transformation
Average incurred incidence cost	Continuous	Log transformation
Average policy holders age	Continuous	
Average vehicle age	Continuous	

for analysis. In this study, SOM clustering is applied in an order from high dimension map setup to low dimension map setup. A trial-and-error process based on the normalized distortion error and quantization error is used to determine the optimal map settings. The final maps with the determined number of clusters were generated after achieving the minimum errors for both normalized distortion and quantization. The distance between datasets on the map indicates the similarity between them. In this study, after some experimentation, three sets of settings have survived and they produced different number of clusters: namely 4, 6, and 10 clusters respectively.

It is a known disadvantage of K-means clustering that the number K used for clustering can result in a large impact on the clustering results and it is normally required that many experiments are run for the same K in order to achieve a reliable and accurate result. Since the optimal numbers of clusters have been selected after applying SOM on the same dataset, the clustering experimentation with K-means algorithm cost less effort in this study. We choose to use all three (4, 6, 10) numbers of clusters using K-means algorithm with a least squares criterion. The results show that 6 clusters produced less informative results compared with the other two options (4 and 10 clusters), due to closer similarities between fewer clusters. Therefore, after the experimentation using both SOM and K-means clustering, 4 clusters and 10 sub-clusters have been confirmed as the final results before further cluster characteristics examination.

Clusters Characteristics Analyses

The cluster membership and risk postcode label have been assigned to each observation in the dataset after the clustering process was completed. The relationship or distribution between 4 clusters and 10 sub- clusters are illustrated in Figure 1 below. There is no child cluster under Cluster 1 (C1), but all other Clusters 2, 3, and 4 (C2, C3, and C4) all have their own sub-clusters. The summarized features for these four clusters are based on primary drivers identified or most influential attributes for discrimination of clusters. They are: the number of business opportunities, the number of units of pure premiums, and the number of vehicles registered within that postcode. A comparison of these three key factors between four clusters and overall average (indicated by Cluster 0 in the illustration) is demonstrated in Figure 2 and 3. To sum up, the characteristics for 4 clusters are: Very high, high, medium, and very low respectively based on the level or value of three primary driving attributes.

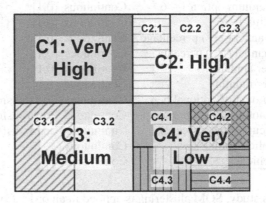

Fig. 1. Overview of 4 clusters

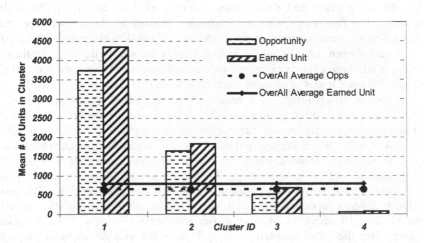

Fig. 2. 'Opportunity' and 'Earned unit' comparison for 4 major clusters identified and the overall average (indicated the trend lines)

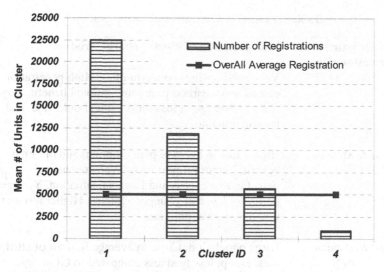

Fig. 3. Number of registration as a key factor for 4 major clusters compared to overall average (indicated by the trend line)

Statistical analysis on 10 sub-clusters was also performed in order to identify their characteristics and key factors or attributes. From the analysis results, 12 factors have been extracted as additional descriptors for each cluster and they are:

1. Average earned premium
2. Average Optional Benefit
3. Proportion of Website channel access
4. Policy Holder Age
5. Insured Vehicle Age
6. Proportion of Flexible excess option
7. Proportion of Vehicle on Financial situation
8. Termination Rate
9. Paid Rate
10. Loss Ratio
11. Average Claim Cost
12. Incident Rate

Based on the descriptors extracted, the characteristics of clusters have been identified. In Table 2 above, 10 featured clusters are illustrated with the population size information. Due to the confidential reason, we cannot disclose further specific or more detailed descriptions of the clusters identified. From the clusters and the average value of their driving attributes, some insights could be gained into the nature pattern of the business and customers for the motor insurance company. The business and customer data have been clustered into different groups and sub-groups on their locality (a postcode level), and these groups are the identified target groups for business strategies design and decision making action.

Table 2. Clusters obtained and their characteristics

Cluster name and population size	Interpretation of cluster characteristics
Favorite (C1) (7%)	Very good business opportunities, High penetration, earned units, earned premium, optional benefit, very close to average in loss ratio, policyholder age, and termination rate
Good & Modern (C2.1) (3%)	Higher rate in average pure premium and optional benefit, Larger proportion of Website channel access, flexible excess and financial situation, Younger policyholders and younger vehicles, Higher termination rate, Lower paid rate
Good & Popular (C2.2) (9%)	Large population, Close to average in most of attributes, except less business compared to C1
Good & Conservative (C2.3) (6%)	Lower rate in average earned premium and optional benefit, Smaller proportion of Website channel access, flexible excess and financial situation, Older policyholders and younger vehicles
High risk & Average (C3.1) (9%)	Higher rate in average earned premium, optional benefit, Younger policyholders and younger vehicles, Higher lose ratio, incident rate, and termination rate, Lower paid rate
Conservative & Average (C3.2) (8%)	Large population in this group, Below average in most of attributes
Business at the Edge (C4.1) (19%)	Less than half in business opportunities compared to Medium group (C3s), Close to average in most of attributes, Older policyholders, Potentials are still there for business
Minor business & High Loss (C4.2) (14%)	Very low volume in business opportunities, Extremely high loss ratio, Older policyholders, Close to average in other attributes
Minor business & Low claims (C4.3) (12%)	Very low volume in business opportunities, Very low claim cost, incident rate, loss ratio and penetration, Close to average in other attributes
No business & Ignore (C4.4) (14%)	Extremely low in business opportunities, can be ignored

Such clustering analysis can also be used as a predictive tool on both macro and macro levels. Taking the analysis as a macro predictive analysis tool, we can form the clusters over different time periods, then the changes of the clusters over time can reveal the dynamic pattern of the business trend. If such clustering process is used on a micro level for prediction, new policy holders can be clustered into the existing cluster groups based on their maximum similarity between their attributes in the profile and the average attributes values of each existing cluster. Such clustering process is an alternative approach for prediction and decision making analysis which is more flexible and data driven compared to traditional classification approaches.

Another advantage of clustering analysis is the explorative feature which can benefit further business knowledge and pattern investigation and hypothesizing. As discussed in this section, clustering can generate data segmentations for the entire business transaction data set, also identifies the most significant or influential factors. The interesting patterns of certain group derived from clustering analysis can help the business analysts to pay more attention to any potential problems or opportunities.

5 Further Business Applications and Discussions

5.1 Marketing Campaign

From the clustering analysis, different groups of policyholders with detailed characteristics have been identified. In common practice, marketing campaigns, such as mailing promotion, are targeted to the entire population of customers within the company or for specific regions. For example, mail promotion for a new insurance product may be sent out for each existing customer in the company's database. The inefficiency of such a campaign has been shown by an extremely low feedback rate. As such, a desire for more focused marketing campaigns on a target group with high potential or more interests has been raised by marketing department. With identified target groups, less expense is required to run a campaign on a smaller size of customer population, however, high return rate can be achieved because the targeted individuals are more likely to take the offer than other potential customers because there is a higher similarity or better match between the products provided in the promotion and their interests.

5.2 Pricing

In the insurance industry, pricing or pure premium is generally formulated based on groups of customers with certain characteristics. These characteristics identified are the most important factors for pricing modeling. For instance, the well-known 'age under 25 male drivers' may share the same premium formula with some other customers who do not belong to 'high risk' group if this specific targeted group was not identified from the overall customer population or was not given enough 'penalty' in the pricing settings. As such, the more accurately the groups and their characteristics are identified, the more accurately the pricing models can be built. In particular, in dynamic industries undergoing rapid change, new segments of the policy holders could emerge and existing segments can also evolve with new features. Therefore, an ongoing clustering analysis based on the most recent data should be performed for target group identification to serve as fundamental analysis for pricing settings and decision making.

6 Conclusions and Discussions

In this chapter, we have demonstrated the potentials of soft computing techniques in insurance industry. More specifically, we have shown the applications (or a case study) of clustering techniques for target group identification on the data set collected from a motor insurance company. The unsupervised soft computing technique, clustering analysis, has been used to reveal the nature patterns of the data, and as such, this data-driven approach also performed a data segmentation task to identify potential target groups and their characteristics. Overall, the benefits of applying clustering techniques on business transactions and customer data sets within insurance industry can be summarized as:

- providing exploratory analysis which serves as a fundemental process for further data investigation and pattern discovery;
- identifying target groups and sub-groups for business stretigies design and plan implementation;
- predicting the overall business trend changes and individual customer group membership;
- building up a flexible process which can be intergrated with other modelling process for various business applications.

However, there are also many issues related to clustering analysis techniques, experimentation, and implementation in real time business environment need to be addressed and tackled, which should be included in the future study.

References

Araya, S., Silva, M., et al.: A methodology for web usage mining and its application to target group identification. Fuzzy Sets and Systems 148(1), 139–152 (2004)

Berry, M., Linoff, G.: Data Mining Techniques: For Marketing, Sales and Customer Support. Wiley, Chichester (1997)

Bocognano, A., Couffinhal, A., et al.: Which coverage for whom? Equity of access to health insurance in France, CREDES, Paris (2000)

Borgatti, S.P.: How to explain hierarchical clustering. Connections 17(2), 78–80 (1994)

Collins, E., Ghosh, S., et al.: An application of a multiple neural network learning system toemulation of mortgage underwriting judgements. In: Neural Networks. IEEE International Conference, pp. 459–466 (1988)

Honkela, T.: Self-Organizing Maps in Natural Language Processing. Unpublished doctoral dissertation, Helsinki University of Technology, Espoo, Finland (1997)

Hsu, W., Auvil, L., et al.: Self-Organizing Systems for Knowledge Discovery in Databases. In: Proceedings of the International Joint Conference on Neural Networks (IJCNN-1999) (1999)

Jain, A.K., Murty, M.N., et al.: Data Clustering: A Review. ACM Computing Surveys 31(3), 265–323 (1999)

Johnson, S.C.: Hierarchical Clustering Schemes. Psychometrika 2, 241–254 (1967)

King, B.: Step-wise clustering procedures. Journal of the American Statistical Association 62(317), 86–101 (1967)

Kohonen, T.: Self-organizing maps. Springer, New York (1997)

MacQueen, J.: Some methods for classification and analysis of multivariate observations. In: Proceedings of the Fifth Berkeley Symposium on Mathematical Statistics and Probability, vol. 1, pp. 281–297 (1967)

Nikolopoulos, C., Duvendack, S.: A hybrid machine learning system and its application to insuranceunderwriting. In: Evolutionary Computation, 1994. IEEE World Congress on Computational Intelligence. Proceedings of the First IEEE Conference, pp. 692–695 (1994)

Smith, K.A.: Introduction to Neural Networks and Data Mining for Business Applications. Eruditions Publishing, Emerald (1999)

Sneath, P.H.A., Sokal, R.R.: Numerical Taxonomy: The principles and practice of numerical classification. W.H. Freeman, San Francisco (1973)

Ward, J.H.J.: Hierarchical grouping to optimize an objective function. Journal of the American Statistical Association 58(301), 236–244 (1963)

Webb, B., C. Harrison, et al.: Insurance Operations, American Institute for Chartered Property Casualty Underwriters Malvern, Pa (1992)

Weber, R.: Customer segmentation for banks and insurance groups with fuzzy clustering techniques. In: Fuzzy Logic. B. JF., pp. 187–196. Wiley, Chichester (1996)

Williams, G., Huang, Z.: Mining the Knowledge Mine: The Hot Spots Methodology for Mining Large Real World Databases. In: Proceedings of the 10th Australian Joint Conference on Artificial Intelligence: Advanced Topics in Artificial Intelligence, pp. 340–348 (1997)

Probabilistic Sales Forecasting for Small and Medium-Size Business Operations

Randall E. Duran

Catena Technologies Pte Ltd, 30 Robinson Road, Robinson Towers #11-04, Singapore 048546, Republic of Singapore

Abstract. One of the most important aspects of operating a business is the forecasting of sales and allocation of resources to fulfill sales. Sales assessments are usually based on mental models that are not well defined, may be biased, and are difficult to refine and improve over time. Defining sales forecasting models for small- and medium-size business operations is especially difficult when the number of sales events is small but the revenue per sales event is large. This chapter reviews the challenges of sales forecasting in this environment and describes how incomplete and potentially suspect information can be used to produce more coherent and adaptable sales forecasts. It outlines an approach for developing sales forecasts based on estimated probability distributions of sales closures. These distributions are then combined with Monte Carlo methods to produce sales forecasts. Distribution estimates are adjusted over time, based on new developments in the sales opportunities. Furthermore, revenue from several types of sources can be combined in the forecast to cater for more complex business environments.

1 Introduction

One of the most important aspects of operating a business is the forecasting of sales and allocation of resources to fulfill sales. While customer relationship management and enterprise resource management systems have helped provide larger organizations with methodologies for forecasting and measuring sales, many small- and medium-size business operations rely on the assessment abilities of one or more sales managers to estimate and manage the sales pipeline. In turn, the outputs of these forecasts as provided to operational mangers are often overly simplistic and do not provide sufficient depth to support resource planning adequately.

The fundamental problem is that sales assessments are usually based on mental models that are not well defined, may be biased, and are difficult to refine and improve over time. Furthermore, defining sales forecasting models for small- and medium-size business operations is often difficult when the number of sales events is small but the revenue per sales event is large. Little or no data may be available to support estimations, and the effect of assumptions and uncertainties is magnified when only a few sales make up the entire revenue stream. Hence, it is not practical to use traditional budgeting processes that set a fixed annual budget when there is a high degree of variability in the expected revenue.

Soft computing approaches, such as Monte Carlo techniques, have been used to determine the general range of possible outcomes where information that is critical to planning is either unknown or unreliable. Savage (2002) focused attention on the problems of using single point estimates in corporate accounting and how Monte Carlo techniques can be used to reflect a more realistic state of corporations' finances. Monte Carlo techniques have also been championed for use in a wide range of

B. Prasad (Ed.): Soft Computing Applications in Business, STUDFUZZ 230, pp. 129–146, 2008.
springerlink.com

applications related to risk modeling and risk analysis (Koller 2000). There has also been research into using adaptive processes and rolling forecasts for budget planning (Hope and Fraser 2003). The objective of the research described in this chapter is to develop an approach that combines computer-based Monte Carlo analysis with adaptive business planning to help provide more accurate business forecast information and support operational resource planning.

This chapter reviews the challenges of sales forecasting in this environment and describes how incomplete and potentially suspect information can be used to produce more coherent and adaptable sales forecasts. It outlines an approach for developing sales forecasts based on estimated probability distributions of sales closures. These distributions are then used with Monte Carlo methods to produce sales forecasts. Distribution estimates are adjusted over time, based on new developments in the sales opportunities. Furthermore, revenue from several types of sources can be combined in the forecast to cater for more complex business environments.

The first section of this chapter presents an overview of the business problem, in context of small- and medium-size business operations. The second section describes the soft computing approach used to address the problem, including an overview of the method and a description of some of the underlying mechanisms. The third section summarizes results of applying this approach in real-world scenarios involving individual, partner, and existing-customer sales situations. The results are then evaluated with regards to the value that they can provide to operational managers, who must base their resourcing decisions on the forecasts. The final section summarizes the benefits of using these techniques and provides suggestions for future areas of research.

2 Overview of the Business Problem

In high value, low volume businesses (HVLVBs) misestimating a single sale can have significant impact on the business. It is also difficult to extrapolate future sales for HVLVBs because the statistical benefits gained through the laws of large numbers are not applicable. There may be few, if any, historical points from which to make inferences, and comparative benchmarks or baselines may not be available. Furthermore, sales forecasts may be affected by events that unfold over time, requiring the forecasts to be changed frequently. Thus, for these types of businesses, planning resource allocation and capital investment according to sales forecasts can be risky. As background information, the following paragraphs provide specific HVLVB examples and describe typical revenue forecasting tools used for planning.

Note that the terms "revenue" and "sales" are used interchangeably, since sales are the primary source of revenue for most small- and medium-size businesses. Likewise, the terms "deal" and "sale" are used interchangeably. Furthermore, a deal or sale "closing" refers to the achievement of contractual commitment that will lead to realization of the projected revenue.

2.1 Business Examples

To help frame the challenges presented when forecasting sales, consider the following two examples. In each example, the businesses' sales are characterized by high value – greater than $100,000 – and low frequency – less than 10 per year – transactions.

They might include business operations that provide highly specialized product and service offerings, such as industry-specific software products, high-end consulting services, and project-focused deliveries.

In the first example, an established company starts a branch office in a foreign country. While the characteristics of product sales may be understood in the domestic market or in other countries, it quite likely that those characteristics will not translate into the new market. Important factors such as branding, cost, and demand may be significantly different in a new market and may not be easily quantified in advance. Hence, uncertainty regarding the expected sales can in turn lead to uncertainty as to how much revenue should be forecast for a new market and how much capital should be invested. As a case in point, many firms have entered new markets, such as China, with unrealistic expectations about how quickly they would be able to generate significant revenue, and they have suffered unexpectedly large losses as a result.

In the second example, consider a startup business that has been operating for only a short period of time. While the general market characteristics may be understood, the willingness of customers to purchase from a newly formed company with no track record may be difficult to quantify. Furthermore, uncontrollable market influences such as natural and man-made disasters may significantly affect the overall economy and, thus, demand for the product. In this situation, deciding when and how much to invest is difficult due to the large number of uncertainties that underlie the assumptions regarding potential revenue.

To simply illustrate the challenge in these examples, imagine that only two sales events are forecast to close within the next six months. In January, one sale – worth $200,000 – is forecast for March, and the other sale – worth $500,000 – is forecast for May. Based on the little information available, the business manager then has to determine how much to invest on presales and sales fulfillment capability (i.e. delivery resources). Table 1 shows the potential outcomes when, using a simplistic approach, a half-year spending budget of $400,000 is allocated and utilized.

Table 1. Results of applying a fixed budget with high sales variance

Possible Outcomes	Resource Investment	Revenue	Financial Effects - P&L	Potential Business Impact
Neither sale closes	$400,000	$0	-$400,000	Need to raise more capital or shut down operation
Opportunity 1 closes, opportunity 2 does not	$400,000	$200,000	-$200,000	Overcapacity: operating losses and reduction of staff
Opportunity 2 closes, opportunity 1 does not	$400,000	$500,000	$100,000	Reasonably matched investment and returns
Both sales close	$400,000	$700,000	$300,000	Under capacity: poor delivery quality, staff turnover, and cancellation of contracts

Note that in this example; only one of the four cases – the third one – produces a fairly positive result. While the fourth case produces the greatest profits, it could have significant negative non-financial results that are likely to impact future revenues. While this scenario is useful for illustrating the fundamental problem, it is important to understand that it has been greatly simplified. The timing of closure, the sale amount, and the likelihood of closing are all estimates that may not be accurate.

Moreover, even businesses that have been operating for several years can also face difficulties forecasting revenue. Consider a mature business whose revenue comes from regular purchases from existing customers, large lump sums from new customer sales, and sales made through external channels such as distribution partners. Different types of uncertainties will underlie each of these revenue sources: sales personnel's estimates may be afflicted by common decision biases (Hammond et al. 2006); repeat business may be inconsistent with respect the timing of orders and amounts; and distribution partners' agreements may be oriented towards annual targets, hence providing little insight as to when revenue, and delivery commitments, will come through. Hence, more mature HVLVBs have similar challenges to new HVLVBs, but with much greater complexity.

2.2 Revenue Forecasting Tools

A common tool used for managing HVLVB sales information is a sales pipeline forecast, which is often embodied as a spreadsheet. This spreadsheet lists potential sales opportunities, their expected closure dates, their percentage chance of closing, the person or entity responsible for closure, and, based on those values, calculates the total expected revenue. While, this technique is helpful for tracking the status of potential sales, its revenue-estimating ability is limited because a simple spreadsheet model does not adequately reflect the true complexity of the sales environment.

With this simple forecasting model, the input estimations related to the percentage chance of closure misrepresents the "all or nothing" nature of sales, especially when the number of sales is small. For example, while ten sales opportunities worth $100,000 each and with 20% likelihood of closing could well (but may not) yield $200,000 revenue, a 20% chance of making a $1,000,000 sale will not. Even the amount forecast is unlikely to be the exact amount realized. The actual dollar amount realized could be greater or smaller, depending on factors such as the agreed final scope of the sale, price negotiations, and fluctuations in foreign exchange rates.

Given the variability of HVLVB sales, it is critical for an experienced sales manager to assess and quantify expected sales revenue. Sales managers will often use the sales pipeline spreadsheet as tracking mechanism, but not as the sole basis for their sales forecast. Instead, they adapt and enhance the calculations provided by the spreadsheet using their own unique estimating methods – i.e. heuristics. Their estimations can be based on knowledge of previous sales results, local purchasing patterns, and the capabilities of individual sales people, and the behavior of customers. Thus, the resulting forecast is highly dependent on the sales manager's personal views and experience, rather than on a system. This relationship is problematic because such personal estimation methods will vary between individual managers, and there is little transparency as to what the individual's methods encompass or are based on. Ten sales mangers could produce significantly different revenue forecasts for the same sales opportunities. Not surprisingly, these highly individual-dependent estimations are not necessarily consistent over time, and can be difficult for others to verify.

Sales forecasts are central to maximizing profits. Profits are achieved by keeping the cost of delivering sales less than actual sales. The cost of delivering sales is largely determined by to the costs of employing resources for sales, presales, and product delivery. It takes time to acquire and develop these resources to the point where they are fully effective. Accurate forecasts are critical to determine how many resources are required and how far in advance they should be secured. Underestimation of sales leads to underinvestment, lower revenue, and lower income at best. Overestimation of sales leads to overinvestment, potential losses, and, in the worst case, business failure. Because the profits of HVLVBs depend on a small number of sales, accurate sales forecasts are especially critical – they can make or break these types of businesses.

2.3 Adaptive Processes

It is difficult, and often not practical, for small- and medium-size businesses to make annual budgets that are fixed and are not affected by the sales volume. Hope and Fraser (2003) argue that this approach is not beneficial for large organizations either, and that more dynamic and adaptive processes are required. Instead of keeping spending budgets fixed, they should be adjusted depending on changes in business over time. As the realization of sales becomes more likely, the budget should be increased, and vice versa.

While dynamic budget adjustment is useful as a concept, its implementation has limitations. Some costs factors, such as travel and capital expenditures, can be increased, reduced or postponed quickly. However, other costs, such as human resource costs, may require significant lead times and may not be easily decreased due to labor laws or overall staff morale considerations. Thus, it is not practical to reevaluate and adjust resourcing on a monthly basis. A more practical approach is to do budget planning based on rolling forecasts – possibly monthly or quarterly – and that may be reassessed on a monthly or quarterly basis, and to take proactive steps to help prepare for the expected outcomes of the forecast. The next section discusses how Monte Carlo techniques can be integrated with such an adaptive forecasting model to yield a Probabilistic Sales Forecasting (PSF) system.

3 A Soft Computing Approach

To improve HVLVB sales forecasting, more complex information must be incorporated into the forecasting model. Partial truths, related to the likelihood of sales completions, must be given better focus. The soft computing approach takes the existing simple sales-pipeline model and enhances it by adding distribution-oriented estimates for the probabilities of sales closures, and allows for adjustment of those distributions over time.

Based on these enhancements to the forecasting model, Monte Carlo simulation techniques are used to assess the potential range of expected sales outcomes. This enhancement transforms the simple pipeline evaluation model from single-value input estimates and a single-value calculated result to a set of inputs and results that reflect the uncertainty of the situation. This approach emulates the implicit mental models that sales managers use to develop their forecasts. But, unlike those implicit estimation methods, it produces much richer forecasting information and provides greater insight into how the estimates of individual sale opportunities relate to the final result.

3.1 Monte Carlo Techniques

In his seminal paper "Risk Analysis in Capital Investment", Hertz (1964) presented a method for creating forecasts and decision models that use range-based approximations of the uncertain variables. This approach has evolved over time as the basis for Monte Carlo simulation techniques. Although software-based tools and computer platforms that implement Monte Carlo simulations have improved tremendously since Hertz's initial work, the overall approach has not changed significantly. The three principal steps that form the basis of Monte Carlo simulations are

1. Estimate the range of values and overall likelihood of those values for each important input variable in the forecasting model, so as to define a probability distribution for each of the variables
2. Simulate random occurrences, based on the defined probability distributions, for each of the input variables estimated in step 1, and measure the resulting value produced by the forecasting model
3. Repeat step 2 many times and record the forecasting results for each simulation. The resulting values represent the likelihood of occurrence for different forecast results, and can be represented as probability distributions for the forecasting model output variables.

These techniques were initially developed for capital investment budgeting purposes and risk analysis and were implemented by custom computer programs. Monte Carlo analysis allows more complex investment scenarios to be modeled more accurately and allows different investments to be compared on their likelihood of different outcomes, rather than just their expected return.

Over the past twenty years, use of Monte Carlo techniques has expanded beyond capital investment, and is used for enterprise risk management, natural resource exploration, and project risk analysis. Furthermore, custom development of computer programs to implement Monte Carlo simulations is no longer necessary. Several tools are available that allow Monte Carlo simulations to be incorporated into spreadsheet models. Mun (2004) provides instruction and examples for applying Monte Carlo analysis using such tools.

The objective of this research is to apply Monte Carlo and adaptive process techniques to the real-world business problems faced by small- and medium-size business operations. Whereas much of the focus of Monte Carlo analysis has been on quantifying risks related to revenue shortfalls, this work focus on the risks of achieving too much as well as too little revenue.

3.2 The Probabilistic Sales Forecasting Model

As shown in Figure 1, the PSF model calculates the expected revenue for the business operation based on multiple sources of sales revenue. The revenue sources considered

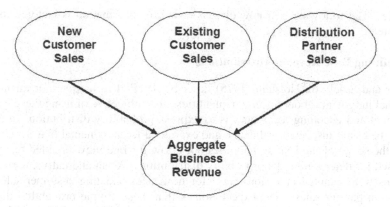

Fig. 1. Relationship of potential revenue sources to overall operating revenue

include new customer sales, existing customer sales, and distribution partner sales. The aggregate business revenue is the sum of revenue from these sources. Depending on the type of business being modeled, it could be the case that one or more of these sources are not relevant; alternatively, other sources of revenue could be applicable. The PSF model is flexible, allowing relevant revenue sources to be included as necessary. Beyond aggregating multiple sources of revenue, the objective of this model is also to factor in the uncertainties and unknowns. Therefore, the likelihood of future sales is also considered.

The key dimensions of the PSF model are the time horizon for revenue acquisition (i.e. sales closure) and the likelihood of achieving various revenue amounts during that time. The PSF model uses a static time horizon of six months. Depending on the type of business, sales closures will often take more than three months to complete, so using a horizon of three months is too short and could easily underestimate actual revenue. On the other hand, it is often difficult to make estimates regarding business with much accuracy beyond six or eight months. Hence, choosing a forecasting horizon of one year would be too speculative as to future events, making the resulting forecast more inaccurate and unreliable.

Six months was chosen as the forecast horizon for the PSF model, as it represents a middle-ground timeframe which is able to capture near-term sale completions at a reasonably accurate level and also allows for sufficient advance notice to support forward planning and implementation of resource adjustments. This strategy is to produce or revise six-month rolling forecasts on a quarterly basis, every three months. However, the forecasts could be generated on a monthly basis, if desired, or updated immediately after major events, such as verbal confirmation of a major new sale, or the sudden and unexpected loss of an existing customer or partner.

The PSF model represents forecast revenue for a six month period with a single output distribution. This result is an aggregation of the potential revenue that could be generated by new sales, existing customers, and distribution partners. The inputs are also probability distributions, and are aggregated using Monte Carlo simulations. The input distributions are constructed based on simple estimates of the sales potential value, specific commitments by customers and partners, and historical data, when it is available. The following will now discuss how this information is factored into the PSF system.

3.3 Defining Base Revenue Distributions

Spetzler and Stael von Holstein (1975) describe detailed techniques for quantifying individual judgments about uncertain quantities. The objective of using these types of assessment and encoding techniques is to express probability distributions in a form that can be easily inspected, validated, and examined for judgmental biases. Unfortunately, the scope of the PSF system does not allow for interview-oriented techniques to be used for determining the probability distributions. As an alternative, normal distributions were adapted in various ways for new sales, existing customer sales, and distribution partner sales. The expectation is that these simple probability distributions would be refined in the future based on experience with the PSF model, or enhanced using interview techniques at a later time. For simplicity, all of the revenue types were modeled as variants of a normal distribution. However, other distribution types may be more appropriate, and this is an area of future research. For the first model implementation, increased precision was traded for reduced complexity. This choice made it easier to implement and maintain, and a simpler model was easier to explain to and review by business stakeholders.

The probability input distributions are estimated somewhat differently for new customer sales, existing customer sales, and distribution partner sales revenue. For new customer sales opportunities, the PSF system models the expected revenue returns as normal distributions that are scaled and adjusted based on four estimated parameters: expected deal size, minimum deal size, maximum deal size, and likelihood of closure. The expected deal size is the "headline" number representing the most likely value of how much the deal is thought to be worth.

Translating the PSF model input parameters into a probability distribution for Monte Carlo simulation is accomplished in two steps. First, the mean of the distribution is defined as the expected deal size. Second, the standard deviation of the distribution is defined as a percentage of the expected deal size, based on the likelihood of closure as shown in Table 2. The method for determining the likelihood of closure is discussed later in this chapter.

As an example, in the early stages a sale the expected size might be of $270,000 with a likelihood of closure at 40%, the baseline distribution would have a mean of $270,000 and a standard deviation of 0.3 x $270,000 = $108,000. As the deal progresses and goes into contractual negotiations – based on refined scope and

Table 2. Rules for determining the standard deviation of new sale input distributions

Expected Likelihood of Success	Standard Deviation
0 – 35%	40%
36-79%	30%
70-90%	20%
90-100%	5%

understanding with the customer – the expected deal size might be reevaluated to be $220,000 and the likelihood of closure adjusted to 80%. In turn, the baseline probability distribution for the sale would be adjusted to have a mean of $220,000 and standard deviation of $220,000 x .20 = $44,000, reflecting increased and more accurate knowledge of the sales situation.

Existing customer sales distributions are estimated based on the mean and standard deviation of historical sales over the previous 18 months. Sale cash flows for consecutive six month periods are calculated, treated as samples of a distribution, and then translated to a distribution to be used for Monte Carlo simulation purposes. Figure 2 shows the monthly sale cash flows, a histogram of their distribution, and the resulting base-probability distribution for an existing customer. Note that the small number of samples is a weakness of this approach. However, using a larger sample – i.e. a longer period of time – may produce even more skewed results if there is a significant growth or reduction trend in progress, which is not uncommon for young businesses. If there are not 18 months of past data available, the customer's potential sales are treated the same way as new customer sales.

For distribution partners the input distributions are determined based on whether the partnership agreement has a revenue target defined. If no revenue target is agreed in advance, expected revenue is estimated based on knowledge of the partners and the opportunities to which they have access. If a target is agreed, the mean is defined as the target amount. The standard deviation to be used for partners, as shown in Table 3, varies depending on the partner maturity and target arrangement. A relatively high standard deviation is used in the case of new partners with no commitment, because of the lack of historical experience and visibility into the potential revenue stream. New partners with a commitment are given a lower standard deviation, to reflect an increased level of confidence in the revenue prediction. Historical sales data is used to determine the standard deviation for more mature partnerships, where there is at least 18 months of historical sales data available.

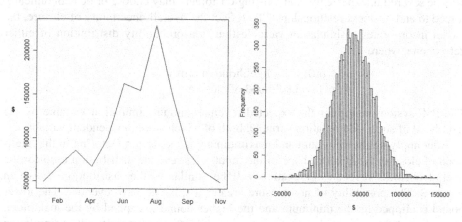

Fig. 2. Existing monthly sales historical data and approximated distribution histogram

Table 3. Rules for determining the standard deviation for partnership revenue

Partnership Characteristics	Standard Deviation
New Partnership, No Commitment	50%
New Partnership, Commitment	25%
Mature Partnership, Historical Data	Based on Historical Data

3.4 Adjustments to the Base Distributions

While normal distributions are a convenient means of modeling the expected amount of revenue that will be generated, actual sale results are not well served by this representation. Therefore, several adjustments to the base input distributions are necessary to make the PSF model more accurate. Specifically, concrete minimum and maximum sale amounts must be incorporated, and the overall likelihood of successfully realizing the revenue must be factored into the simulation model.

Within the PSF model the normal distributions are adjusted by expected minimum and maximum expected revenue potential. The minimum and maximums sale amounts are estimates based on factors such as the minimum size of a sale that is cost-effective for the business to undertake. The maximum may be related to the overall budget or sign-off authority of the purchasing party or, alternatively, the largest size project that the business could deliver with its current or near-term resource capability. During the PSF system simulations, if an input variable fell below the minimum amount, it was treated as a zero value case. If the simulation variable was greater than the maximum amount, it was capped and set as the maximum amount for that simulation instance.

The normal distributions are also adjusted by the sale's likelihood of success. There are many reasons for revenue to not to be realized at all: new customer sales may not close due to competition or change of plans by the customer; partnerships may be severed unexpectedly, and existing customers may choose or be economically forced to end business relationships. To reflect the overall uncertainty of closure, the model incorporates a simulation variable that is a probability distribution of either zero or one, where

$$p(0) = 1 - \text{likelihood of closure}$$
$$p(1) = \text{likelihood of closure}$$

The PSF system multiplies the expected revenue amount simulation variable by the likelihood of success simulation variable, both of which were independent variables.

After applying the minimum and maximum adjustments and factoring in the likelihood of closure, the general shape of the input sales revenue distribution used for the overall business revenue aggregation will be similar to the distribution shown in Figure 3. The possibility of non-closure creates a spike at zero revenue. The lower bound is clipped by the minimum and the higher bound is capped by the maximum. A small spike occurs at the maximum expected revenue amount, since simulation cases that exceed that amount are rounded down to the maximum amount.

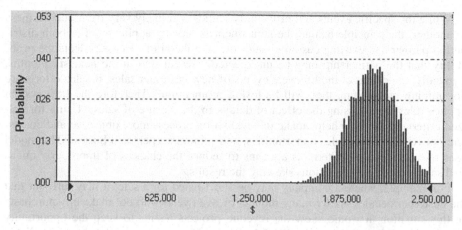

Fig. 3. Adjusted new customer sale probability input distribution

3.5 Likelihood of Success Estimation and Adjustment

While allowing inexact information to be factored into the forecasting process, the PSF system links the likelihood of success to measurable events in the sales cycle. As a new customer sales opportunity progresses, there are different events that will occur that affect the likelihood of its revenue being realized. A company may initially be competing with many others for a particular sale. Being short-listed (or not) for a sale will significantly affect the likelihood of the revenue being realized in the future. Thus, the model's likelihood-of-closure variable is adjusted in accordance with events that affect the sale. Table 4 shows the event-based rules that are applied within the model. While, initially, personal biases may affect the likelihood of closure assigned to various stages of the sales lifecycle, the intention is that the comparison of model performance to actual performance over time can be used to refine the values and eliminate these biases.

Table 4. Rules for determining the likelihood of success for new customer sales

Level	Event	Likelihood of Closure
0	Discussions with Customer	0%
0	Selection of competitor	0%
0	Notification of non-purchase	0%
1	Budget Amount and Decision Timeframe Confirmed	10%
2	Short-listed	40%
3	Pre-engagement Funded	55%
4	Verbal Confirmation of Selection	60%
5	Written Confirmation of Selection	70%
6	Contractual Negotiations Begin	80%
7	Contractual Negotiations Completed	95%

While the specific events and probability values will likely vary from one business to another, the principle behind the adjustment is broadly applicable. For both distribution partner and existing customer sales the likelihood of success reflects the probability that the partnership may be unexpectedly terminated in the next six months. Typically, in context of the lifecycle events of new customer sales, if sales efforts are not gaining momentum, they will be losing momentum. Therefore the PSF system applies rules for assessing the effect of delays in the closure of sales. Using formalized criteria in this way helps make the evaluation process more objective and consistent between different sales opportunities, and minimizes personal biases. The model uses these time-based criteria as a means to reduce the chances of improperly qualified sales or changing situations skewing the results.

For example, where a company may be short-listed for a sale, it may turn out that the customer decides – after many months or even years – to not make any purchase. In this situation, it would skew the revenue projects results to keep the opportunity projected at 40% probability for all of that time. Hence, when one of the events listed in Table 4 occurs, an opportunity closure date is either initially estimated or redefined. If the closure date is past and the sale is not closed, the likelihood of closure factor is reduced to the next lower event level and the closure date is reset to be three months later. This approach allows for the likelihood-of-closure to be increased while the sale is progressing in a positive direction, but reduces the closure likelihood once momentum stops and no progress is being made.

Continuing with the same example, consider that in January the company was short-listed for an opportunity and the closure date was April. If, in April, when the sale was expected to be completed, the company was still at the short-listed stage, then the time-based criteria would cause the likelihood of closure to be reduced from 40% to 10%. This adjustment reflects that the initial estimations regarding the customer decision-making process for closure were overly optimistic, and should be discounted until another positive event occurs. If verbal confirmation of selection was provided in May, the likelihood of closure would be reset as 60% and a new closure date would be estimated. However, if the verbal confirmation does not occur by the end of July – three months after the initially predicted closure date – no further progress would be deemed to have been made and the likelihood of success would be again reduced from 10% to 0%.

The time-based adjustments are applied only to new sales, since in most HVLVBs they have the highest value and the greatest transparency. Adjustments related to changing business conditions over time for existing customer sales are incorporated into the model by using the previous 18 month sales history as the basis of the base probability distribution. Distribution-partner sales estimates are not adjusted on a rolling basis; rather, they are expected to be readjusted on an annual basis based on the previous year's performance and coming year's targets.

4 Results

Having defined the basis of how the PSF system, with the help of sales personnel and managers, determines the input distributions for the business's potential revenue streams, Monte Carlo simulations are then used to calculate the expected revenue for a six month period, producing a aggregated return distribution of revenue. The results of the simulation, which are non-normal, enable managers perform analysis in more depth and answer questions like: "what is the likelihood that the total business unit revenues will be less than $200,000 this quarter?" A sales person could also use the model to get an objective assessment of the total revenue they are likely to achieve at a 50% or 80% level of confidence.

Application of this approach is illustrated using three real-world scenarios related to individual, partner, and existing customer sales in a HVLVB. The first scenario forecasts the sales pipeline of a new sales person, who has great potential, but a limited track record. The second scenario estimates sales that will be generated by a new distribution partner. The third scenario relates to estimating follow-on business with an existing customer. Finally, all three scenarios are combined in the context of a business operation that has its own sales people, distribution partners, and existing sales account managers, who all contribute to the business operation's revenue.

4.1 Estimating New Sales

The first scenario forecasts the sales pipeline of a new sales person, who has great potential, but a limited track record. The model's moderately conservative likelihood-of-closure assumptions are used and then adjusted as sales opportunities materialize or fail to be achieved. This model helps objectively measure sales people's progress over time and predict their overall success. It also can help improve their ability to estimate sales closures based on their historical accuracy.

In June, they have forecast two sales to close in the next six months. The parameters of the two deals are shown in Table 5.

Table 5. Sale parameter determining revenue probability distributions for two new customer sale opportunities

Sale Parameters	New Customer 1	New Customer 2
Min	$600,000	$1,300,000
Expected	$1,000,000	$2,100,000
Max	$1,500,000	$2,500,000
Stdev	$400,000	$420,000
Closure Date	October	November
Event Level	1	5
Event Description	Budget and Decision Timeframe Confirmed	Written Confirmation of Selection
Likelihood of Success	10%	70%

Figure 4 shows a cumulative chart of the expected combined revenue from both opportunities. From the resulting distribution we can see that there is about a 30% chance that no revenue will be generated by either sale and about a 50% chance that the sales will generate more than $1,800,000 of revenue. A simple non-probabilistic model could have estimated of the maximum revenue as the sum of the maximum amounts of the two potential deals, $4,000,000. However, the PSF model shows that there is less than a 5% chance of making more than $2,500,000 from new customer sales.

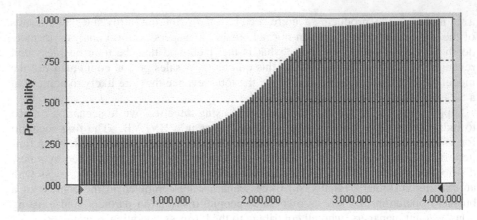

Fig. 4. Cumulative probability chart of expected revenue from two combined new customer sales

4.2 Estimating Partner Sales

The second scenario estimates sales that will be generated by two distribution partners. Compared to solely relying on the partners' commitment, the PSF technique can help account for the potential variance in a partner's revenue contribution and can

Table 6. Sale parameters determining revenue probability distributions for two distribution partners

Sale Parameters	Partner 1	Partner 2
Commitment	No	Yes
Expected	$250,000	$150,000
Stdev	$125,000	$37,500
Likelihood of Success	95%	95%

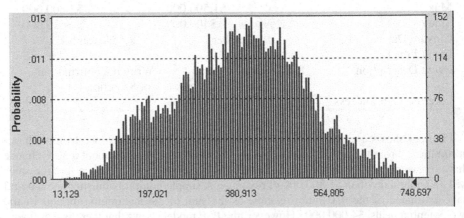

Fig. 5. Histogram of expected combined sales of two partners

help assess whether a partnership should receive more or less investment. Figure 5 shows the probability distribution of the expected combined sales of two distribution partners.

4.3 Estimating Existing Customer Sales

The third scenario relates to estimating follow-on business with two existing customers.

Table 7. Sale parameters determining revenue probability distributions for two existing customers

Sale Parameters	Existing Customer 1	Existing Customer 2
Expected	$277,000	$623,000
Stdev	$75,000	$155,000
Likelihood of Success	95%	75%

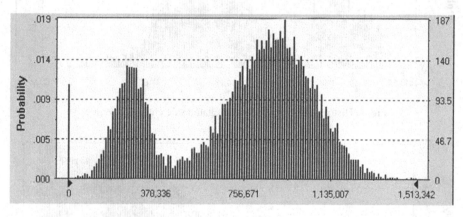

Fig. 6. Histogram of expected combined sales of two partners

The likelihood of success reflects the likelihood that the relationship may be unexpectedly terminated in the next six months. Whereas the business with customer 1 has somewhat higher variance in percentage terms, the relationship is quite solid. In comparison, the relationship with customer two may not be as stable or the overall commitment period may be much shorter, reflecting a higher possibility of the entire revenue stream dropping off during the six month period.

Figure 6 shows the expected revenue from both customers for the six-month period. The result is a distribution that is bimodal and also has a spike at zero, representing the potential loss of both customers.

4.4 Aggregated Forecasts

While the PSF system can be used to generate aggregated sales estimates for each of individual scenarios identified, for operational and resource planning purposes it is more important to look at a forecast that the model would generate for the aggregate business produced by these different scenarios. Therefore, the three scenarios have been combined in the context of a business operation that has its own sales people, distribution partners, and existing sales account managers, who all contribute to the business operation's revenue. By combining the underlying probabilities for each of the sales participants' opportunities within a single Monte Carlo simulation, an aggregate distribution of expected sales for the entire business operation is forecasted. Figures 7 and 8 show the histogram and cumulative probability views of the aggregate revenue forecast.

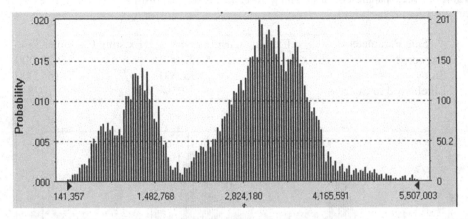

Fig. 7. Histogram of expected aggregated sale channel revenue

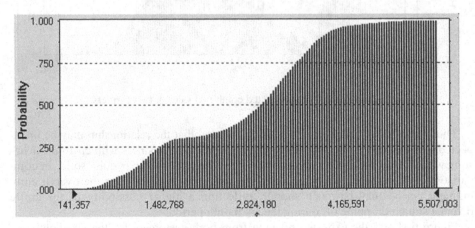

Fig. 8. Cumulative probability chart of expected aggregated sale channel revenue

While the mean, approximately $263,000, could be used as a simple, single-value revenue forecast, the distribution provides a better picture of all the possible outcomes. Using this distribution, further analysis can be easily performed to determine that: while the maximum possible revenue is over $5,500,000, there is less than a 10% chance of realizing more than $3,850,000; there is a 75% chance of achieving at least $1,460,000 in sales revenue; and there is only about a 10% chance that the total revenue will fall between $1,500,000 and $2,500,000. All of these conclusions can be determined by assessing the point at which the curve, as shown in Figure 8, intersects with the respective dollar amounts. This depth of information enables business managers to understand the range of possibilities better and, in turn, more effectively manage the costs and investments that are required to deliver sales.

4.5 Utilizing Probabilistic Forecasts for Resource Planning

Staffing and resourcing is a significant challenge, even when customer demand is well understood. There are many trade offs in balancing operating costs versus readiness for delivery, hiring and development of permanent staff versus the use of contractors, and setting prices to manage the volume of business, and thus, resources required. Adding customer demand as another predominant variable makes resource planning all the more difficult.

The objective of producing sales forecasts in the form of distributions is to help provide more transparency into possible range of outcomes in customer demand, so that resource planners can better determine how best to adjust the other variables under their control. An underlying assumption is that sales revenue can be, more or less, directly translated into resource requirements to deliver the product. This translation often can be roughly calculated as revenue per employee. For example, if the historical revenue per employee is $100,000 and aggregate distribution shows that there is a 50% chance of achieving between $2,700,000 and $3,800,000, then one interpretation could be that an operation manager would need to plan to have between 27-38 staff ready to deliver that business.

There are potentially many ways to utilize these forecasts. Likewise, how the other resourcing variables are adjusted in response will depend on the business and its resource supply options. In the case of the expected revenue shown in Figure 7, if managers were taking a cautious approach they might hire permanent staff to be able to fulfill $2,000,000 of business – taking care of the "small hump" – and begin interviewing and identify potential contractors to use in case the revenues looked like they would fall into the case of the "large hump" part of the curve. Alternatively, if mangers were bullish on future business or were concerned about the ability to acquire additional resources in a tight labor market, they might choose to hire to fulfill $3,500,000 to ensure that they can deliver in at least 75% of the cases.

One of the greatest benefits of the PSF system is being able to regenerate the forecasts quickly and adapt plans as events unfold. Better yet, managers can plan ahead to see how the revenue picture will change if they do unfold and take low-cost steps to help prepare for those outcomes. As events unfold, their overall effect can be better gauged. Relying on emotional instincts when good or bad events occur can be dangerous. When a big sale closes, there is a temptation to over hire, and not consider the potential failed initiatives that might also occur in the future. Likewise, when an expected sale or existing customer is lost, there is the protective reaction to cut costs, possibly to the detriment of other future business. A dynamic model can help reduce the speculative aspects and ensure that decisions are based on the aggregate situation rather than one-time events. To help smooth the effects of periodic revaluations of the forecasts, it could also be useful to adjust the resourcing based on a moving average, so as not to over react to short term changes.

5 Conclusion

By incorporating inexact and approximated information, it is possible to transform a simple, one-dimensional sales estimation technique into a rich and dynamic forecasting tool. Enhancing the forecasting model can help reduce dependencies on hard-to-measure mental models that are used by, and are unique to, individual sales managers. Using a formal model, forecast results can be easily shared, questioned, and further analyzed. Formal models are also more consistent and are less liable to physiological biases, such as sunk costs and the influences of interpersonal relationships. This approach is not expected to deliver precise financial forecasts, but rather to serve as an aid in operational planning. While there is a risk that garbage in may produce garbage out, quantifying the uncertainty of the forecast's inputs is an important step in improving the quality and consistency of the forecast over time. It allows the forecaster's assumptions to be more easily reviewed and questioned by others.

As with all predicative models of complex systems, this approach is not foolproof and should be combined with other information that is not captured within the model to make business decisions. While the PSF model is believed to be sufficiently general and applicable to many types of businesses, it is expected that many of the underlying parameters will need to be adjusted to match specific characteristics of different businesses. Likewise, the model parameters may need to be tuned over time, based on experience using the model and changes to the business' characteristics.

This project has identified a number of areas for extension and future research. One obvious modification would adapt the model to focus on profits instead of sales revenues. Another area of work would to be use more formal encoding techniques for capturing and modeling the uncertainty of sales based on individual's experience, instead of using a one-size-fits-all approach. Beyond optimizing resources, these tools could also be used to perform "what if" analysis on the effects of specific success or failures, and, in turn, determine the appropriate level of aggression when pricing deals, according to their potential impact on resource allocation. Finally, there are many adaptations that could be made to the model with respect to the probability distributions: non-normal distributions – such as triangular or Weibull – could be used as the basis of the variable inputs; alternatively, correlations could be introduced between different probability distributions to reflect interrelationships between different sale opportunities.

References

Hammond, J., Keeney, R., Raffia, H.: The Hidden Traps in Decision Making, Harvard Business Review, January 2006, pp. 118–126 (2006)

Hertz, D.: Risk Analysis in Capital Investment. Harvard Business Review 42, 95–106 (1964)

Hope, J., Fraser, R.: Beyond Budgeting. Harvard Business School Press, Boston (2003)

Koller, G.: Risk modeling for determining value and decision making. Chapman & Hall, CRC, Boca Raton (2000)

Mun, J.: Applied risk analysis: Moving beyond uncertainty in business. John Wiley & Sons, New Jersey (2004)

Savage, S.: Accounting for uncertainty. J Portfolio Management 29(1), 31–39 (2002)

Spetzler, C., von Stael Holstein, C.: Probability encoding in decision analysis. Management Science 22(3) (November 1975)

User Relevance Feedback Analysis in Text Information Retrieval: A Rough Set Approach

Shailendra Singh[1] and Bhanu Prasad[2]

[1] Samsung India Software Center, D-5, Sector-59, Noida -201301, India
[2] Department of Computer and Information Sciences, Florida A&M University, Tallahassee, Florida, 32307, USA

Abstract. User relevance feedback plays an important role in the development of efficient and successful business strategies for several online domains such as: modeling user preferences for information retrieval, personalized recommender systems, automatic categorization of emails, online advertising, online auctions, etc. To achieve success, the business models should have some kind of interactive interface to receive user feedback and also a mechanism for user relevance feedback analyis to extract relevant information from large information repositories such as WWW. We present a rough set based discernibility approach to expand the user preferences by including the relevant conceptual terms extracted from the collection of documents rated by the users. In addition, a rough membership based ranking methodology is proposed to filter out the irrelevant documents retrieved from the information repositories, using an extended set of conceptual terms. This paper provides a detailed implementation of the proposed approach as well as its advantages in the context of user relevance feedback analysis based text information retrieval.

1 Introduction

User (i.e., customer) relevance feedback plays an important role in the development of efficient and successful business strategies for several domains. There are enormous applications that are based on user relevance feedback. This includes online advertising, online auctions, modeling user preferences for information retrieval, personalized recommender systems, automatic categorization of emails etc. (Web 2.0 1997). Successful business models should have an interactive interface to receive user feedback and should comprise a mechanism for user relevance feedback analysis. These features are needed in order to extract relevant information from large information repositories, such as World Wide Web, to fulfill the user needs. In the following sections, we provide some details on the user relevance feedback analysis for the domain of Text Information Retrieval (TIR).

1.1 Text Information Retrieval

The World Wide Web, with its large collection of documents, is a storehouse of information for any user. Search engines help the users to locate information from this vast repository but these search engines usually retrieve a huge list of URLs which are ordered according to a general relevance computation function. Most of the time, the users may find a large proportion of these documents as irrelevant.

However, user queries are usually very short and contain only two words (i.e., terms) in many cases (Gauch et al. 1999). The words, when considered in isolation,

B. Prasad (Ed.): Soft Computing Applications in Business, STUDFUZZ 230, pp. 147–178, 2008.
springerlink.com

cannot convey the content information effectively. The context, in which a word is used, needs to be considered to compute the relevance of a document. Therefore two documents with a high degree of overlap in terms of words may be significantly different from each other in the content. However, the context of natural language texts is very difficult to model mathematically. Since the web documents contain information as a free-form text written in HTML, there is only a visual clue about the document presentation but not much information about its content.

The problems that affect the users and search engine designers are: dealing with the synonyms, dealing with ambiguous terms, finding the exact query and finding associated information (Deerwester et al. 1990). In conclusion, it is very difficult to come up with query-term (or word-based) generic relevance computation functions that can work for all users and domains.

1.2   Improving the Performance of Text Information Retrieval

Text retrieval research is focused on finding some effective solutions that can improve the retrieval precision. One of the mechanisms to improve the functioning of information servers has been to consider user relevance feedback on a training set of documents (Korfhage 1997). The feedback is analyzed to learn the preferences of the user. These preferences may be used to build a modified query that better represents the user's bias. This method was originally proposed by Rocchio (1971). It has undergone many changes (Salton et al. 1985; Crestani 1993; Fuhr and Buckley 1993; Allan et al. 1995) since then. The systems based on these methods are called information-filtering systems.

The paper presented by Meng and Chen (2005) discusses an intelligent web search system that can find the desired information for a user by accepting as little relevance feedback as possible. It also proposed the *websail*, *yarrow* and *features* search engines to achieve significant precision in the search process.

User preferences may also be maintained in the form of user profiles (Roldano 1999). In the context of web search, user profiling can provide valuable guidelines for organizing the way in which the information is being presented to the user. The systems based on this approach are called content-based recommender systems (Pazzani et al. 1996; Jochem et al. 1999; Shytkh and Jin 2007) and they use relevance feedback mechanism to build the personalized profiles inside a search system. This approach is designed to capture the user search intentions derived from explicit and implicit feedback. It also helps to keep the analyzed data in layered profiles for further query augmentation, document recommendation and contribution personalization in the form of subjective index.

Relevance feedback approach has been used for web page clustering in the work presented by Anagnostopoulos et al. (2006). This paper proposes a *balanced relevance weighting mechanism* and uses the proportion of the already relevant categorized information amount for feature classification. Experimental measurements over an e-commerce framework, which describes the fundamental phases of web commercial transactions, are used to verify the robustness of the mechanism on real data.

For effective search, search engines offer user-friendly interfaces. Wilson et al. (2007) presented a formative framework for the evaluation of search interfaces in handling the user tactics and varying user conditions.

1.3 Reasoning Paradigms Used by Text Retrieval Systems

Different reasoning paradigms have been applied to improve the performance of text retrieval systems. Initially, the document retrieval problem was treated within a Boolean model (Korfhage 1997) that considers whether a word was present or not in a document. However, because of the restrictions of the model, a probabilistic model (Maron and Kuhns 1960) was found to be more appropriate. In this approach, every word has a probability of occurrence associated with it and is computed from a set of training documents. The semantic model proposed by Deerwester et al. (1990) reduces the dimensionality of the document space by identifying the conceptual terms within the document collection. This method can handle the *synonymy* and *polysemy* problems but it is computationally complex.

Since natural language documents are inherently based on context-dependent grammar, crisp bi-valued logic is incapable of handling the nuances of natural language. As a result, the bi-valued logic is not always suitable for the text-information retrieval tasks. Szczepaniak and Gil (2003) have argued that new methods based on the fuzzy reasoning model (Zadeh 1965) provide a better platform for handling the uncertainties of text-based reasoning. Another reasoning paradigm that has been found to be suitable for handling uncertainty of knowledge is the rough set based reasoning technique proposed by Pawlak (1982). More details on rough sets are provided later on.

1.4 Scope of the Proposed Work

It is clear from the earlier discussion that the web documents, which have to be retrieved on the basis of words present in them, cannot have a unique relevance factor associated with them because "relevance" is a function of individual user preferences. Two documents may contain the same set of words in entirely different contexts. It is also possible that two completely different sets of words may actually convey the same information. We consider that rough sets would be the ideal reasoning tool for user-centric document classification because it can handle non-unique classificatory analysis effectively.

In this paper, we have proposed a complete framework to show how rough set based reasoning can be used to build a user relevance feedback analysis system. We observed from the literature that the concept of discernibility has not been explored in the context of user relevance-feedback analysis. Discernibility can be the key for building a relevance-feedback based text retrieval system because it can be used to store a user's preference by analyzing the user's interests. It is observed that the precision of retrieval is greatly improved by using the modified query. Thus, the modified query can be used to develop a customized text filtering system. In order to develop a complete customized information filtering system, the documents fetched by the modified query have to be ranked by the system and only the relevant ones should be presented to the user. We have proposed rough-membership functions for computing document relevance in the concept approximation space. These functions are capable of handling document relevance computation better then traditional rough membership computing functions. Although rough set based analysis had been proposed earlier to state how approximations for sets can be constructed, to the best of our knowledge, its potential in building a complete relevance-feedback based document filtering system was not reported earlier.

2 A Review of Current Trends in User Relevance Feedback Analysis

Most of the information available on the web is in an unstructured format. One of the key aspects of text information retrieval systems is to design some efficient strategies to locate and retrieve the information (in response to a user query) from the repository. The repository is usually indexed by some means. Some of these systems integrate a user relevance-feedback mechanism in order to improve the performance. This kind of system is of primary interest to us. We also present a detailed review of the literature. We start the review process with an introduction to the search engines and a review of various text retrieval systems. We have traced the history of the development of these systems and present the salient features of the reasoning paradigms that have been employed by the text retrieval systems.

2.1 Search Engines

Search engines maintain very large databases of web documents. Query words entered by the users are used to retrieve relevant documents from this database. Search engines update the databases regularly through a special class of programs called "robots" or "spiders" that constantly scan the web for new content. Some of the most popular search engines are Yahoo!, AltaVista, WebCrawler, Excite and Google. Each search engine has its own relevance computation function to compute the relevance of all the indexed documents against a user query. The efficiency of a search engine is dependent on its indexing, storage and retrieval mechanisms. The performance of a search engine is dependent on its computation technique(s).

Yahoo (1994) is the web's oldest "directory", in which, initially, human editors organized the web sites into categories. However, since October 2002, Yahoo is making use of Google's crawler-based listings for its main results. Yahoo! does not support Boolean operators or nested searching. It does allow a user to indicate terms to be definitely included or definitely excluded, while searching for a document.

AltaVista (1995) is the oldest crawler-based search engine on the web and is good for images and news search. It allows two distinct search modes: simple search and advanced search. AltaVista supports full Boolean searching with the operators AND, OR and NOT. Searching can be nested using parentheses. Phrase searching is also possible. It also provides a free translation service into some languages.

Google is the most popular search engine on the Internet. It originated in Stanford University in 1998 (Google 1998). The success of Google lies in its computation of relevance of a page on the basis of the page rank. The page rank for a page is computed as a function of the number of links that goes out from the page and the number of links that are directed to the page and also the page-ranks of the linked pages. This mechanism enables the retrieval of a page even when it does not contain the query words explicitly. The searching mechanism of Google also considers the relative positions of words in the query and in the documents. As a result, the search returns more intelligent results than other search engines. Google maintains both forward indexing from the documents to the words and backward index from the words to the documents to compute the relevance efficiently. The Google database maintains a compressed version of the web page it rates, and this web page accounts for the Google's efficiency.

Though search engines are the most popular mechanisms for locating and retrieving relevant information, they return a huge list of URLs in response to any user query. This is because the simple word based pattern matching can find a lot of match in any repository. It is also due to the fact that no two users usually have the same perspective of searching and hence there is a lot of redundancy in the search results. An effective way to tackle this problem is by considering the user-related parameters also, while searching. Search engines in general do not entertain any interaction with the user other than the initial query. Though many popular search engines today accept feedback from the user about the quality of the results and how to improve them, these feedbacks are mostly used offline to tune the system parameters rather than providing better results immediately to the user.

2.1.1 Relevance Feedback

Relevance feedback from the users provides an assessment of the actual (as opposed to predicted) relevance of a document to a query. This can be utilized effectively to improve the performance of the retrieval mechanisms (Korfhage 1997).

A relevance feedback method describes the characteristic of the process that ultimately improves the retrieval performance of the ranked-output Information Retrieval (IR) systems, by including the user in the overall process. Relevance feedback based IR systems are designed as the learning systems that can perform the following steps:

- User poses a query.
- The system returns a set of documents.
- The user provides his/her judgments about the relevance of a subset of the documents.
- The relevance feedback is then used to improve the precision of search for the user.

The relevance feedback based methods tune the system parameters to provide the best performance after analyzing the user's feedback. These methods have been classified as implicit methods (Mitchel 1997). These methods perform parameter adjustment that either leads to a changed document representation or a changed query representation or both. These methods are further categorized as document-oriented, query-oriented or a unified mechanism oriented, depending on how they function. Their definitions are as follows:

Document Oriented View

In the document oriented view, system parameters are modified in such a way that the documents move closer to the relevant queries. Relevance feedback from the user is also used to tune the documents' representations. Thus the system's internal representation of the document changes with time.

Brauen et al. (1968) described an approach for document vector modification. Salton (1971) further described the advantages of accumulating the relevance feedback and making permanent changes to the document representation. However, the problem with this approach is that no unique representation of the document can be permanently maintained. The performance suffers severely if the new queries are not significantly similar to the old ones.

Query Oriented View

Most of the modern IR systems are built by using the query-oriented view that was initially proposed by Rocchio (1971). The major steps in this method are:

- Using the relevance feedback, the initial query is expanded and used to retrieve a new set of documents.
- Irrelevant documents are filtered out using a user-specific relevance function.

Crestani (1993) proposed a neural-network based query modification method. Allan et al. (1995) presented an overview of Text Retrieval Conference (TREC) experiments related to the query expansion. TREC collection contains topics, documents and relevance assessments in the topic X document space, where a topic can be considered as a more structured and extensive version of a query. Potential queries were generated from this collection and then expanded.

2.2 Interactive Information Retrieval Systems

All the models and algorithms presented earlier consider the users primarily as feedback providers. Once the feedback is received, the models process the feedback without interacting with the user again. The user usually has no role in the final retrieval results. More interaction among the users and the retrieval system was sought and a new class of systems was thereby designed. Koeneman and Belkin (1996) reported that a handshaking between the user and the system does improve the results of purely automated query expansion mechanisms reported by Salton and Buckley (1990).

There have been several studies to identify the benefits of interactive information retrieval systems. Belkin et al. (2001) investigated the impacts of query length on task retrieval performance and/or interaction. InQuery (Croft 1995) is a text retrieval system based on an interactive relevance feedback technique. It uses Bayesian reasoning to combine evidence from the corpus (i.e., from the documents themselves) and the users. The automatic query expansion is accomplished by using a corpus-analysis technique. InQuery has obtained excellent results with TREC collection. WebPlanner (Jochem 1999) is another system that guides users towards their needs by using structured domain-specific queries. However, it is not feasible to have the domain-specific structured queries to be stored for all possible domains. Hence a better way to provide individual satisfaction is to use a dynamic representation for the users' preferences. ProspectMiner (Intarka 2000) is another retrieval system that learns the users' interests based on user rating. The retrieval system suggests better queries, which will fetch more relevant pages.

2.2.1 Modeling User Behavior

As interactive information retrieval systems gained popularity, modeling the user behavior became one of the key research areas. Relevance-feedback based methods accept the relevance feedback from the users in a query-specific manner but did not perform any general user behavior modeling. User modeling for information retrieval was introduced by Widyantoro et al. (2001). User modeling accumulates information about users' preferences from relevance feedback. At the same time, this information is usually made permanent. The user behavior model is adapted over time and is used to study the long time preferences and also enables real-time learning of short-time interests.

One of the important advances in this area of research is the consideration of user profiles. *User profiling* aims at *determining a representation of the user preferences so that the stored values may serve as input parameters for a filtering action operating on the available offer* (Roldano 1999). In the context of web search, user profiling can provide long-term valuable guidelines for organizing the way in which the information is being presented to the user. User profiling is often coupled with learning from the user feedback. User modeling is heavily used in recommender-systems introduced by Goldberg (1992). "Syskill & Webert" (Pazzani et al. 1996) is a software agent that learns a user's interest profile and uses the profile to identify the interesting web pages for the user. It learns a separate profile for each topic of each user. Various classification algorithms like the Bayesian classifier, nearest neighbor classifier (Mitchel 1997), PEBLS classifier (Cost and Salzberg 1993), etc. have been used to classify the new documents and reported their performance analyses.

2.3 Reasoning Models for Text Retrieval

In the earlier sections, the design of different kinds of text retrieval systems according to their functional characteristics is presented. We will now present an overview of various reasoning paradigms for text retrieval. The reasoning paradigm dictates the document and query representation schemes and the logical design of the text retrieval systems. The most popular TIR models are the Boolean model, vector-based model, probabilistic model and the fuzzy model. A brief overview of these models is presented next.

2.3.1 Boolean Model

The Boolean model was the first text retrieval model to be developed. The queries are logical Boolean constructs or functions linking the search terms with AND, OR, or NOT connectives. Queries are logical constructs and hence they cannot be directly compared for similarity with documents. Documents are usually represented as vectors of 1s and 0s. The value 1 at position k in a vector V representing the document D indicates that the term k is present in document D. The value 0 indicates the absence of the term. Though this model is simple and ensures fast retrieval, there are few disadvantages with it (Verhoeff et al. 1961; Korfhage 1997). Extended Boolean models such as fuzzy, infinite-norm and P-One Boolean models were developed by Lee (1994).

2.3.2 Vector-Based Model

The Vector-based model allows the system to have the same representation for the documents and the queries. Queries and documents are represented as N-order vectors. N is the number of terms that characterizes the whole document collection. Usually the vectors contain normalized frequency-based weights like Term Frequency (TF) / Inverse Document Frequency (IDF) that are normalized over a document or the entire collection. Many similarity measures were presented in the literature. Most of them are *lexical similarity measures* such as distance-based and angle-based measures (Korfhage 1997) and are used to assess the relevance. Based on the similarity measures, it is possible to have a ranked output.

2.3.3 Probabilistic Model

The probabilistic model was first introduced by Maron and Kuhns (1960). It uses a membership-like function (of a document to the retrieved set of relevant documents) but the function is based on probabilities. The probabilistic model also allows ranked-output. In response to a query, the system retrieves the documents whose probability of being relevant to the query is more than a certain threshold value.

Probabilistic models have shown good retrieval performance, starting with simple experiments as reported by Fuhr (1986). More complex and effective retrieval models are also presented (Losee 1988; Fuhr and Buckley 1991). The probabilistic model has also been applied in combination with Boolean and fuzzy models to create better hybrid models (Losee and Bookestein 1988; Fuhr and Buckley 1993; Wong and Yao 1995). The disadvantage of the probabilistic models is that they check the simple occurrences or co-occurrences of words but cannot handle the problems related to the uncertainties (uncertainties are inherent in natural language texts).

2.3.4 Fuzzy TIR Systems

Fuzzy set theory (Zadeh 1965) allows different degrees of membership strength to a set. The inherent flexibility of this reasoning paradigm allows all documents to have a certain degree of membership to the set of retrieved documents, which is based on the set of terms from the query.

Several information retrieval applications have been developed using the fuzzy model. HP Laboratories developed a help-desk retrieval system based on fuzzy theory (Piccinelli and Mont 1998). Szczepaniak and Niewiadimski (2003) have argued that new methods are essential to establish textual fuzzy similarity among various documents to provide retrieval results that are more acceptable to the human user. Several fuzzy similarity measures for comparing documents and their effectiveness are also mentioned in that work. Details of other related work can be found in Subtil et al. (1996), Glockner and Knoll (1999) and Kim and Lee (2001).

3 Discernibility for User Relevance Feedback Analysis

It is observed from the overview that the research in text retrieval has gradually shifted from general-purpose search mechanisms, as implemented by search engines, to the design of user-specific and more recently domain-specific retrieval systems. The reasoning paradigm has also shifted from crisp logic based reasoning to fuzzy and rough set based models since the later models can handle the domain uncertainties more elegantly.

The rough set based reasoning technique proposed by Pawlak (1982) provides a granular approach for reasoning. Rough set is a tool to deal with inexact, uncertain or vague knowledge. Miyamoto (1998) discussed the applicability of rough sets for information retrieval. Stefanowski and Tsoukias (2001) have shown how rough reasoning can be applied to classify imprecise information. Specifically, this approach provides a mechanism to represent the approximations of concepts in terms of overlapping concepts. Since words do not provide a unique classification of documents when considered in isolation, a granular approach that can extend the document or the query space to include other conceptually related words is ideally suited for TIR. Hybrid models that integrate the granularity of the rough set based approach with the uncertainty handling principles of fuzzy reasoning are also becoming popular.

Some researchers (Srinivasan et al. 2001; Das-Gupta 1988) have proposed the use of rough-approximation techniques for query expansion that is based on a rough set model. Bao et al. (2001) have developed a hybrid system for document categorization using the latent semantic indexing and rough set based methods. This system extracts a minimal set of co-ordinate keywords to distinguish between classes of documents. Chouchoulas and Shen (2001) have demonstrated the applicability of rough set theory to the information filtering by categorizing e-mails. Jensen and Shen (2001) have used rough set theory for automatic classification of WWW bookmarks. Menasalvas et al. (2002) have provided a rough set based analysis of affability of web pages. They have also used rough set based approaches to compute user interest parameters for web usage mining.

In this section, we have presented a rough set based framework for designing a user-relevance feedback analysis system.

3.1 Rough Sets and Discernibility

As opposed to fuzzy set theory, the focus of rough set theory is to model the uncertainty within a domain, which may arise due to indiscernibility or limited discernibility of objects in the domain of discourse. The idea is to approximate a concept by a pair of subsets called lower and upper approximations. Since the concepts themselves may be imprecise, integration of rough sets and fuzzy sets can lead to the development of other reasoning models. These models can deal with the uncertainties of the real world better than any other reasoning paradigms (Pal and Skowron 1999).

Rough set theory provides a framework to perform discernibility analysis for a pre-classified data set. Discernibility based data analysis focuses on finding the defining properties that can discern maximally between objects of different categories. *Discernibility Matrix* (Komorowski et al. 1999) provides an effective way of finding the minimal set of attributes and their values, which can distinguish between the maximum number of objects of different classes. Therefore, discernibility analysis paves the way for the following:

* *Dimensionality reduction for large data sets* – attributes which are not in the discerning set are usually redundant for the task of classification.
* *Representation of categories* – the set of discerning attributes and the values obtained through discernibility analysis can be used to design the classifiers. Classification rules can then be framed to categorize new data.

Other known approaches for reducing the dimensionality include Singular Value Decomposition (SVD) and Principal Component Analysis (PCA). SVD reduces the dimension and also takes care of the synonymy and polysemy (Deerwester et al. 1990). However, the complexity of this method is very high and hence it is not very much suitable for large data sets. In addition, this method does not specify how the reduced dimensions are to be chosen. PCA (Jolliffe 2002) is not suitable for document categorization.

Since rough set theory introduces the concept of non-unique classification and granular approach for reasoning, we have explored the possibility of using this form of reasoning for designing user-relevance feedback based document filtering mechanism. Relevance feedback by the user provides an assessment of the actual relevance of a document from the user's perspective. Since the user queries are usually short, the same set of documents retrieved (as a result of a single query) may receive very different relevance feedback from different users. Adding more words arbitrarily to the query does not help because some existing documents may be missed out while trying to judge the relevance for all the words. Though Boolean combinations help, it becomes practically infeasible for the user to think a priori of all the synonyms that the author might have used, or all the words that are defining his or her particular perspective. Rather, we propose that it is possible to analyze the user response to a set of documents retrieved by the broad query and then automatically extract a modified query, which will be useful for retrieving more relevant documents in the future. The basis for modifying the initial query will be based on the discernibility analysis. The analysis can detect the optimal set of features and values that can represent the user's categorization of the documents, as relevant or irrelevant, effectively. While query-modification is not a new concept in user-relevance feedback based document retrieval, the concept of discernibility has not been explored earlier.

3.2 Using Discernibility to Represent User Preference

Since document categorization is a non-unique classificatory analysis task, we propose to use the rough-theoretic concept of *discernibility* to track the user preferences. The aim is to extract a minimal set of words that can distinguish between relevant and irrelevant classes of documents for a user. Using this approach, we can distinguish between two categories of discerning words - one set, which by their presence, usually makes a document relevant to the user; the other set, which by their presence, makes a document irrelevant to the user. We define these two sets of discerning words as follows:

- Words which are present with a high relative importance in the good documents and are not present or have low importance in the bad documents are *positively discerning* words. Such words are desirable in a document in the future.
- Words which are present with a high relative importance in the bad documents and have low importance in the good documents are *negatively discerning* words. Such words need to be avoided while searching for relevant documents.

The concept of good and bad documents is explained in the next section. We use the positive and negative discerning words to formulate an improved search query for the user. When this modified query is fed to the search engine, it is observed that the returned documents are better in general. Since discernibility analysis yields the minimal set of attributes along with their values, these values can be used to build the classifiers for a decision system. We will show that, in our case, the classifiers can be used to imbibe the user's preferences. Some mathematical preliminaries of rough set theory and the concept of discernibility analysis are presented by Komorowski et al. (1999).

3.3 Synthesizing Concept Approximation

One of the main goals of the information system analysis in rough set theory (Komorowski et al. 1999; Pawlak 1982) is to synthesize the approximations of target concepts such as decision classes in terms of the defining attributes. A decision system expresses all the knowledge about the model. The decision classes constructed above can be characterized by the attribute values of the elements lying in the class. There can be many possible ways to characterize the decision since there can be inherent redundancy in the data or some attributes may actually be insignificant. As a result, the minimality of decision representation is a crucial issue. While the equivalence relation introduced earlier is one of the ways of dimension reduction, the main objective of a decision system is to identify the significant features that can help in inducing the decision boundaries and help the decision making process in the future.

The study of discernibility aims to find the optimally discerning attributes and their values for various decision classes. One of the ways of finding the significant attributes in a decision system is to observe the effects of removing these features on the overall decision system. Rough set based discernibility analysis plays an important role in finding the minimal set of discerning features for the class classification problems.

3.4 Designing a Discernibility Based Query Modifier

The remaining sections of this paper illustrate how we have modeled the problem of user-relevance feedback based TIR as a decision theoretic problem to be analyzed using the rough set theory.

The decision system is comprised of documents which are retrieved by a search engine in response to a user query and are rated by a user on the basis of its relevance to the query. Each document is represented as a vector of weighted terms, as explained in the next section. Initially, the user provides ratings as "good", "bad" or "average", for a set of training documents. The ratings reflect a particular user's bias and have nothing to do with the underlying search engine's relevance computation. A decision table is then constructed using the words as features and their weights as values and the user's rating as the decision. This is a numeric valued table from which we construct the discernibility matrix. In the original analysis of discernibility matrix, all decision differences are considered as identical. The Minimal Degree - Heuristic (MD-Heuristic) algorithm (Komorowski et al. 1999) cannot be applied straight away to learn the user's preferences because this is a graded decision space. The error committed in judging a "good" document as "average" and the error committed in judging a "bad" document as "good" are definitely not equal. Hence we have presented on how to conduct a discernibility analysis that is more suitable to the problem of relevance feedback based text retrieval problem. We have also presented an algorithm which is built upon the original MD-Heuristic algorithm, but it is modified to suit the requirements of the text retrieval problem. This representation is better suited to extract target concepts representing the user's preference. We present the details of the discernibility analysis for user preference extraction in the next few sections. The results obtained using the procedure for different domains are also presented.

3.4.1 Representation of a Document

Let x be an *information item* in a finite universal set U. The *information item* x is an object used for *information retrieval*. For our text filtering system, x can be an HTML document, a text document or an XML document. Let U be defined as

$$U = \{x_1, x_2, x_3, \ldots, x_n\}.$$

Let A be the set of information features used to define an item and it is defined as

$$A = \{a_1, a_2, a_3, \ldots, a_m\}.$$

For defining the text documents, A can be a collection of keywords or query terms or simply the set of all the words that occur in a document. For web document analysis in general, the set A is extended to include image features, icons, captions, video-frame characteristics and so on. Each information item has a value associated with each attribute. The domain of definition of an attribute restricts the set of admissible values for the attribute.

The feature set used in our text filtering system is comprised of the non-HTML words occurring in a text. The value of a word is computed as a function of its frequency of occurrence in the document. Since there are several words like "is", "was", "among", etc., which are an essential part of the English language grammar but convey little information about the content of a document, are called the stop words. A list of stop words is maintained by our system. The words that are not part of the stop word list are considered as the features. Each document x_i in our universe is represented as a set of weighted words. To calculate the weights of the words, we use the HTML source code of the pages. Documents authored in HTML make use of a number of tags to give special emphasis to the visual presentation of the document. The special emphasis enables the author to convey his/her intentions to the reader, in a more effective way. Thus, a document with a good representative set of keywords can provide more information about its contents than a document whose keywords have been chosen rather haphazardly. Similarly, when a word is emphasized within a running text, it stands out and conveys more about the contents of the paragraph. Thus HTML tags, in a way, help the contextual analysis of words. Search engines like Google also make use of this concept.

Since each tag in an HTML document has a special significance, we designate separate weights to each one of them. We have used tag weights in the range of 1 to 10 with an interval of 2, where 10 for <TITLE – indicates title of the document>, 8 for <META -contains all possible keywords >, 6 for <BOLD> and <I - emphasized words>, 4 for <H1> & <H2>, 2 for <H3> & <H4> headings tag of different font sizes respectively. Words occurring without any tag have weight 1. The weight of a particular word a is then computed as follows:

$$W(a) = \sum_{i=1}^{m} w_i \times l_i \tag{3.1}$$

where w_i represents the weight of tag i and l_i represents the number of times the word a appears within tag i, and m is the total number of tags (including no tag) considered.

Finally, for every document, the weight of all the words are divided by the maximum weight to normalize all weights within the range of 0 to 1.Thus the weight of the most important word in the document is always 1.0. The normalization takes care of the fact that longer documents do not get unfair advantage over shorter documents during relevance computation. The words are then arranged according to decreasing weights.

3.4.2 Building the Decision Table for Discernibility Analysis

To build our initial decision table, from which the system learns the client's preference, the user is asked to rate a set of training documents fetched by a standard search engine in response to a user query. The user is requested to rate on a scale of 1-3 with 3 denoting "Good", 2 denoting "Average" and 1 denoting "Bad". We did work with finer divisions of rating on a 1-5 scale also.

Each training document is converted to its weighted word representation as described in the earlier section. Since different documents may have different number of words, we use a fixed size word vector of size n, where $n \geq 30$, to represent each document. All our results have been obtained with $n = 50$, as no significant improvement was observed with $n > 50$. Thus, each training document is represented in terms of its top 50 most important words. The set of information features in our decision table is then constructed as a collection of all distinct words occurring in the training set. Each document x_i is now represented in terms of the information features as follows: $x_i = \{a_1: nw_1, a_2: nw_2, a_3: nw_3, \ldots, a_k: nw_q\}$

Here, nw_j $(1 \leq j \leq q)$ represents the normalized weight of the word a_j in a document. It is 0.0 if it does not occur in the document; else it has a weight between 0.0 and 1.0. k denotes the total number of distinct words in the training set.

A decision table is then built as follows:

$$D_T = \left(W_M \cup \{d_i\} \right) \tag{3.2}$$

where d_i is the decision of user about training document x_i and

$$W_M = \left[n\, w_{ij} \right]_{N \times k} \tag{3.3}$$

where N is the total number of documents in the training set and nw_{ij} represents the weight of the word w_j in the training document x_i.

3.4.3 Discernibility Analysis for Learning the User Preference

The motivation for using discernibility is explained through a hypothetical example. Let us assume that an user has rated four documents D_1, D_2, D_3, and D_4 as bad (1), average (2), good (3) and good (3) respectively. Further, suppose that D_1, D_2, D_3, and D_4 have words W_1, W_2, W_3, and W_4 with weights as shown in the Table 1. This shows that the word W_3 doesn't have the capacity to distinguish between the "good", "bad" or "average" documents because it has a high weight for all of them. On the other hand, the word W_1 has the potential to distinguish a "bad" document from a "good" one. Thus we can say that W_1 may be a negatively discerning word. Similarly it may be argued that W_2 and W_4 are positively discerning words.

If we can extract the most discerning words using the decision Table 1, these can be used to formulate a modified query as follows. The modified query is constructed as a Boolean function of all the positive and negative discerning words. Positive

Table 1. A decision table D_T for documents

Documents	W_1	W_2	W_3	W_4	Decision
D_1	1.0	0.1	0.9	0.2	1 (bad)
D_2	0.5	0.5	0.9	0.75	2 (average)
D_3	0.0	1.0	1.0	0.9	3 (good)
D_4	0.2	0.9	0.9	0.9	3 (good)

discerning words are indicated as the desirable words in the documents while negatively discerning words are indicated as undesirable.

For example, a query "$W_2 + W_4 - W_1$" would indicate that we want the documents which contain W_2 and W_4 but do not contain W_1. Now we will explain how the most discerning words can be extracted from the decision table D_T.

3.4.4 Constructing the Discernibility Matrix
Since the word weights stored in the decision table are continuous, it is not ideally suited for finding significant words. Thus we first *discretize* the decision table D_T. The discretization technique returns a partition of the continuous value set for each information feature, by dividing the entire range into intervals (Komorowski et al. 1999).

Let us suppose the number of distinct documents in the training set is N and the number of distinct words in the entire training set is k. We will now show how the *discernibility matrix* for this set is constructed. For each distinct word in the domain, its weights in different documents are arranged in an ascending order. An interval set P_a is then constructed for the word a as follows: $P_a = \{[I_0, I_1), [I_1, I_2), \ldots, [I_r, I_{r+1})\}$ where $I_a = I_0 < I_1 < I_2 < \ldots < I_r < I_{r+1} = L_a$.

For each interval in the interval set, the mid point of the interval is called a **cut**. Each distinct word a is thus associated with a set of cuts defined as $\{[a, c_1], [a, c_2], \ldots, [a, c_r]\} \subset A \times \Re$ where c_i is the mid point of $[I_{i-1}, I_i]$.

Since each word may not be present in all the documents, the number of intervals and therefore the number of cuts may be different for different words. Let us suppose that the word a_i has p_i cuts. Then the total number of cuts for the entire set of words is $\Sigma_i p_i$ where $1 \le i \le k$.

Table 2. Discernibility matrix $D_T{}^*$

Document Pairs	$W_1, 0.1$	$W_1, 0.35$	$W_1, 0.75$	$W_2, 0.3$	$W_2, 0.7$	$W_2, 0.95$	$W_3, 0.95$	$W_4, 0.47$	$W_4, 0.82$
(D_1, D_2)	0	0	-1	1	0	0	0	1	0
(D_1, D_3)	- 2	- 2	- 2	2	2	2	2	2	2
(D_1, D_4)	0	- 2	- 2	2	2	0	0	2	2
(D_2, D_3)	-1	-1	0	0	1	1	0	0	1
(D_2, D_4)	0	-1	0	0	1	0	0	0	0

Let D_T* denotes the discernibility matrix. D_T* is constructed with the help of the decision table and the cuts. D_T* has one column for each cut induced over D_T and one row for each pair of documents (D_i, D_j) where D_i and D_j have different user categorizations (i.e., different decisions).

An entry v_{ij}^k in D_T^* is decided as follows:

- $v_{ij}^k = 0$ in D_T^* if the document pair D_i and D_j have different decisions but the weights of the word i in both the documents are on the same side of the cut.
- $v_{ij}^k = d_i - d_j$ if the weight of the word k in the document i is more than the cut and the weight of the word in the document j is less than the cut, and the documents have different decisions d_i and d_j respectively.
- $v_{ij}^k = d_j - d_i$ otherwise.

Thus, a non-zero entry corresponding to a word a in D_T* indicates that the word has two different significance levels in two documents of different decisions. The absolute value of the entry determines the power of the word to distinguish between the two different categories. A negative value indicates that the word has a higher weight in a bad document than in a good document. This means that the word may be a negatively discerning word. Table 2 denotes the discernibility matrix D_T* constructed from Table 1.

3.4.5 Analyzing the Discernibility Matrix

Here we explain how D_T^* is analyzed to get the most discerning words and the corresponding values of the cuts. Since, theoretically, there can be an infinite number of such cuts possible, we can apply the MD-Heuristic algorithm presented by Komorowski et al. (1999) to obtain the minimal set of maximal discerning cuts.

However, the MD-Heuristic algorithm works with a discernibility matrix in which all the decision differences are considered as identical. We have modified this algorithm (Komorowski et al. 1999) to handle a graded classification scheme like ours, where the classes "good" and "average" are definitely more close to each other in terms of quality than "good" and "bad. Thus we need to give more emphasis on the words that can distinguish between the documents of two extreme categories like good and bad rather than on the words that are less distinctive.

To obtain this, we first consider the columns (in the discernibility matrix) that induce the highest degree of difference in decision, followed by the next highest and so on, until there are no more discerning words in the set. This is summarized in the following algorithm.

Modified MD-Heuristic Algorithm *(to find the minimal set of words to discern maximal number of documents)*

Input: Discernibility Matrix D_T^*.
Output: Discerning features in the form of cuts and cut values.
Step 1: *Let W denotes the set of most discerning words. Initialize W to NULL. Initialize T = r, where r is the maximum difference possible in the decision (in our case it is 2);*
Step 2: *For each entry in D_T^*, consider the absolute value of the decision-difference stored there. If none of the absolute values are equal to T then set T = T − 1. If T=0 then stop else go to Step 3;*

Step 3: *By considering the absolute values of decision difference, choose a column with the maximal number of occurrences of T's – this column contains the word and the cut that is discerning the maximum number of documents corresponding to the current level of discernibility;*

Step 4: *Select the word a* and cut c* corresponding to this column. In case of a tie, the leftmost column is chosen. Delete the column from D_T^*. Delete all the rows marked T in this column since this decision-discernibility is already considered. Delete all the columns for a* from D_T^*;*

Step 5: *If majority of the decision differences for this column are negative then the word is tagged with a negative (-) sign to indicate that it is a negatively discerning word. Otherwise it is tagged with a positive sign (+) to indicate that it is a positively discerning word.*

Step 6: *Add the tagged word a* and cut c* to W.*

Step 7: *If there are more rows left then go to Step 2. Else stop.*

Discussion on the Algorithm - This algorithm outputs a list of words along with their cut-values, which collectively discern all pairs of documents rated by the user. A pair (a^*, c^*) denotes a word and its weight such that the documents having the word a^* with a weight less than c^* is likely to have one categorization while the document having the word a^* with a weight greater than c^* is likely to have a different categorization. If the same word occurs more than once with different cut values then it indicates the fact that the documents having the word with different weights can be classified into different categories with respect to the word. The categorization is depending on which side of the cut values the weight falls in. Each word has an associated tag to indicate whether it is positively discerning or negatively discerning as determined in Step 5 of the previous algorithm. The presence of the positively discerning words and the absence of the negatively discerning words are desirable in good documents. Bad documents can be identified by high weights of negatively discerning words. Since positively discerning words are desired in good documents, these words can be used to represent a modified query. For search engines which support Boolean queries, one can build a Boolean expression with the positive and negative discerning words appropriately. The weights of the cut values are not used for the query modification. These values will be used in the automatic grading of documents as presented in Section 4.

We will now explain the whole methodology with an example.

Example 1 - One of the queries we fed to the Google search engine was *"alcohol addiction"*. This is one of the queries mentioned in TREC list of topics. In this domain, we were particularly interested in authoritative documents, on alcohol addiction, that discussed the causes and remedies of alcohol addiction (rather than the list of remedial centers, etc.). We have rated the top 50 documents returned by Google and used them as our training set. After analyzing the HTML source codes for these documents and by considering the top 50 most important words from each document, we compiled the total collection of information features as a set of 700 words, some of which are shown below.

A = [research, information, project, addictions, alcohol, drugs, abuse, help, collection, description, dedicated, contains, alcoholism…promises, centre, enquirer].

The decision table D_T of the entire set contains 50 rows and 700 columns where each row represents one document and each column represents one word. Apart from this, the last column represents the user's rank for the document. The entries in this table reflect the weight of a particular word in a document.

$$D_T = \left(W_M \cup \{d_i\} \right)$$

$$\begin{bmatrix} 0 & 0.25 & 0 & 1 & 0.38 & 0.38 & 0.37 & 0.128 & 0.127...0 & 0 \\ 0 & 0.003 & 0 & 0 & 0.076 & 0.075 & 0.075 & 0 & 0...0 & 0 \\ ... & & & & & & & & & \\ 0 & 0 & 0 & 0.40 & 0.44 & 0 & 0 & 0 & 0...0.209 & 0.209 \end{bmatrix} \cup \begin{bmatrix} 2 \\ 1 \\ ... \\ 1 \end{bmatrix} \qquad (3.4)$$

The decision table D_T is constructed from the training documents and then the discernibility matrix is constructed using D_T as explained in the Sections 3.4.4 and 3.4.5, in order to find the most discerning words with a minimal set of cuts c^*.

The set W of most discerning words for alcohol addiction domain is obtained as follows:

$W = \{a_1^* = $ Alcohol, $a_2^* = $ Addictions, $a_3^* = $ Abuse, $a_4^* = $ Drugs, $a_5^* = $ Treatment, $a_6^* = $ Health, $a_7^* = $ Description, $a_8^* = $ Rehabilitation, $a_9^* = $ Help, $a_{10}^* = $ Revised $(-)\}$

The words a_1 to a_9 are positively discerning words, while a_{10} is a negatively discerning word. A *modified query* can be constructed using these discerning words W. The modified query can be fed to the search engine again. In general, we have observed that fewer documents are retrieved with the modified query. This is expected since the modified query becomes more focused. The total number of documents retrieved by Google with the initial query is 52800 while it is only 3270 with the modified query. Since Google does not treat negative (-) operators, we deleted the negative words while constructing the modified query.

The list of URLs for the initial and modified queries need not be the same. Certain documents which were not retrieved with the original query may be retrieved in the modified query. Similarly certain documents which were obtained with the initial query may not be retrieved with the modified one. It is observed that the relevance of the documents retrieved with the modified query is also significantly better from the user's perspective. While the user rated 48% of the top 50 documents as "bad" with the initial query "alcohol addiction", with the modified query, the same user rated only 4% of the top 50 documents as "bad". The decrease in the percentage of bad

Table 3. User ratings for the training sets of 3 URLs with query *"alcohol addiction"*

No.	List of top 5 retrieved URLs out of 50 documents collection with the initial query "alcohol addiction"	User ratings
1.	http://center.butler.brown.edu/	1
2.	http://www.well.com/user/woa/	2
3.	http://www.thirteen.org/edonline/lessons/alcohol/alcoholov.html	1

Table 4. User ratings for top 3 retrieved documents with modified query *"alcohol, addiction, abuse, drugs, treatment, health, description, rehabilitation, help"*

No.	List of top 3 retrieved URLs out of 50 documents collection with the modified query "alcohol, addictions, abuse, drugs, treatment, health, description, rehabilitation, help"	User ratings
1.	http://dmoz.org/Health/Addictions/Substance_Abuse/Treatment/Alternative/	3
2.	http://dmoz.org/Society/Issues/Health/Mental_Health/Substance_Abuse/	3
3.	http://uk.dir.yahoo.com/health/diseases_and_conditions/addiction_and_recovery/	3

documents with the modified query proves the effectiveness of the modified query. Table 3 lists the top 3 URLs retrieved with the initial query *"alcohol addiction"*. Table 4 lists the top 3 URLs with the modified query *"alcohol, addictions, abuse, drugs, treatment, health, description, rehabilitation, help"*.

URL number 1 *"http://center.butler.brown.edu/"* of the initial query of Table 3, which was rated "bad" by the user, is not retrieved in the top 50 web documents fetched using the modified query. In the same way, many URLs like URL no. 1 *"http://dmoz.org/Health/Addictions/Substance_Abuse/Treatment/Alternative/"* of Table 4 are rated "good" by the user due to useful information as well as many good links. This is not retrieved by the original query.

The most discerning words with their minimal cuts for *"alcohol addiction"* are as follows: $P = [\{(alcohol, 0.25), (alcohol, 0.37)\}, \{(addictions, 0.5)\}, \{(abuse, 0.07), (abuse, 0.29)\}, \{(drugs, 0.0038)\}, \{(treatment, 0.039)\}, \{(health, 0.24)\}, \{(description, 0.034)\}, \{(rehabilitation, 0.15)\}, \{(help, 0.038)\}, \{(revised, -0.02)\}]$.

The 1st and 2nd cuts of the word *'alcohol'* are 0.25 and 0.37 respectively. Similarly we have cuts for other discerning words. These cuts value can be used to categorize new documents retrieved through the modified query.

3.5 Results

We have experimented on a variety of domains to judge the performance of the query modification scheme. Some of the query topics like "HIV", "Antarctica Expedition" were chosen from TREC's topic list. The user perspectives for these topics were also kept according to TREC requirements. One of the topics we worked with is "Blood cancer", as opposed to the TREC mentioned topic "Brain Cancer". Similarly, we chose "Indian Tourism" as a domain. We chose some new queries like "Alternative medicine" and "Air pollution" because different kinds of users are likely to have interest in these topics (and hence these could be rated from different user perspectives).

Different groups of users were asked to evaluate different domains depending on their interest in the topics. For each domain, the same user(s) has rated the initial and final set of documents to maintain uniform standards of rating. In all the cases, usually the authoritative pages containing good documentation about the topic were

Table 5. The percentage of bad documents in the top 50 URLs correspond to the initial and modified query. The Boolean operator "-" indicates the negatively discerning words which need to be avoided to construct the modified query. Negatively discerning wording are not fed to Google while querying.

Initial query and number of retrieved URLs using Google	% of bad documents in top 50 from the initial query	Modified query and total number of retrieved URLs using Google search engine with the modified query	% of bad documents in top 50 from the modified query
Alcohol Addiction 528,000	48	"Alcohol + addictions + abuse + drugs + treatment + health + description + rehabilitation + help – revised" 3270	4
Alternative Medicine 1,910,000	44	"Health + medicine + alternative + therapy + yoga + Acupuncture + stress + diet + disease – agriculture - altvetmed – dosha" 3940	12
Blood Cancer 196,000	44	"Cancer + health + medical + information + blood + leukaemi + help + myeloma + alive + symptom – companion - safety - poison" 2410	0
Air Pollution 2,180,000	30	"Air + pollution + health + carbon + environmental + research + smog + quality + rain + clean" 8810	6
HIV 6,360,000	44	"AIDS + treatment + HIV + epidemic + health + description + information + service + virus – details" 6150	14
Indian Tourism 695,000	76	"India + tourism + information + Indian + Pradesh + indiaworld" 190	18
Antarctic Exploration 134,000	36	"Antarctic + exploration + expedition + history + polar + science + discovery" 8380	32

rated higher than the hub pages containing links to other pages. We will now produce some results obtained for various domains.

As already mentioned, it is generally observed that the documents fetched with the modified query are more relevant to the user. Table 5 authenticates this claim. Columns 2 and 4 of Table 5 show the percentage of bad documents among the top 50 URLs obtained with the initial and the modified queries respectively. Table 5 also shows both the initial and modified queries mentioned above for each domain. The table clearly shows that there is a substantial reduction in bad documents among the top 50 URLs for all the domains. While the user had rated 48% of the documents retrieved using the query "Alcohol Addiction" as "bad", only 4% of those retrieved with the modified query was rated as "bad" by the same user. For the query "blood cancer", the percentage of bad documents in the top 50 URLs was brought down substantially from 44% to 0%. A similar decrease in the bad documents was observed for the query "Air Pollution" where the percentage of bad documents among the top

50 came down from 30% to 6%. For the HIV domain, even with the modified query, 14% of the top 50 documents were rated as "bad" by the user. While this is not a very good sign, we may observe that initially 44% of the documents were bad. Based on an investigation, we found that this is because a huge number of HIV patient case histories are available over the web and all the search terms are usually present in all these case histories. As a result, such domains are difficult to discern.

3.6 Modeling Different Perspectives for the Same Domain

We now present the results for some domains which have a lot of natural sub-topic divisions and usually a user is interested in only one of them at a time. But since the user initially does not know how many documents are there on the web or their relative organizations, the user starts with a query on a broad topic of interest and selects some documents of his/her choice by rating them along the way. We will show how our discernibility analysis based system learns the subtopic selection criterion effectively. We present the results for two domains whose initial queries to Google were *"alternative medicine"* and *"Indian tourism"* respectively.

With alternative medicine as the query, a lot of documents containing information on "aromatherapy", "yoga", "homeopathy", etc. were retrieved. The user was particularly interested in authoritative documents on the topic of "acupuncture". The modified query in this case is *"Health, Medicine, Alternative, Therapy, Yoga, Acupuncture, Stress, Diet, Disease"*.

In case of "Indian Tourism", the user first showed interest in the pages that gave information about all the Indian states and their historical buildings. The modified query in this case was *"forts, palaces"*. Interestingly, even without the word *"India"*, the top 10 documents retrieved by Google were on historical places in India only. In the second case, the user showed interest in hill resorts. The modified query was *"hill, mountains, India"*. These results show that it may be possible to apply discernibility analysis for developing an automatic document classification system, which can be trained to learn the rules of classification in a supervised manner.

In this section, we have presented a discernibility matrix approach to extract most discerning words along with their cut values. Most discerning words are used as a modified query to retrieve documents while cut values are not used. In the next section, we will explain how these cut values can be used to design a user-preference based text filtering system that can filter out bad documents for the user (Singh and Dey 2005).

4 Rough Membership Based Document Grading Scheme

In the earlier sections, we saw how rough set theory based discernibility analysis can be used to learn user preferences. We showed how a modified query can be obtained from the initial query, to increase the precision of document retrieval for a particular user. However, we observed that the retrieved list of URLs were still ordered by the underlying search engine's relevance computation preferences and did not match with the user's perspective. To improve the system's retrieval precision, the newly retrieved documents have to be graded by the system and only those documents which are not "bad" according to the system rating, should be passed on to the user. Hence we have proposed a new set of rough membership functions for computing the membership of a document for various user classification categories. This is because the traditional rough membership tools cannot be directly applied for classifying the documents (equivalence classes with respect to a subset of words can very often be "NULL" for an arbitrary set of words). We have also shown how a complete document filtering system can be built using the proposed concepts. The filtering system initially analyzes a set of training documents which are evaluated by the user as mentioned in the earlier section. The filtering agent created through the discernibility analysis is called a *sieve*. The sieve resides at the client's side and grades new documents for the user by using the rough membership functions.

4.1 Construction of Sieve to Represent User Preference

In this section, we will show how the sieve is constructed as a reduced decision table D_T containing the most discerning words of W along with the associated cut values that represent the user's preference.

Let P denotes the partition of most discerning words W using the minimal set of cuts. P is represented as follows:

$$P = \left[\left\{ \left(a_1^*, c_1^{1*} \right), ..., \left(a_1^*, c_1^{r*} \right) \right\}, \left\{ \left(a_2^*, c_2^{1*} \right), ..., \left(a_1^*, c_2^{r*} \right) \right\}, \left\{ \left(a_m^*, c_m^{1*} \right), ..., \left(a_m^*, c_m^{r*} \right) \right\} \right]$$

(4.1)

where $c_i^{1*}, ..., c_i^{r*}$ are the 1st, ...r^{th} cut for a_1^*.

Since this set can effectively discern between all pairs of classification categories, we can use this reduced set P to represent the earlier decision table D_T with a reduced dimensionality. Using P, we can also convert the numeric valued decision table D_T into a discrete decision table. The discrete valued decision table is called the *symbolic decision table S*. This is obtained from D_T as follows.

- All information features which are not contained in the minimal cut set are removed from D_T.

- Assign interval name 0 to all values of a_1^* less than c_1^{1*} for the interval $\left[a_1^*, c_1^{1*} \right)$, interval name 1 to all the values of a_1^* lying in the interval $\left[c_1^{1*}, c_1^{2*} \right)$, interval name 2 to all the values a_1^* lying in the interval $\left[c_1^{2*}, c_1^{3*} \right)$, and so on.

- An analogous assignment is done for each of the other discerning words $a_2^* ... a_m^*$.

- For each discerning word, interval value -1 indicates the absence of the word in the document.

After this assignment, the decision table D_T is now converted to a symbolic decision table S. In other words, the symbolic decision table S contains only the most discerning words for all the documents and their ratings by a user. The interval number corresponding to the weight of the discerning word in the document replaces its actual weight in the symbolic decision table. A weight -1 indicates the absence of the word in the document. S is thus described as follows:

$$S = \left(SW_M \cup \{d_i\} \right) \text{ where } d_i = \{1,2,3\} \text{ and } SW_M = \left[Sv_{ij} \right]_{N \times k}$$

where N is the number of training documents and

k is the number of discerning words and

(4.2)

$$Sv_{ij} = \left\{ \begin{array}{l} e : e \in \text{interval no. corresponding to the minimal} \\ \text{cuts of } j^{th} \text{ word of } i^{th} \text{ document} \end{array} \right\}$$

$e = -1$, if word is absent in that document

The set of most discerning words and their cut values for the query "*alternative medicine*" are shown as follows: $W = \{a_1^* = \text{Health}, a_2^* = \text{Medicine}, a_3^* = \text{Alternative}, a_4^* = \text{Therapy}, a_5^* = \text{Stress}, a_6^* = \text{Diet}, a_7^* = \text{Acupuncture}, a_8^* = \text{Yoga}, a_9^* = \text{Disease} \}$.

Table 6. Symbolic decision table consisting of most discerning words for the query "*alternative medicine*"

$$S = \left(SW_M \cup \{d_i\} \right)$$

	a_1^*	a_2^*	a_3^*	a_4^*	a_5^*	a_6^*	a_7^*	a_8^*	a_9^*		d
x_1	2	2	1	-1	-1	-1	-1	-1	-1		2
x_2	-1	0	1	-1	-1	-1	-1	-1	-1		2
x_3	-1	0	0	-1	-1	-1	-1	-1	-1		1
x_4	-1	2	-1	-1	-1	-1	-1	-1	-1		1
x_5	-1	2	-1	1	1	1	1	-1	-1	$\cup$	3
x_6	2	2	1	-1	-1	-1	-1	1	-1		2
x_7	-1	2	1	-1	-1	-1	-1	-1	-1		1
x_8	2	0	1	-1	-1	-1	-1	1	1		3
x_9	2	-1	0	1	0	-1	-1	1	-1		3
x_{10}	2	2	2	1	-1	1	-1	1	-1		3

SW_M d_i

$P = [\{(\text{Health}, 0.41), (\text{Health}, 0.63)\}, \{(\text{Medicine}, 0.20), (\text{Medicine}, 0.29)\}, \{(\text{Al-}$
ternative, 0.125), (Alternative, 0.16)}, { (Therapy, 0.20)}, { (Stress, 0.35)}, { (Diet,
0.49)}, { (Acupuncture, 0.3)}, { (Yoga, 0.19)}, { (Disease, 0.27)}}].

We assign the interval name 0 to all the values of "health" that are less than 0.41,
the interval 1 to all the values that are within the range [0.41, 0.63) and 2 to [0.63,
max. value corresponding to health). Other discerning words are also appropriately
treated. The symbolic decision table S obtained is shown in Table 6. Again, a cursory
glance through S shows that the good and average documents usually have words
$a_1^*(\text{health})$ and $a_8^*(\text{yoga})$ with a high frequency, while in bad documents the words
$a_1^*(\text{health})$ and a_8^* (yoga) are occurring with a low frequency.

4.2 Grading of New Documents

In this section, we will present a new rough membership function to compute the
grades of new documents. We observed in the earlier section that the symbolic interval, to which the frequency of a discerning word belongs, can be an indicative of the
relevance of a document to the user. Hence we will use these values for computing
the degree of overlap and thereby the rough memberships of a new document to the
categories "good", "bad" and "average".

4.2.1 Classical Rough Membership Computation

Classically, *rough membership* function quantifies the degree of relative membership
of an element into a given category. The degree of membership of an element x to a
category d, denoted by μ_d^A, is given by the degree of overlap between the set of
documents in category d (denoted by C^d) in the given information table and the
equivalence class $[x]_A$ to which x belongs (Komorowski et al., 1999). Here A is the set
of description attributes. The memberships are defined as follows:

$$\mu_d^A : U \rightarrow [0,1] \quad \text{and} \quad \mu_d^A(x) = \frac{|[x]_A \cap C^d|}{|[x]_A|} \tag{4.3}$$

The traditional rough set based membership function defined above does not work
well for the text documents because the decision table that is obtained for a set of
documents is usually sparse. Thus, for most of the subset of words, the equivalence
class for a document may be very small or NULL. Also, different words when taken
in isolation may indicate different kinds of categorization for the same document.
This motivated us to propose some new rough membership computation functions,
which are more suitable for the document grading tasks.

4.2.2 A New Rough Membership Function for Document Grading

We preferred to design a membership function which can take into account the contribution of all the words together. The proposed function takes into account the relative degree of membership of a document into different categories with respect to
each discerning word and then provides a final categorization of the new document as
a function of all these memberships.

To categorize a new document, we consider only those words of the new document which are presenting in the set W of most discerning words. The weight of the word in the document is determined and normalized, as explained in Section 3.4.2. Then, on the basis of minimal cuts described in the section 4.1, we determine the interval to which the weight of the word belongs. Let us take an example document "http://altmedicine.about.com/mlibrary.htm" from the domain of "*Alternative Medicine*", which was not retrieved originally but is retrieved with the modified query. The interval values for the different discerning words are shown in Equation 4.4. As can be seen from Table 6, the good words appear with a very high frequency.

$$New\ Symbolic\ W_M = \begin{bmatrix} & a_1^* & a_2^* & a_3^* & a_4^* & a_5^* & a_6^* & a_7^* & a_8^* & a_9^* \\ nx_1 & 2 & 2 & 1 & -1 & -1 & -1 & -1 & -1 & -1 \end{bmatrix} \qquad (4.4)$$

The equivalence class for a word is determined as a function of the word weight in the document as shown in Equation 4.5. Let $e_a(x)$ denotes the weight of a in document x. The equivalence class of each discerning word a^* is a collection of those training documents that have the same interval number.

$$\left[a^* \right] = \left\{ \left(x, x' \right) \in U^2 \mid \forall a^* \in W, e_{a^*}(x) = e_{a^*}\left(x' \right) \right\} \qquad (4.5)$$

The new document's rough membership to each category d is computed by taking into account the relevance of each discerning word a^* for that category. The membership to category d corresponding to the discerning word a^* is denoted by $\mu_d^{a^*}(x)$, and is computed using Equation 4.6.

$$\mu_d^{a^*}(x) = \frac{\left| \left[a_j^* \right] \cap C^d \right|}{\left| \left[a_j^* \right] \right|},\ where\ d = \{1,2,3\},\ \forall a_j^* \in W,\ \&\ C^d\ is\ a\ collection$$

$$of\ documents\ of\ category\ d,\ provided\ \left| \left[a_j^* \right] \right| \neq \phi,\ \mu_d^{a^*}(x) \in [0,\ 1], \qquad (4.6)$$

$where\ (1 \le j \le k),\ k = no.of$ most discerning words

The total membership value of a document x to decision category d is computed by $\mu_d(x)$ as shown in the Equation 4.7. This function computes the membership with respect to the complete set of discerning words.

$$\mu_d(x) = \left[\sum_j \mu_d^{a_j^*}(x) \right] \qquad (4.7)$$

The final membership of the document x to a category d is given by $\mu_d^*(x)$ and is provided in Equation 4.8. This equation computes the final membership value of a document x to a category d, by normalizing it against all the membership values. Finally, the decision category with the maximum weight is assigned as the category of the document.

$$\mu_d^*(x) = \left[\mu_d(x) \Big/ \Sigma \mu_d(x) \right] \quad \& \quad \text{Decision} = \left\{ d : \max\left(\mu_d^*(x) \right) \right\} \tag{4.8}$$

4.2.3 An Example Membership Computation

An example is provided to illustrate how new documents are retrieved for the domain *"alternative medicine"*. The symbolic decision table for this domain was presented in Table 6. The equivalence classes for each of the discerning words can be computed from the training set. Using the Equations 4.6 to 4.8 presented above, we show how the grade of a new document nx_1 can be computed.

This document contains the discerning words *"alcohol, addictions, abuse, drugs, treatment, health, description, rehabilitation, help"*. The weights of these words in the new document are replaced by the interval numbers to which they belong.

We will now illustrate the calculations using the symbolic decision table obtained from a small sample training set of 10 documents, presented in Table 6. The set of training documents having a decision class d is denoted by C^d where $d = \{1, 2, 3\}$. C^1 shows the collection of training documents which falls under decision 1. The other collections are also similarly computed as follows: $C^1 = \{x_3, x_4, x_7\}$, $C^2 = \{x_1, x_2, x_6\}$ and $C^3 = \{x_5, x_8, x_9, x_{10}\}$.

The equivalence class of each discerning word present in nx_1 document, constructed using the small sample training set, is given as follows: [health (a_1^*)] = $\{x_1, x_6, x_8, x_9, x_{10}\}$, [medicine (a_2^*)] = $\{x_1, x_4, x_5, x_6, x_7, x_{10}\}$ and [alternative (a_3^*)] = $\{x_1, x_2, x_6, x_7, x_8\}$.

We find that the equivalence classes of health (a_1^*), medicine (a_2^*) and alternative (a_3^*) are non-empty while equivalence classes of others words are null. We now show the rough membership computations for different decisions, corresponding to the word "health (a_1^*)" using Equation 4.6.

$$\mu_1^{a_1^*} = \frac{\left| \left[a_1^* \right] \cap C^1 \right|}{\left| \left[a_1^* \right] \right|} = \frac{|\{\phi\}|}{|\{x_1, x_6, x_8, x_9, x_{10}\}|} = \frac{0}{5} = 0.0$$

$$\mu_2^{a_1^*} = \frac{\left| \left[a_1^* \right] \cap C^2 \right|}{\left| \left[a_1^* \right] \right|} = \frac{|\{x_1, x_6\}|}{|\{x_1, x_6, x_8, x_9, x_{10}\}|} = \frac{2}{5} = 0.40 \tag{4.9}$$

$$\mu_3^{a_1^*} = \frac{\left| \left[a_1^* \right] \cap C^3 \right|}{\left| \left[a_1^* \right] \right|} = \frac{|\{x_8, x_9, x_{10}\}|}{|\{x_1, x_6, x_8, x_9, x_{10}\}|} = \frac{3}{5} = 0.60$$

Rough membership computation for each discerning word medicine (a_2^*) and alternative (a_3^*) will be done in a similar way. The integrated membership to each decision category for each word is then computed using Equation 4.7. The normalized rough membership value for each decision category is calculated as explained in Equation 4.8. $\mu_2^*(nx_1)$ has a maximum rough membership value (i.e., 0.44) and it indicates that the document belongs to the decision category 2 (i.e., it is an "average" document). It is found that this document is not an authoritative one but has very good links to other web pages. Hence, according to the user's perspective, this is categorized as "average" or grade 2.

4.3 Designing a User Preference Based Document Filtering System

In this section, we present the outline of a complete document filtering system that is designed based on the methodologies proposed earlier. Fig. 1 presents a schematic view of the system.

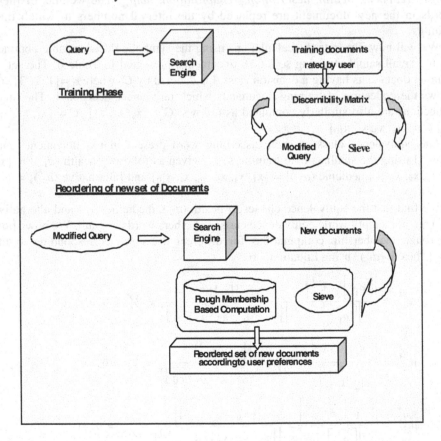

Fig. 1. A schematic view of user relevance feedback based document retrieval system

The proposed filtering system is designed to work in collaboration with an information collector at a user site. The information collector is an agent, which interacts with a backend search engine to fetch the documents. The system goes through two phases. In the first phase it accepts a user query and sends it to a backend search engine, which in our case is Google. The top 50 retrieved documents are presented to the user who rates them. The system then builds the decision table and the discernibility matrix. The discernibility matrix is scanned for the set of most discerning words which is used to form a modified query and the sieve.

In the second phase, the agent constantly monitors the Internet for new documents, using the modified query derived in the first phase. However, before presenting all these documents to the user, the documents are passed through the sieve and the rough membership of each document is computed for each category. In this phase, the sieve rates the new documents fetched for the user. It filters out the bad documents and presents only the good and average documents to the user after reordering them.

4.4 Results and Discussion

In this section, we have presented a rough-membership based document categorization scheme which can be used to determine the relevance of a retrieved document before presenting it to the user. Thus, for each document that is retrieved with the modified query, we computed the rough membership of the document to various decision categories. In order to do that, we made use of the cut-values obtained along with the discerning set of words.

Each new document was first converted into its word vector. Thereafter, for each domain, we constructed a symbolic decision matrix consisting of the discerning words alone. Depending on the cut-values for a discerning word, each word weight in the word vector of a document was replaced by the interval name in which the weight lay. The rough membership computation for each category was then performed according to the equations. The final categorization decision is done according to the Equations 4.6 to 4.8. Thus, each document in the list retrieved with the modified query has a decision attached to it. We now present the results of the document grading scheme for *"alternative medicine"*.

Query "Alternative Medicine" – Since there are various forms of Alternative Medicine, the user can prefer one topic over another and mark all those documents which are relevant to that particular topic as good and the irrelevant ones as bad. The system is expected to learn the preference and perform a similar categorization for the new documents.

In the results presented, the user had marked the documents related to "acupuncture and yoga" as good while those on ayurveda, aromatherapy, homeopathy, etc. as bad. After analyzing the entire training list of 50 documents and their ratings, the most discerning words were found and these were used to build a modified query. The modified query comprising of the most discerning words, in this case, *Health, medicine, alternative, therapy, yoga, acupuncture, stress, diet, and disease*. Table 7 presents the rough membership evaluation based misclassification matrix for the set of 50 retrieved URLs through the modified query corresponding to *Alternative Medicine*. It shows that the misclassification rate out of top 50 documents is 80%.

Table 7. Misclassification matrix for system rating versus user rating for the domain *"alternative medicine"*

System→ User↓	Rated 1	Rated 2	Rated 3
Rated 1	4	2	0
Rated 2	4	32	1
Rated 3	1	2	4

4.4.1 Average Performance Analysis

Pazzani et al. (1996) had reported the average classification accuracy for different classifiers used by *Syskill and Webert* in terms of 20 documents rated from different domains. We also performed a similar study in order to judge the accuracy of the rough membership based grading scheme. We compared the system-generated grades with feedback taken from the user for the top 20 documents retrieved with the modified query. We have also shown the accuracy for the top 50 documents. The overall accuracy of the system is defined as

$$\text{Accuracy} = \frac{\text{no. of matches in system rating and user rating}}{\text{Total no. of documents rated by system}} \times 100 \quad (4.10)$$

Table 8 summarizes the accuracy of the system in various domains. The average accuracy of the rating scheme for the top 20 test documents and top 50 test documents are 81% and 77.6% respectively. This establishes the effectiveness of the grading scheme.

Since bad documents can be identified by the system, these can be eliminated from the list presented to the user. Average accuracies reported by Syskill and Webert across various domains on 20 test documents also show similar precision values for various classifiers. The average accuracy values reported are 77.1% with the Bayesian classifier, 75.2% with the PEBLS classifier, 75.0% with the back propagation network, 75.0% with the nearest neighbor approach and 70.6% with the ID3 (Mitchel, 1997) approach.

Table 8. Accuracy of system evaluation for various domains

Domain	Accuracy (for 20 documents)	Accuracy (for 50 documents)
Alcohol Addiction	90%	92%
HIV	85%	74%
Air Pollution	80%	80%
Antarctic Exploration	80%	72%
Blood Cancer	70%	70%

5 Conclusions

In this paper, we have presented a discernibility matrix based approach to learn the user preferences. We have shown how the rough set theoretic concept of discernibility can be extended to learn a user's bias for rating of text documents, by analyzing a relevance feedback table. A modified implementation of the MD-Heuristic algorithm is proposed, which outputs the minimal list of words that can collectively discern all pairs of documents rated by the user. We have introduced the concepts of positively discerning words and negatively discerning words. We have also shown how a *modified query* can be constructed using the discerning words, along with Boolean operators.

When the modified query is fed to the search engine again, it is observed that the quality of retrieval improves significantly. Some good documents which were not retrieved during the original query may be retrieved by the modified query. We have provided some experimental details to show that the percentage of bad documents reduces substantially in almost all the domains. However, some irrelevant documents are still retrieved and the list is not ordered according to the user preference. This is due to the fact that the documents are still ordered according to the relevance computation function of the underlying search engines. Therefore, we have presented a rough membership based document grading mechanism. We have shown how this grading scheme in combination with the discernibility analysis presented earlier can be used to build a complete user relevance feedback based document retrieval system. The rough membership functions rate a new document on the basis of the weights of the discerning words present in them. We have shown, through example queries, that the performance of the system is quite satisfactory. The system is found to be capable of eliminating bad documents quite effectively.

References

Altavista (1995), http://www.altavista.com

Allan, J., Ballesteros, L., et al.: Recent Experiments with INQUERY. In: Proceedings of the Fourth Text Retrieval Conference (TREC-4), NIST Special Publication, pp. 49–63 (1995)

Anagnostopoulos, I., Anagnostopoulos, C., Vergados, D.D., Maglogiannis, I.: BRWM: A Relevance Feedback Mechanism for Webpage Clustering. AIAI, 44–52 (2006)

Brauen T.L., Holt R.C., et al.: Document Indexing Based on Relevance Feedback. Report ISR 14 to the National Science Foundation, Section XI, Department of Computer Science. Cornell University, Ithaca, NY (1968)

Bao, Y., Aoyama, S., Yamada, K., Ishii, N., Du, X.: A Rough Set Based Hybrid Method to Text Categorization. In: Second International Conference on Web Information Systems Engineering (WISE 2001), vol. 1, p. 294 (2001)

Belkin, N.J., Cool, C., et al.: Rutgers TREC 2001 Interactive Track Experience. In: The Tenth Text Retrieval Conference (TREC 2001) (2001)

Cost, S., Salzberg, S.: A Weighted Nearest Neighbor Algorithm for Learning with Symbolic Features. Machine Learning 10(1), 57–78 (1993)

Crestani, F.: Learning Strategies for an Adaptive Information Retrieval System using Neural Networks. In: Proceedings of the IEEE International Conference on Neural Networks, San Francisco, CA (1993)

Croft, W.B.: Effective Text Retrieval Based on Combining Evidence from the Corpus and Users. IEEE Expert: Intelligent Systems and Their Applications 10(6), 59–63 (1995)

Chouchoulas, A., Shen, Q.: Rough Set-Aided Keyword Reduction for Text Categorisation. J of Applied Artificial Intelligence 15(9), 843–873 (2001)

Das-Gupta, P.: Rough Sets and Information Retrieval. In: Proc. of the Eleventh Annual International ACM SIGIR Conference on Research and Development in Information Retrieval, Set Oriented Models, pp. 567–581 (1988)

Deerwester, S., Dumais, S., et al.: Indexing by Latent Semantic Analysis. J. of the American Society for Information Science 41(6), 391–407 (1990)

Fuhr, N.: Two Models of Retrieval with Probabilistic Indexing. In: Proceedings of 1986 ACM conference on Research and development in information retrieval, Palazzo dei Congressi, Pisa, Italy, pp. 249–257 (1986)

Fuhr, N., Buckley, C.: A Probabilistic Learning Approach for Document Indexing. ACM Transactions on Information Systems (TOIS) 9(3), 223–248 (1991)

Fuhr, N., Buckley, C.: Optimizing Document Indexing and Search Term Weighting Based on Probabilistic Models. In: The First Retrieval Conference (TREC-1) NIST Special Publication, pp. 89–99 (1993)

Goldberg, D., Nichols, D., Oki, B., Terry, D.: Using Collaborative Filtering to Weave an Information Tapestry. Communications of the ACM 35(12), 61–70 (1992)

Google (1998), http://www.google.com

Gauch, S.J., Wang, et al.: A Corpus Analysis Approach for Automatic Query Expansion and its Extension to Multiple Databases. ACM Transactions on Information Systems 17(3), 250–269 (1999)

Glöckner, I., Knoll, A.: Natural Language Navigation in Multimedia Archives: An Integrated Approach. In: Proceedings of the seventh ACM international conference on Multimedia (Part 1), Orlando, Florida, United States, pp. 312–322 (1999)

Intarka Inc.: ProspectMiner (2000), http://www.intarka.com

Jochem H., Ralph B., Frank W.: WebPlan: Dynamic Planning for Domain Specific Search in the Internet (1999), http://wwwagr.informatik.uni-kl.de/~webplan/PAPER/Paper.html

Jensen, R., Shen, Q.: A Rough Set-Aided system for sorting WWW Bookmarks. In: Proc. 1st Asia-Pacific Conference, Web Intelligence, pp. 95–105 (2001)

Jolliffe I.T.: Principal Component Analysis. Springer Series in Statistics (2002)

Koenemann, J., Belkin, N.J.: A Case for Interaction: A Study of Interactive Information Retrieval Behavior and Effectiveness. In: Conference proceedings on Human factors in computing systems, Vancouver, Canada, pp. 205–212 (1996)

Korfhage, R.R.: Information Storage and Retrieval, pp. 221–232. Wiley Computer Publishing, Chichester (1997)

Komorowski J., Polkowski L., Andrzej S.: Rough Sets: A Tutorial (1999), http://www.let.uu.nl/esslli/Courses/skowron/skowron.ps

Kim, D.W., Lee, K.H.: A New Fuzzy Information Retrieval System Based on User Preference Model. In: The 10th IEEE International Conference on Fuzzy Systems (2001)

Losee, R.M.: Parameter Estimation for Probabilistic Document-Retrieval Models. J. of the American Society for Information Science 39(1), 8–16 (1988)

Losee, R.M., Bookstein, A.: Integrating Boolean Queries in Conjunctive Normal Form with Probabilistic Retrieval Models. Information Processing and Management 24(3), 315–321 (1988)

Lee, J.H.: Properties of Extended Boolean Models in information Retrieval. In: ACM Annual Conference on Research and Development in Information Retrieval, pp. 182–190 (1994)

Maron, M.E., Kuhns, J.L.: On Relevance, Probabilistic Indexing and Information Retrieval. J. of ACM 7(3), 216–244 (1960)

Mitchel, T.: Machine Learning. McGraw-Hill, New York, USA (1997)

Miyamoto, S.: Application of Rough Sets to Information Retrieval. JASIS 49(3), 195–205 (1998)

Menasalvas, E., Millan, S., Hochsztain, E.: A Granular Approach for Analyzing the Degree of Afability of a Website. In: Alpigini, J.J., Peters, J.F., Skowron, A., Zhong, N. (eds.) RSCTC 2002. LNCS (LNAI), vol. 2475. Springer, Heidelberg (2002)

Meng, X., Chen, Z.: Intelligent Web Search Through Adaptive Learning from Relevance Feedback. In: Khosrow-Pour, M. (ed.) Encyclopedia of Information Science and Technology, pp. 3060–3065. Idea Group Publishing (2005)

Pawlak, Z.: Rough Sets. Int. J. of Computer and Information Sciences 11(5), 341–356 (1982)

Pazzani, M., Muramatsu, J., Billsus, D.: Syskill & Webert: Identifying Interesting Web Sites. In: Proc. of the National Conference on Artificial Intelligence, Portland, OR, USA (1996)

Piccinelli, G., Mont, M.C.: Fuzzy-Set Based Information Retrieval for Advanced Help Desk. HP Laboratories, Bristol, UK (1998)

Pal, S.K., Skowron, A.: Rough Fuzzy Hybridization- A New Trend in Decision-making. Springer, U.S.A. (1999)

Rocchio, J.J.: Relevance Feedback in Information Retrieval. In: G. Salton - The SMART Retrieval system: Experiments in automatic document processing, pp. 313–323. Prentice Hall, Englewood Cliffs (1971)

Roldano, C.: UserProfiling with Bayesian Belief Networks (1999), http://www.labs.bt.com/profsoc/facts/workshop/abstract/BBN.html

Salton, G.: The SMART Retrieval System: Relevance Feedback and the Optimization of Retrieval Effectiveness. Prentice Hall, Englewood Cliffs, USA (1971)

Salton, G., Fox, E.A., et al.: Advanced Feedback Methods in Information Retrieval. J. of the American Society for Information Science 36(3), 200–210 (1985)

Salton, G., Buckley, C.: Improving Retrieval Performance by Relevance Feedback. J of the American Society for Information Science 41(4), 288–297 (1990)

Subtil, P., Mouaddib, N., et al.: A fuzzy information retrieval and management system and its applications. In: Proceedings of the 1996 ACM symposium on Applied Computing, Philadelphia, Pennsylvania, United States, pp. 537–541 (1996)

Stefanowski, J., Tsoukias, A.: Incomplete Information Tables and Rough Classification. Computational Intelligence 17, 454–466 (2001)

Srinivasan, P., Ruiz, M.E., Kraft, D.H., Chen, J.: Vocabulary Mining for Information Retrieval: Rough Sets and Fuzzy Sets. Information Processing and Management 37, 15–38 (2001)

Szczepaniak, P.S., Niewiadimski, A.: Internet Search Based on text intuitionistic Fuzzy Similarity. In: Intelligent Exploration of the Web, pp. 96–102. Physica –Verlag, Springer, Heidelberg, New York (2003)

Szczepaniak, P.S., Gil, M.: Practical Evaluation of textual fuzzy Similarity as a Tool for Information retrieval. In: Menasalvas, E., Segovia, J., Szczepaniak, P.S. (eds.) AWIC 2003. LNCS (LNAI), vol. 2663, pp. 250–257. Springer, Heidelberg (2003)

Singh, S., Dey, L.: A new Customized Document Categorization scheme using Rough Membership. Int. J. of Applied Soft-Computing, 373–390 (2005)

Shtykh, R.Y., Jin, O.: Enhancing IR with User-Centric Integrated Approach of Interest Change Driven Layered Profiling and User Contributions. In: 21st International Conference on Advanced Information Networking and Applications Workshops (AINAW 2007), pp. 240–245 (2007)

Verhoeff, J., Goffman, W., et al.: Inefficiency of the use of Boolean Functions for Information Retrieval Systems. CACM 4(12), 557–594 (1961)

Wong, S.K.M., Yao, Y.Y.: On Modeling Information Retrieval with Probabilistic Inference. ACM Transactions on Information Systems 13(1), 38–68 (1995)

Widyantoro, D.H., Ioerger, T.R., et al.: Learning User Interest Dynamics with a Three Descriptor Representation. J. of the American Society for Information Science and Technology 52(3), 212–225 (2001)

Wilson, M.L., Schraefel, M.C., White, R.W.: Evaluating Advanced Search Interfaces using Established Information-Seeking Models. J. of the American Society for Information Science and Technology (Submitted 2007)

Web 2.0 Business Models (2007),
    http://www.deitel.com/eBook/Web20BusinessModels/tabid/2498/Default.aspx

Yahoo! (1994), http://www.yahoo.com

Zadeh, L.A.: Fuzzy Sets. Inform. and Control 8, 338–365 (1965)

Opportunistic Scheduling and Pricing Strategies for Automated Contracting in Supply Chains

Sabyasachi Saha* and Sandip Sen

Department of Mathematical & Computer Sciences
University of Tulsa
Tulsa, Oklahoma, USA
sabyasachi.saha@gmail.com, sandip@utulsa.edu

Summary. Supply chains form an integral cornerstone of the daily operations of today's business enterprises. The effectiveness of the processes underlying the supply chain determine the steady-flow of raw material and finished products between entities in the marketplace. Electronic institutions facilitate free and fair competition between providers and suppliers vying for contracts from customers and manufacturers. We assume that contracts are awarded via competitive auction-based protocols. We believe that the efficiency of the scheduling and pricing strategies of the suppliers play an important role in their profitability in such a competitive supply chain. The suppliers can decide their scheduling strategy for completing the contracted tasks depending on their capacity, the nature of the contracts, the profit margins and other commitments and expectations about future contracts. Such decision mechanisms can incorporate task features including length of the task, priority of the task, scheduling windows, estimated future load, and profit margins. Robust, opportunistic task scheduling strategies can significantly improve the competitiveness of suppliers by identifying market niches and strategically positioning available resources to exploit such opportunities. Effective price adjustment mechanisms are also required to maximally exploit such market opportunities.

1 Introduction

With the increasing popularity of e-markets, business-to-business ($B2B$) trading has received an unprecedented boost. To stay ahead of the competition in this scenario, complexity of business strategies is increasing proportionally. New mechanisms and models are needed to enable modern businesses to adapt to this rapidly changing and evolving marketplace that present new possibilities and new risks. Agent-based systems have been proposed to optimize the functional modules and business strategies of an enterprise [1]. Stable and profitable $B2B$ relationships are predicated on the effectiveness of the supply chains linking these organizations. The effectiveness of the processes underlying the supply chain determine the steady-flow of raw material and finished products between entities in the marketplace. Electronic institutions facilitate free and fair competition between providers and suppliers vying for contracts from customers

---

* Currently at Toyota-IT Center, USA at Palo Alto, CA.

B. Prasad (Ed.): Soft Computing Applications in Business, STUDFUZZ 230, pp. 179–199, 2008.
springerlink.com   © Springer-Verlag Berlin Heidelberg 2008

and manufacturers. We believe that optimizing supply chain management is a key step towards developing more profitable businesses. Efficient supply chain management seeks to

1. Reduce waste by minimizing duplications and inefficient productions
2. Reduce inventory costs by efficient planning of future demand
3. Reduce lead time through optimal allocation of the subtasks
4. Improve product quality
5. Develop strong and long term partnerships
6. Increase profitability

From the multiagent systems perspective, a supply chain is often modeled using a decentralized network of agents where each agent can do a part of a task. In $B2B$, dependability of the business entities determine the stability of the inter-business relationships. The net productivity of a manufacturing enterprise, for example, depends on the performance of the downstream components of its supply chain, which in turn depends on the performance index of the supplier entities. The tasks which the manufacturers have to complete differ in priority or the deadline before which they must be completed. Tasks of different priorities represent the dynamics of market demand. In this paper, we have considered a three-level supply chain with primary and secondary manufacturers and a group of suppliers. At each level there is a group of competitive agents. The functionalities of agents in different layers are different. In the uppermost level there is a group of primary manufacturers, in the second level there is a group of secondary manufacturers who are linked with a group of suppliers in the lowest level in the hierarchy. We use a contracting framework for connecting the suppliers to the manufacturers: manufacturers announce contracts for tasks with given specifications (deadline and processing time); suppliers bid on these tasks with prices; and the contract is allocated by an auction to a supplier who fulfills all task constraints [8]. We present different task scheduling mechanisms of the suppliers and experimentally investigate their effectiveness on the profitability of the suppliers under varying task mixes and group composition. Suppliers who are responsive to accommodate dynamically arriving tasks incurs higher profit [8]. We also investigate the effect of different pricing strategies by suppliers who can create market niches by smart, predictive scheduling algorithms [9]. A supplier who can meet deadlines that few others can meet will be able to demand more from the manufacturer for their services. Also when competition is very high, a supplier reduces its price in order to win the contract.

2   Responsive Supply Chains

In the present-day markets, business enterprises have integrated the functional groups within a supply chain [12]. Additionally, the success of a modern enterprise depends, in turn, on its success to maintain stable and profitable relationships with other enterprises that it is functionally coupled to. In [2] we find how software agents following *structured conversations* achieve effective coordination in the context of a supply chain. However, in the present age of globalization, achieving this goal is a daunting task. Today's supply-chain managers have to consider supplier enterprises who are geographically distributed. They have to manage the flow of materials from the distributed suppliers to global manufacturing facilities. Emphasis should be given on keeping low inventory, minimizing operations cost, having flexibility and providing efficient customer service. Effective decision support tools on modeling the dynamics of the supply chain offer an edge of performance to managers in achieving these goals [14].

While a company might adopt a myopic policy of adjusting the logistics activities individually to attain overall excellence, they will fail to visualize the interdependency-tradeoffs present among the activities. It is acknowledged that *integrated logistics* is the key to better business. Equally crucial is the issue of operational flexibility to accommodate the fluctuating demand of the market. In a successful *B2B* trading scenario, *integrated logistics* coupled with this flexibility can guarantee business growth. Adoption of a suitable *supply-chain strategy* as well as an effective *supply-chain planning* is deemed necessary. Supply-chain strategy refers to the policy tradeoff between cost and service that the company would adopt (at what point on the cost-service curve does the company wishes to place itself?). Supply-chain planning is the company's working policy to achieve the adopted strategy (what operations dynamics should the company adopt to move itself from the current position to the target position?).

The motivation behind our work has been this aspect of a supply chain being responsive to dynamically arriving tasks which is hinged upon the performance of the suppliers. We believe that a distributed approach where each supplier tries to maximize its profit will enable a robust, scalable solution to the responsiveness issue. We use simulations to evaluate various scheduling algorithms used by suppliers to schedule the tasks allocated to them. We have designed a utility based strategy which uses a model of periodic variations in the distribution of arriving tasks. Tasks with varying priorities (deadline) are stochastically generated, and arrive sequentially. Suppliers bid for tasks in auctions to win contracts. Suppliers with a more flexible scheduling policy can expect to win larger number of more profitable contracts. Each suppliers goal is to maximize local profit. Suppliers with more flexible schedules can meet the deadline requirements of tasks which few others can meet and hence hike their winning bids, leading to increased profitability.

3   Related Work

More recently the problem of supply chain management has drawn the attention of multiagent systems (MAS) researchers. In MAS, a supply chain is conceptualized as a group of collaborative autonomous software agents [5]. It is argued that managers can better coordinate and schedule processes by distributing the organization-wide business management system into autonomous problem solving agents. This approach is more appropriate when the participating enterprises are geographically distributed and manually controlling *B2B* trading is not possible.

Tackling coordination in supply chains using partial constraint satisfaction problems by mediating agents is investigated by [3]. Negotiation-based supply chain management has been proposed in [4], where a *virtual* supply chain is established by the negotiating software agents when an order arrives. Using *blackboard architecture*, which is a proven methodology to integrate multiple knowledge sources for problem solving, a mixed-initiative agent-based architecture for supply chain planning and scheduling is implemented in MASCOT [7].

Swaminathan *et al.* has provided a framework for efficient supply chain formation [13]. Labarthe *et al.* [6] has presented a heterogeneous agent based simulation to model supply chains. Multiagent Systems research has emphasized on the emergence of the optimal configuration of the supply chain. Walsh *et al.* has shown the optimal dynamic task allocation in a supply chain using combinatorial auction [15]. They have proved that, given a task, composed of a group of subtasks, the dynamic formation of the supply chain that will produce maximum profit.

Recently, trading agent competition (TAC) [11] introduced a realistic supply chain framework. The customers requires some computers with different configurations. The generation of this tasks are stochastic. The agents bid for the contracts to the customers. They also contract with different suppliers for the supply of different raw materials. A supplier manufactures in its own factory when it has all the required raw materials and then deliver the order to the customers. Agent's challenge is to design strategies for efficient handling of the factory schedules, efficient contracting with the suppliers, competitive and profitable bidding to the customers, reduce the inventory cost, reduce penalty for late delivery of cancel the orders. It needs to find out efficient bidding strategy to increase its own profit.

4   Scheduling Strategies

We evaluate four scheduling strategies using which the suppliers will schedule assigned tasks. The goal is to allocate a task t of length $l(t)$ and deadline $dl(t)$ (the deadline determines priority $\mathcal{P}(t)$: we have assumed that a task with immediate deadline is considered to be of high priority or priority 1 and a task of normal deadline is of ordinary priority or priority 0.) on the calendar $\mathcal{C}$ given the current set of scheduled tasks on in $\mathcal{C}$. The following scheduling strategies have distinct motivations:

First fit: This strategy searches forward from the current time and assigns t to the first empty interval of length $l(t)$ on the calendar. This produces a compact, front-loaded schedule.

Best fit: This strategy searches the entire feasible part of the schedule (between the current time and deadline $dl(t)$) and assigns t to one of the intervals such that there is a minimal number of empty slots around it. This strategy produces clumped schedules, but the clumps need not be at the front of the schedule.

Worst fit: This strategy searches the entire feasible part of the schedule (between the current time and deadline $dl(t)$) and assigns t to one of the intervals such that there is maximum empty slots around it.

Expected Utility Based (EU): All the strategies discussed above schedule the tasks in a greedy way. They bid for a task whenever it can schedule it. In reality, there are periodic patterns in the task arrival distribution. Hence, it might be prudent at times to not bid for a task with the expectation of keeping the option open to bid for a more profitable task in the future. In the EU strategy, suppliers use the knowledge of the task arrival pattern and whenever a new task arrives a decision is made whether it is worthwhile to bid on it. If the task arrival pattern is not known *a priori*, it can be learned over time provided the pattern does not change drastically in a short time. Unfortunately, there is a risk involve in such opportunistic scheduling as some production slots may remain unused if the expected high-profit tasks never materialize. According to this algorithm, it will bid for a task if its expected utility is positive. This will be negative when the expectation that it can make more profit later overpowers the risk of not bidding now. So, it will not bid for this task. We will discuss the expected utility based bidding strategy in the next section.

5   Utility of Scheduling a Task

For each arriving task $t \in \Upsilon$, where Υ is the set of all tasks, the utility for the supplier for scheduling that task is given by the function $u(l(t), \mathcal{P}(t))$, where $l(t)$ is the length and $\mathcal{P}(t)$ is the priority of that task. Let $esd(t)$ and $dl(t)$ be the earliest start date and the deadline for processing the task. A supplier calculates the number of empty slots, $fs(d)$, for each day $d \in D$, where D is the set of all days on the calendar $\mathcal{C}$ between and including $est(t)$ and $dl(t)$. For each of these days, the supplier generates two sets of task combinations $\mathcal{T}^k_{fs(d)}$ and $\mathcal{T}^k_{fs(d)-l(t)}$, where, $\mathcal{T}^k_\alpha = \{T | T \subset \Upsilon, \alpha = \sum_{t \in T} l(t), \mathcal{P}(t) > k \ \forall t \in T\}$ is the set of all those task combinations where the lengths of the tasks in each combination adds up to α and each task has a priority higher than k. We compare the utility of scheduling this task now and scheduling the remaining slots later with all possible ways of scheduling the currently empty slots. We choose to schedule the task now if the corresponding utility dominates all other ways of filling up the empty slots without scheduling the current task.

Let $n^T_{i,j}$ to be the number of tasks in T of length i and priority j, $Pr(i, n^T_{i,j}, d, \mathcal{H})$ is the probability that at least $n^T_{i,j}$ number of tasks of length i and priority j will arrive at a later time on day d, where $\mathcal{H}$ is the history of tasks that have already arrived. Given the task distribution (discussed in the Section 7), the probabilities $Pr(i, n^T_{i,j}, d, \mathcal{H})$ can be calculated by using multinomial cumulative probability function.

The expected utility of scheduling the current task t on day d on calendar $\mathcal{C}$ given the history of task arrivals $\mathcal{H}$ is

$$EU(t, d, \mathcal{H}, \mathcal{C}) = u(l(t), \mathcal{P}(t)) + den(\mathcal{C}, est(t)) \times$$
$$(AU(t, d, \mathcal{H}, fs(d) - l(t))$$
$$- AU(t, d, \mathcal{H}, fs(d))),$$

where

$$AU(t, d, \mathcal{H}, h) = \frac{1}{|T_h^{\mathcal{P}(t)}|} \sum_{T \in T_h^{\mathcal{P}(t)}} \sum_{k > \mathcal{P}(t)} \sum_{i=1}^{l_{max}}$$
$$Pr(i, n_{i,k}^T, d, \mathcal{H}) n_{i,k}^T u(i, k),$$

is the average expected utility of scheduling h hours on day d with tasks of higher priority than task t given the history of task arrivals $\mathcal{H}$. l_{max} is the maximum possible length of a task, and the function den returns the density of the calendar, i.e., the percentage of slots that have been scheduled, up to a given date.

For a given day on the calendar, the expected utility expression adds the sum of the utility of the current task to the difference of the average utility of all possible ways of filling up the calendar with higher priority tasks with or without the current task being scheduled. We schedule the task provided that the expected utility is positive for at least one of the days considered and the day chosen is the one which provides the maximum expected utility, i.e., $\arg\max_{d \in D} EU(t, d, \mathcal{H}, \mathcal{C})$.

6 Pricing Strategies

We investigate the ability of suppliers to use adaptive pricing schemes to exploit market opportunities. A supplier may find it beneficial to hike prices when it believes there is less competition for contracts and vice versa. As current information about competitors may not be available, such schemes may also backfire. While aggressive exploitation of sleeping market opportunities may be beneficial, a more conservative approach might be prudent in the face of competition.

We have defined three pricing strategies for the suppliers. In all the three strategies, a supplier will increase the price of the bid when it is winning contracts continuously. A supplier, however, cannot bid more than the reserve price of the secondary manufacturer. Suppose a supplier s is asked to bid for a task t by a secondary manufacturer m, whose reservation price is R_m. The length of the task is $l(t)$ and deadline $dl(t)$.

Linear strategy: If the supplier won previous contract for this type of task t, it will increase its bid, by a constant α, from its previous bid, provided the new bid will not exceed the reserve price of the secondary manufacturer it is bidding to. Though here, for a fair comparison, we have used the same α for all suppliers. Similarly, if the supplier looses its previous contract for the same type of task, it will reduce its bid by α from its previous bid, provided the bid is not less than its own reserve price R_s. So, if b was its bid for the last contract for the of same length and priority as that of task t, then its bid for task t will be,

$$bid = \min(b + \alpha, R_m), \text{ if it won the last contract}$$
$$= \max(b - \alpha, R_s), \text{ if it lose the last contract} \qquad (1)$$

Defensive strategy: This is a cautious strategy. It increases (or decreases) its bid when it is winning (or loosing) with a caution. If the supplier with this strategy won k contract at a row for the similar type of tasks, it will increase its bid from its previous bid by

$$incr = \alpha - (k \times \delta), \text{ if, } \alpha > k \times \delta$$
$$= 0, \text{ otherwise}$$

provided the increased bid will not be more than the reserve price of the secondary manufacturer it is bidding to. Here, δ is another constant parameter of the supplier. This supplier will decrease its bid by the same amount, i.e. $incr$, from its previous bid if the supplier looses k contracts at a row of similar type tasks provided the decreased bid will not be less than its own reserve price. If the last bid of the supplier for the same type of task was b then new bid can be found out from Equation 1 using $\alpha - (k \times \delta)$ in stead of α.

Impatient strategy: Suppliers following this pricing strategy increases and decreases the bid sharply as a consequence of winning and loosing a bid, respectively. If the supplier with this strategy won k contracts at a row for the similar type of tasks, it will increase its bid from its previous bid by $\alpha + (k \times \delta)$, provided the new bid does not exceed the secondary manufacturer's reservation price. Similarly, if the supplier looses k contracts at row, it will decrease its bid by $\alpha + (k \times \delta)$, provided the decreased bid will not be less than its own reserve price. If it's last bid was b the new bid can be found from Equation1 using $\alpha + (k \times \delta)$ in stead of α.

We have assumed that at the start of the simulation, each supplier has the same bid for a particular type of task.

7 Experimental Framework

Our experimental framework is designed to facilitate the evaluation of the relative effectiveness of alternative scheduling and pricing schemes, described above, under varying environmental conditions. In our simulations, each period consists of a five day work week, each day having six slots. We vary the arrival rate of different task types and vary the percentage of priority tasks over different simulations. Each of the tasks is generated and allocated to a manufacturer depending on whether the manufacturer can accomplish the task by using one of its suppliers. Selection of one supplier over another by a manufacturer is performed through a first-price sealed-bid auction protocol.

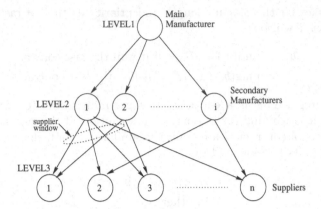

Fig. 1. Supply chain structure

The supply chain that we have used for our simulations consists of 3 levels, each being populated by one or more enterprises having similar functional capabilities. There is one *main manufacturer* in level one, six *secondary manufacturers* in level two, and twelve *suppliers* at level three. The supply chain structure is shown in Figure 1. In our simulation we have consider a whole task as a combination of some parts. We can think of a whole task T as $L = L_m + L_{sm} + L_{su}$, where, L_m, L_{sm} and L_{su} are the part of the whole task to be done by main manufacturer, secondary manufacturer and suppliers respectively. The main manufacturer will contract a part of each task that it has to complete to one of the secondary manufacturers. Each secondary manufacturer, in turn, contracts part of each task it wins to a supplier in some subset of all suppliers. We call this subset of suppliers the *supplier window* of the secondary manufacturer. In our experiments, we have chosen the number of secondary manufacturers and suppliers such that each supplier can receive contracts from exactly two different secondary manufacturers.

A task assigned to the main manufacturer has two properties: (i) *priority*, which can have a value of either 0 (indicating an ordinary task) or 1 (indicating a high priority task). High priority tasks have a *deadline* of one day and low priority tasks have a *deadline* of one week, and (ii) *length*, which is the number of time units (slots) the task requires to be completed. A task must be scheduled on consecutive slots on the same day. Higher priority tasks are more profitable.

In our simulation, the tasks are generated stochastically. We use different types of task distributions to verify the robustness of the different algorithms. In the first task distribution type we use, the high priority tasks arrive in the later part of the week. Most of the tasks arrive in the early part of the week and these are mostly of low priority with a deadline of the whole week. Fewer, high priority, tasks are generated in the last part of the week and have a short deadline. We present an example of such a task distribution in table 1. The number in the jth row and $(2i-1)^{th}$ $(2i^{th})$ column of this table presents the probability of

Table 1. Probabilities of Generating Tasks: the row number gives the length of the task; the column $(2i - 1)$ shows the probability of generating high priority task in the ith day and the column $2i$ shows the probability of generating a low priority task

	1	2	3	4	5	6	7	8	9	10
1	.001	.14	.005	.005	.002	.01	.03	.01	.04	.01
2	.001	.14	.005	.005	.002	.01	.03	.01	.04	.01
3	.001	.15	.005	.005	.001	.005	.02	.02	.03	.01
4	.001	.15	.005	.005	.001	.005	.02	.02	.03	.01

generating a *high* (*low*) priority task of length j in the ith day of the week. In a different task distribution type that we have experimented with, all tasks, of different lengths and arrival dates, are generated with equal probability.

Suppliers at level 3 of the supply chain can employ one of the three pricing strategies mentioned in Section 6 and any one of the four scheduling algorithms mentioned in Section 4. Preemptive scheduling is not allowed in our simulation which precludes use of leveled commitment protocols [10]. Hence, a commitment to a contract by a supplier cannot be undone by any task that arrives at a later time to the supplier. All suppliers use the same initial bid prices: this value is equal to the length of the task for an ordinary task and is twice the length of the task for a high priority task. Subsequently, based on their win/loss record on past contracts, suppliers adjust their bid prices according to one of the price adjustment schemes discussed in Section 6. Only suppliers who can accommodate, within their current schedule, a task announced by a secondary manufacturer and are within the supplier window of that manufacturer, can bid for that task. The supplier with the minimum bid price wins the contract for the task from the secondary manufacturer, i.e., contracts are awarded by a first-price-sealed-bid auction. When a supplier wins a bid, it increases its *wealth* by an amount equal to its bid price minus the cost it incurs to accomplish the task. Initially, wealth of all suppliers is set to zero.

8   Experimental Results

To describe our experimental results we use the following conventions: first fit, best fit, worst fit and expected utility based scheduling algorithms are represented by FF, BF, WF and EU respectively. Suppliers' different pricing strategies *viz.* linear strategy, defensive strategy and impatient strategy are referred to as LIN, DEF and $IMPT$ respectively. The naming convention we will use is as follows: if a supplier is using scheduling strategy x and pricing strategy y, we call that supplier as x_y. For example a supplier using EU scheduling algorithm and LIN pricing strategy will be referred to as EU_LIN. Now we will discuss our experimental results in the following two subsections.

8.1   Experiments with Different Scheduling Strategies

In this subsection, we investigate the performance of different scheduling strategies, discussed in Section 4. Here we assume that all the suppliers use same pricing mechanism. All of them have the same initial price for tasks of same type and then if a supplier wins a contract, it will increase its asking price for that type of task by a constant amount.

In the first set of experiments, we have a total of 12 suppliers, each scheduling strategy is followed by three suppliers and the task distribution is as mentioned in the Table 1. A run consists of 20 periods (weeks) of 5 day each. The number of tasks per week is 150. We study the variation in the wealth earned by suppliers using different types of scheduling algorithms. The result is shown in Figure 2. The average wealth earned by EUs are much better compared to the other three scheduling types. Since the EU strategy is not myopic, the suppliers using it wait for more profitable high priority tasks, ignoring low-priority tasks that arrived earlier, and make more profits throughout the period of execution of 100 days. The EUs benefit from earning more than the other types of suppliers by winning the bids for these high-profit tasks. Other, myopic suppliers, do not consider passing on arriving tasks, and hence fill up slots by committing to tasks arriving early in the run, and therefore fail to accommodate and bid for high priority tasks that arrive later. The EUs are followed in performance by WFs. Since the worst fit algorithm produces an evenly-loaded schedule, suppliers using this type of scheduling can better accommodate the high priority tasks (and ordinary tasks too) that arrive more frequently later in the contracting period. We note that at the end of the run, the prices for the high priority task charged by EU suppliers increase considerably (see Figure 3). Interestingly, Figure 4 shows that the EU scheduling strategy not only garners more profit, but also accommodates more tasks compared to the other scheduling strategies. The reason for this is that in the task distribution, more high priority tasks are of smaller length. In this metric too, the WF strategy comes in second: they accommodate more tasks compared to BF and FF.

In the second set of experiments, we alter the task distribution and generate the tasks such that all task types and their arrival dates are equally likely. Here, tasks per week is decreased to 100, to see if lower task loads will allow the other strategies to compete more effectively with the EU strategy. We present the wealth distribution from this set of experiments in Figure 5. Here also the EU suppliers do better than all the other types of suppliers. The reason is that though all of the suppliers is accommodating some high priority tasks, the other strategies also contract a number of ordinary tasks. The EUs, however, focus more on the high priority tasks. Since, it is highly probable that different high priority tasks will come everyday, they will try to contract only the high priority tasks and increase their profit. Here the bid price for all supplier types increase significantly, but the increment is much more for EUs. So, we see that the EU suppliers are robust and can outperform suppliers using other strategies under different task distributions.

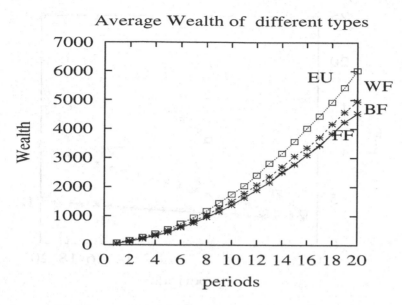

Fig. 2. Average Wealth earned by different supplier types. There are 3 suppliers using each of four scheduling strategies. Task distribution is as shown in Table 1.

In the third set of experiments, we assume that the *EU* suppliers are not knowledgeable about the task distribution. So, they try to estimate the distribution in the first two periods and bid for all tasks and uses a first fit strategy during this time. From the third period onward, however, they use the expected utility strategy based on the estimated task distributions. In Figure 6 we can see that *EU*s utility is less in the first two periods. But the *EU*s dominate the other strategies in the later periods. Similar observations hold for the number of tasks contracted by each type of suppliers shown in Figure 7. In this experiment set the *EU* continues to estimate, in every period, the task arrival probability distribution with the task arrival data of last two periods using relative frequency. Unlike the earlier experiments they use this estimate of task distribution instead of the real task distribution for calculating expected utility. So, this *EU* scheduling strategy works well even when the task generating distribution is not known *a priori*.

In the fourth set of experiments, we investigate the effect of increasing the number of EU suppliers in the population. In this experiment we increase the number of the *EU*s and keep number of supplier under a secondary manufacturer equal and change tasks proportionately. From Figure 8 we observe that if the number of *EU* suppliers increase, their individual utility decreases. This decrease is due to the presence of too many "smart" people (*EU*s) in the population. The competitive advantage of EU suppliers is predicated on the absence of similar strategic schedulers who keep their options until late in the run with the goal of accommodating late-arriving high-payoff tasks. In absence of

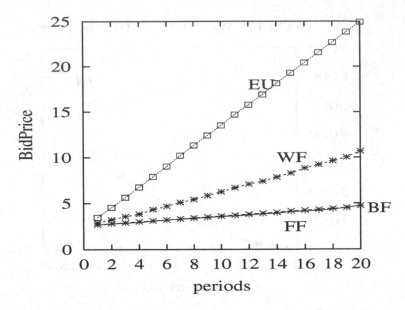

Fig. 3. Increase in bid prices for high priority tasks over the course of a run

competition it can hike the price and garner larger profits. But, presence of many strategic bidders increases competition for late-arriving high-priority tasks and eliminates such opportunistic profit making. Though EUs still perform better than other strategies, the differences in performances decreases because they now start bidding for a larger number of low-priority tasks after the first period. This happens because the density of the of the calendar is lower compared to the previous cases.

We note that, in these experiments, we have supplied a relatively large number of tasks. We use at least a total of 100 tasks for 12 suppliers per weak (which means 360 empty slots per week). We have noticed that if the number of tasks per period is very small, the profits of different suppliers or the number of tasks they do do not differ significantly. This is because then scheduling strategies do not have a significant effect on accommodating late-arriving tasks as most of the slots remain empty.

8.2   Experiments with Different Pricing Strategies

In this section our goal is to investigate the effectiveness of different pricing strategies. In the first experiment, there is a set of 12 suppliers, where we have 3 suppliers each of FF_LIN, BF_LIN, WF_LIN and EU_LIN types. All the suppliers, therefore, are using the same pricing strategy. We wanted to evaluate the relative merits of the *linear* pricing strategy in exploiting any market niches created by different scheduling strategies. The experiment covers a period of one week with 200 tasks. In Figure 9, we present the wealth generated by these

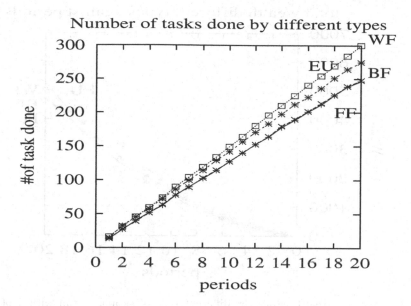

Fig. 4. Number of contracts won by different supplier types

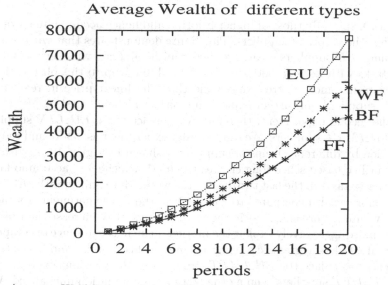

Fig. 5. Average Wealth earned by different supplier types with uniform task distributions

strategies in a week. In the initial part of the week, FF, BF, and WF suppliers generated more wealth by winning more contracts while the EU strategy passed over some of these tasks. In the fourth and fifth day, EU accommodates

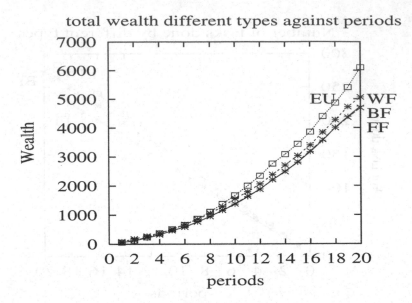

Fig. 6. Average Wealth earned by different type of suppliers, run with 3 of each suppliers. Task distribution is learned by EU suppliers.

more tasks, specially those of higher priority, and hence accumulate more wealth than the other types of suppliers. This figure demonstrates that with responsive scheduling, *EU* suppliers created a market niche and can win contracts for profitable tasks with less competition. The logical followup to this observation is to investigate if a more aggressive scheme than the linear price increase used by the *EU* suppliers can further exploit the market niche created.

In the second experiment, therefore, we replace three *EU_LIN* suppliers by three *EU_IMPT* suppliers. We ran a similar experiment as before and the wealth generation by different groups of suppliers is shown in Figure 10. As before, expected utility based scheduling creates the market niche and accommodates the high priority tasks in the last two days of the week. More importantly, *EU_IMPT* creates more wealth compared to *EU_LIN* in the last experiment. As they succeed in winning more contracts in the last two days, they increase their bid prices more to more aggressively exploit the market niche in the absence of competition.

To evaluate less aggressive pricing schemes in this context, and in the third experiment, we replace the *EU_IMPT* suppliers in the previous experiment with three *EU_DEF* suppliers keeping the rest of the population unchanged. We find that suppliers with *EU* scheduling strategy generate more wealth than others but because of their defensive or cautious pricing strategy they can not extract as much wealth as the *EU_IMPT* suppliers did given the same market niche (see Figure 11).

In the last three experiments, we can see that the *EU* suppliers generate maximum wealth using impatient pricing as this strategy takes full advantage of

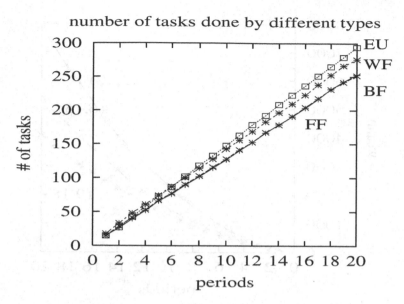

Fig. 7. Number of contracts won by different suppliers. Task distribution is learned by EU suppliers.

the opportunity to bid very high when there is very little or no competition in the population, *i.e.* only few suppliers can meet the requirements for scheduling the high priority tasks generated in the last two days of the week. We derive the following observation from this result: *if the scheduling strategy distribution is somewhat even and if high priority tasks come in bursts with imminent deadlines, aggressive pricing schemes, similar to the impatient strategy used in this study, can produce significant wealth gains for predictive schedulers.*

Next, we evaluate whether the impatient pricing strategy would benefit suppliers using the *FF*, *BF* and *WF* scheduling strategies as well. In Figure 12, we have compared the results in two different experiments. In both experiments, we have 12 suppliers, where each scheduling algorithms have 3 suppliers. In one experiment, *FF*, *BF* and *WF* suppliers follow *IMPT* pricing strategy and *EU* suppliers follow *LIN* pricing strategy. In the other experiment, all of the suppliers use *LIN* pricing strategy. In the figure, we have shown that *FF*, *BF* and *WF* suppliers are better off when they are using *IMPT* pricing in stead of *LIN* pricing. It appears then that *with uniform scheduling strategy distributions, all suppliers can benefit from aggressive pricing as market niches, of differing size and wealth, are produced for all scheduling strategies.*

Now, we want to evaluate the relative performance of the pricing schemes when all suppliers are using the *EU* scheduling strategy. This is a direct head-to-head comparison when market niches may not be that pronounced as everyone is using predictive scheduling. We again use 12 suppliers with four suppliers each using *LIN*, *IMPT*, and *DEF* pricing strategies. We have reduced the number

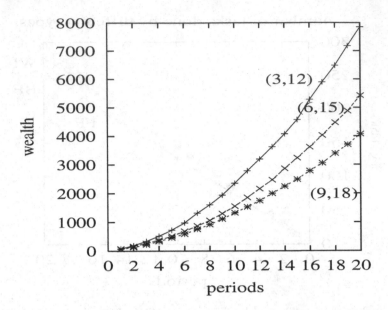

Fig. 8. Average wealth varies as the number of *EU* increases, in the figure the number of *EU* and total number of suppliers are mentioned in the brackets

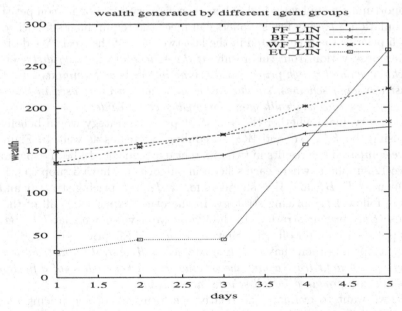

Fig. 9. Average Wealth earned by different supplier types. There are 3 suppliers using each of four scheduling strategies each using linear pricing strategy.

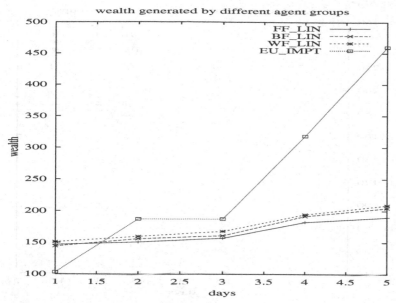

Fig. 10. Average Wealth earned by different supplier types. There are 3 suppliers using each of four scheduling strategies. FF, BF and WF are using linear price strategy while EU is using impatient pricing.

of tasks to 100 per week. In Figure 13, we see that the $IMPT$ supplier performs the best followed by LIN and then DEF. The reason for the impatient strategy doing well here is somewhat opposite to why it does well in the previous experiments. Since all suppliers are opting for similar high priority jobs, competition is high, leading to higher failure rate in winning contracts. $IMPT$ strategy users lower their bids sharply and hence wins more auctions by underbidding the suppliers with other pricing strategies. Hence, *aggressive pricing can be dominant even when all suppliers are using the same scheduling strategy.*

Next, we investigate whether aggressive pricing is preferable when everyone is using it. We have conducted three different experiments, in each case keeping only suppliers of one pricing strategy. Thus we have run three experiments with each of the twelve suppliers using either EU_LIN, EU_DEF and EU_IMPT pricing scheme. In Figure 14, we have shown that in such settings suppliers with DEF pricing strategy generates maximum wealth while suppliers with $IMPT$ pricing performs the worst. Because of the lack of clear market niches, impatient suppliers get into a downward spiral of price wars which significantly erodes everyone's profitability. DEF suppliers are less aggressive about reducing their bids when they lose contracts and hence this more patient attitude rakes in more profit in the long run. In essence this is the opposite scenario of more wealth generated by aggressive pricing in the presence of market niches. We can conclude that *in the absence of market niches, less aggressive pricing mechanisms will produce more profits.*

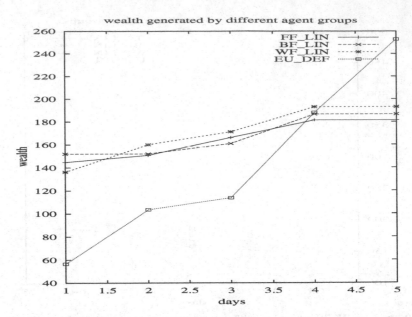

Fig. 11. Average Wealth earned by different supplier types. There are 3 suppliers using each of four scheduling strategies. FF, BF and WF are using linear price strategy while EU is using defensive pricing.

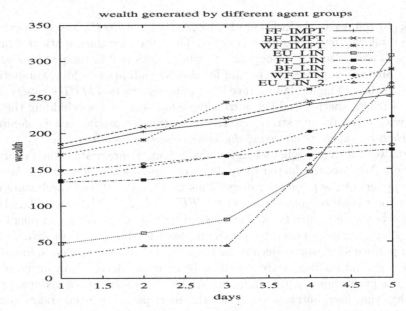

Fig. 12. Comparison of the average wealth generated by different group of suppliers in two different experiments. In one experiment FF, BF and WF follows linear pricing and in the other they follow impatient pricing. EU supplier used linear pricing in both experiments.

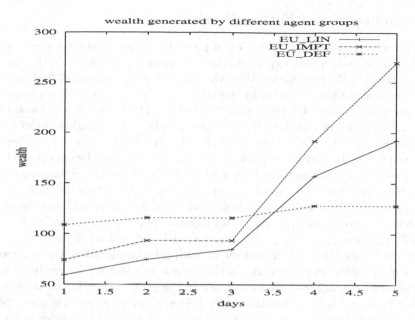

Fig. 13. Average Wealth earned by different supplier types all using EU scheduling and different pricing strategies.

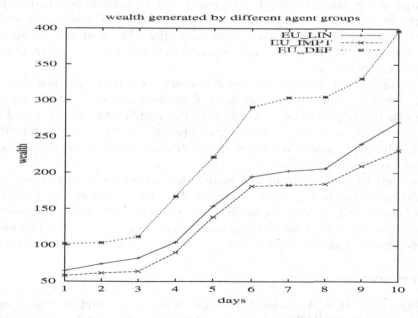

Fig. 14. Average Wealth earned by different supplier using EU scheduling and different pricing mechanism when they are the only type of suppliers in the population.

9   Conclusion and Future Work

In this work, we evaluate combination of predictive scheduling strategies and aggressive pricing schemes to exploit market niches in supply chains in the context of $B2B$ trading. We have shown that when a significant number of high priority tasks are present, the expected utility based strategy and the worst-fit strategy provides greater flexibility. On the contrary, when such tasks are rare and also there are longer tasks in the mix, best-fit strategies may be good enough. Opportunistic scheduling heuristics, like EU, are more effective in identifying market niches and effective pricing strategy are extremely efficient in exploiting the opportunities to generate more profit in an environment where tasks are arriving dynamically.

We have also discussed the advantages and disadvantages of different pricing strategies under different environmental conditions. Based on our observations, one can identify the dominant pricing strategy given a population mix and task generation distribution. Interestingly we have observed that both when the competition is fierce and when there is a market niche, with little competition, the aggressive, impatient pricing scheme dominate. In the former case, this is achieved by winning more contracts by aggressively underbidding the competition. In the latter case, profit is increased by significantly hiking prices. Another important observation is that aggressive bidding can benefit even simple scheduling strategies as long as there is a good mix of scheduling and pricing strategies in the population.

But when there is a scarcity of tasks and most of the suppliers use the same pricing and scheduling scheme, less aggressive pricing schemes dominate aggressive pricing schemes. The latter produces profit-reducing price wars that undermines everyone's profitability. Less aggressive price adjustment mechanisms are more tolerant to failure to win contracts and results in more cumulative profits in such circumstances.

Our approach provides a novel combination of opportunistic scheduling and aggressive pricing to improve profitability of suppliers in the context of a dynamic supply chain. We plan to build on this work by providing an analytical model to predict and choose the best scheduling and pricing strategy combination either given, or having formed, expectations of the strategy choices of competing agents and task arrival distributions.

We need to incorporate in our model the concept of decommitment and penalty in the contracting process which will be more representative of real-life supply chains. We are planning to investigate the effect of trust in this context. We also plan to study swapping of resources based on side-payments. This allows for effective coalitions to develop to better respond to uncertain market needs.

References

1. Arpinar, S., Dogac, A., Tatbul, N.: An open electronic marketplace through agent-based workflows: MOPPET. International Journal on Digital Libraries 3(1), 36–59 (2000)
2. Barbuceanu, M., Fox, M.S.: Coordinating multiple agents in the supply chain. In: Proceedings of the Fifth Workshop on Enabling Technologies: Infrastructure for Collaborative Enterprises, pp. 134–142 (1996)

3. Christopher Beck, J., Fox, M.S.: Supply chain coordination via mediated constraint relaxation. In: Proceedings of the First Canadian Workshop on Distributed Artificial Intelligence (May 1994)
4. Chen, Y., Peng, Y., Finin, T., Labrou, Y., Cost, R., Chu, B., Sun, R., Willhelm, R.: A negotiation-based multi-agent system for supply chain management. In: Working Notes of the ACM Autonomous Agents Workshop on Agent-based Decision-support for Managing the Internet-enabled Supply-chain (May 1999)
5. Jennings, N.R., Faratin, P., Norman, T.J., O'Brien, P., Odgers, B.: Autonomous agents for business process management. International Journal of Applied Artificial Intelligence 14(2), 145–189 (2000)
6. Labarthe, O., Tranvouez, E., Ferrarini, A., Espinasse, B., Montreuil, B.: A heterogeneous multi-agent modelling for distributed simulation of supply chains. In: Mařík, V., McFarlane, D.C., Valckenaers, P. (eds.) HoloMAS 2003. LNCS (LNAI), vol. 2744. Springer, Heidelberg (2003)
7. Sadeh, N., Hildum, D., Kjenstad, D., Tseng, A.: Mascot: An agent-based architecture for coordinated mixed-initiative supply chain planning and scheduling. In: Working notes of the ACM Autonomous Agents Workshop on Agent-Based Decision Support for Managing the Internet-Enabled Supply Chain (May 1999)
8. Saha, S., Sen, S.: Opportunistic scheduling and pricing in supply chains. KI-Zeitschrift fur Kunstliche Intelligenz (AI - Journal of Artificial Intelligence) 18(2), 17–22 (2004)
9. Saha, S., Sen, S.: Aggressive pricing to exploit market niches in supply chains. IEEE Intelligent Systems 20(1), 36–41 (2005)
10. Sandholm, T., Lesser, V.: Leveled commitment contracting: A backtracking instrument for multiagent systems. AI Magazine 23(3), 89–100 (2002)
11. Tac, supply chain management, http://www.sics.se/tac/
12. Shen, W., Norrie, D.: An agent-based approach for manufacturing enterprise integration and supply chain management. In: Jacucci, G., et al. (eds.) Globalization of Manufacturing in the Digital Communications Era of the 21st Century: Innovation, Agility, and the Virtual Enterprise, pp. 579–590. Kluwer Academic Publishers, Dordrecht (1998)
13. Swaminathan, J.M., Smith, S.F., Sadeh, N.M.: Modeling supply chain dynamics: A multiagent approach. Decision Sciences 29(3), 607–632 (1998)
14. Swaminathan, J.M., Smith, S., Sadeh, N.: Modeling the dynamics of supply chains. In: Proceedings the AAAI-94 Workshop on Reasoning About the Shop Floor (August 1994)
15. Walsh, W.E., Wellman, M.P.: Decentralized supply chain formation: A market protocol and competitive equilibrium analysis. Journal of Artificial Intelligence Research 19, 513–567 (2003)

Soft Computing in Intelligent Tutoring Systems and Educational Assessment

Rodney D. Nielsen[1], Wayne Ward[2], and James H. Martin[3]

[1,2,3] Center for Spoken Language Research ([2] Director)
[1,2,3] Institute of Cognitive Science
[1,2,3] Department of Computer Science
[3] Department of Linguistics
University of Colorado, Boulder
Rodney.Nielsen@Colorado.edu

Abstract. The need for soft computing technologies to facilitate effective automated tutoring is pervasive – from machine learning techniques to predict content significance and generate appropriate questions, to interpretation of noisy spoken responses and statistical assessment of the response quality, through user modeling and determining how best to respond to the learner in order to optimize learning gains. This chapter focuses primarily on the domain-independent semantic analysis of learner responses, reviewing prior work in intelligent tutoring systems and educational assessment. We present a new framework for assessing the semantics of learner responses and the results of our initial implementation of a machine learning approach based on this framework.

1 Introduction

This chapter reviews work toward the goal of improving the state of intelligent tutoring systems to the point where they are comparable to human tutors, as measured by increases in the associated student learning gains. In order to make this goal a reality, many soft computing techniques are required. Dynamic interaction will require more flexible techniques than permitted by the conventional scripted Finite State Machine dialogs that exist today. Automatic question generation requires that the system learn good predictors of content significance. Speech recognition, which already relies heavily on soft computing techniques such as machine learning, is fundamental to the intelligent tutoring system's interface. Semantic analysis of children's responses to the tutor requires numerous soft computing technologies, including statistical methods to assess lexical similarity and machine learning methods to learn how to classify the structural relationships within the student's answer as well as the semantic relationships between the student's answer and the desired answer. Determining how best to respond to a child's answer requires probabilistic reasoning to assess the value of requesting further elaboration, to assume they understand the subject, or to recognize with high confidence that they have contradicted the reference answer and that remediation is required. Optimizing student learning requires dynamic and probabilistic user modeling in addition to learning strategies that effectively engage students and produce meaningful learning effects.

B. Prasad (Ed.): Soft Computing Applications in Business, STUDFUZZ 230, pp. 201–230, 2008.
springerlink.com

This chapter reviews prior work and outlines initial steps toward developing this natural, engaging, domain-independent tutoring system, particularly in the area of semantic analysis of children's responses to questions. We build on prior work in lexical semantics (e.g., Banerjee and Pedersen 2003; Landauer and Dumais 1997; Turney et al. 2001, 2003), paraphrasing and entailment (e.g., Barzilay and Lee 2003; Barzilay and McKeown 2001; Glickman et al. 2005; Hickl et al. 2006, 2007; Lin and Pantel 2001; MacCartney et al. 2006; Nielsen et al. 2006), shallow semantic parsing (e.g., Nielsen and Pradhan 2004; Pradhan et al 2005), probabilistic classification (Nielsen 2004; Platt 2000), and to some extent cognitive modeling (Kintsch 1998). We describe a framework and method for assessing learner responses to tutor questions (Nielsen and Ward 2007) and a machine learning based system to automatically generate this assessment. Another significant contribution of this work is that the semantic assessment of answers is completely question independent – the system does not need to be retrained for new questions or even new subject areas. No prior work in the area of tutoring systems or answer verification has attempted to build such question-independent systems. They have virtually all required hundreds of examples of learner answers for each new question in order to train aspects of their systems or to hand-craft information extraction templates.

Imagine a time in the future when, in addition to instructor-led group interaction in the classroom, children get one-on-one or small group (two or three to one) tutoring with a subject matter expert. At a time when funding for education seems to be continually cut, this seems impossible to imagine and almost surely never will come to be, at least not with human tutors, but what about computers? Might it be possible for computers to engage students in this same form of natural face-to-face conversation? Might they even be more capable of patiently tailoring their interactions to each child's learning style or knowing just the right question to ask at just the right time to maximize learning outcomes? What capabilities must the system possess to carry out this feat? While there are many unsolved problems between today and the future just described, this chapter reviews several soft computing strategies that take steps in the direction of solving these problems, with an emphasis on systems that strive to understand a child's utterance in the context of a tutor's question.

Consider the question in example (1a), the desired answer in (2a), and the child's answer shown in (3a).

(1a)   *Kate dijo: "Un objeto tiene que moverse para producir sonido." ¿Estás de acuerdo con ella? ¿Por qué sí o por qué no?*

(2a)   *De acuerdo. Las vibraciones son movimientos y las vibraciones producen sonido.*

(3a)   *Sí, porque los sonidos vibran, chocan con el objeto y se mueve.*

What is required for the computer to understand the relationship between the student's answer and the reference answer? The relationship seems obvious to us, well, at least if we understand the language, but what would we do if, like the computer, we were not fluent in the language? One approach utilized by many is to examine the lexical similarity between the answers. We might recognize in this case that the student used other derivational forms of nearly all of the right words and hence, assume that they answered the question correctly. Unfortunately, the problem is not that easy. In fact, in this case, the student got the cause and effect completely backwards, as can be seen in the original English versions (1b), (2b) and (3b). The student was on topic, used virtually all the right words, but clearly does not understand the concepts being tested. It is critical that the computer be able to assess the interplay and relations between the words. This chapter describes the evolution of computing strategies, from early conventional approaches through the latest machine learning based systems, and describes our current efforts toward this goal of inferring which concepts the student does and does not understand.

(1b)   *Kate said: "An object has to move to produce sound." Do you agree with her? Why or why not?*

(2b)   *Agree. Vibrations are movements and vibrations produce sound.*

(3b)   *Yes, because sounds vibrate and hit the object and it moves.*

We begin with a brief motivation for why intelligent tutoring systems (ITSs) are an important component of future educational settings. Then we describe early, more conventional efforts in this domain along with their shortcomings. We describe the progression of soft computing strategies and discuss how they improve over earlier efforts and indicate where they still have room for improvement. Soft computing approaches to large-scale assessment are then reviewed and contrasted with automated tutoring technology. We also describe the relevance of some current active areas of natural language processing research such as paraphrase recognition and textual entailment. We then highlight our current efforts to achieve more robust semantic understanding by applying machine learning to detect the relationships between phrases that entail an understanding of a tutored concept and those that do not. In closing, we briefly touch on related areas of open research in tutoring that will require further advances in soft computing.

2 Why Are Intelligent Tutoring Systems Important?

Improving reading comprehension is a national priority, particularly in the area of science education. Recent estimates suggest that as many as 36% of fourth grade readers and 27% of 8th and 12th grade readers cannot extract the general meaning nor even make simple inferences from grade level text (NAEP 2005). While many students may appear to learn to read and understand text by third grade, evidence shows that their apparent competence is often an illusion—as texts become more challenging in fourth grade, many students cannot read nor understand them (Meichenbaum and Biemiller 1998; Sweet and Snow 2003). There is thus a critical need for programs that engage beginning readers in a way that supports comprehension. The lack of sufficient comprehension of texts seems likely to be a major contributing factor leading to the current state of U.S. science education:

> Current levels of mathematics and science achievement at the elementary and secondary levels suggest that the United States is neither preparing the general population with levels of mathematics and science knowledge necessary for the 21st century workplace, nor producing an adequate pipeline to meet national needs for domestic scientists. In the 2000 National Assessment of Educational Progress (NAEP), only two percent of U.S. students attained advanced levels of mathematics or science achievement by Grade 12. 34 percent of Grade 4

students, 39 percent of Grade 8 students, and 47 percent of Grade 12 students scored below the "basic" level in science.

The Institute of Educational Sciences
http://www.ed.gov/programs/mathresearch/2006-305k.pdf

Current levels of mathematics and science achievement at the elementary and secondary levels suggest that the United States is neither preparing the general population with levels of mathematics and science knowledge necessary for the 21st century workplace, nor producing an adequate pipeline to meet national needs for domestic scientists. In the 2000 National Assessment of Educational Progress (NAEP), only two percent of U.S. students attained advanced levels of mathematics or science achievement by Grade 12. 34 percent of Grade 4 students, 39 percent of Grade 8 students, and 47 percent of Grade 12 students scored below the "basic" level in science.

How do we tailor instruction to accommodate differing learner needs and address these comprehension deficits? In 1984, Benjamin Bloom determined that the difference between the amount and quality of learning for a class of thirty students and those who received individualized tutoring was 2 standard deviations (Bloom 1984). The significant differences in proficiency between those children who enjoy one-on-one tutoring versus those who have little or no individualized support is testament to the need for further exploration of the individualized tutoring model.

In the two decades since Bloom reported a two sigma advantage of one-on-one tutoring over classroom instruction in several subjects, evidence that tutoring works has been obtained from dozens of well designed research studies, meta-analyses of research studies (e.g., Cohen, Kulik and Kulik 1982) and positive outcomes obtained in large scale tutoring programs in Britain (e.g., Topping and Whitley 1990) and the U.S. (Madden and Slavin 1989). Effective intelligent tutoring systems exist in the laboratory that produce learning gains with high-school, college, and adult subjects through text-based dialog interaction (e.g., Graesser et al. 2001; Peters et al. 2004; VanLehn et al. 2005), some demonstrating up to a one sigma gain relative to classroom instruction (Anderson et al. 1995; Koedinger et al. 1997). Therefore, intelligent tutoring systems show promise in approaching the effectiveness of human tutoring and systems that are accessible, inexpensive, scalable and above all effective would provide one critical component of an overall educational solution.

From a cost-benefit perspective, computers and associated learning software have the potential to provide a relatively inexpensive solution in today's education system. The cost of training and employing additional teachers to provide the level of individualized attention that many students need is prohibitively expensive, whereas interactive computer systems that use advanced human communication and interface technologies to provide individualized tutoring could inexpensively provide focused, individualized, adaptive, scientifically-based instruction. In short, advances in soft computing technologies, communications, and language technologies combined with advances in cognitive science and the science of reading and learning, provide a powerful and timely potential solution to our nation's education crises.

3 Prior Work on Intelligent Tutoring Systems

Early automated tutors largely followed conventional computing approaches, for example, utilizing multiple choice questions. Current research shows that, if properly designed, multiple choice questions can be a very effective tool for assessing comprehension. However, one of the most successful means of improving learning gains is to force students to articulate their beliefs, leading them to a better understanding of what they do and do not know and strengthening the causal relations between the bits of knowledge they have acquired (Chi 1996; VanLehn et al. 2003). Therefore, much research has moved away from multiple choice questions and toward free response questions, requiring students to express their deeper understanding of the concepts in the text.

The first most obvious step in this direction was the use of scripted, domain-specific dialog techniques, often implemented utilizing Finite State Machines (FSM). For example, SCoT, a Spoken Conversational Tutor (Peters et al. 2004), uses *activity recipes* to specify what information is available to the recipe, which parts of the information state are used in determining how to execute the recipe, and how to decompose an activity into other activities and low-level actions. The Phoenix semantic parser (Ward 1991) defines semantic frames which include patterns to be matched and extracted from the dialog. The dialog is then driven by what frame elements have not yet been addressed and the specific system turns attached to those frame elements. The advantage of this technology lies in its precision – it is generally very accurate in the classification of relevant text fragments, but this is typically at the cost of poorer performance in recall – finding all of the relevant information nuggets (see Table 1 for a description of precision and recall).

We conducted a pilot experiment to determine how well an automatic system based on domain-specific shallow semantic parsing (the process of recognizing predicate-argument structure or semantic relationships in text) could grade young children's summaries of stories. We collected spoken summaries of a single story, "Racer the Dog", from 22 third and fourth grade students. We divided the summaries into a training set consisting of fifteen summaries and a test set of seven summaries. Multiple researchers generated reference summaries, which we distilled to consist of the key points that should exist in a good summary of the story. We utilized the Phoenix semantic parser to map summaries to semantic frames (see Fig. 1 and Fig. 2 for examples).[1] We first wrote an initial parser grammar that could extract the points in the reference summary and then utilized the fifteen training summaries to expand the coverage of the grammar. The seven summaries comprising the test set were of course not examined during system development. After system development, we manually parsed the test set to create gold standard reference parses for evaluation purposes. The manual parse identified a total of 36 points that addressed content from the reference summaries. Compared to this gold standard, automatic parses of the test set had a recall of 97% (35/36) and a precision of 100% (35/35) – the parser found all but one of the relevant points and produced no erroneous ones. (In order to consider a parse correct, the concept from the child's summary had to have the same semantic roles as one in the reference summary and the roles had to be filled by references to the same entities.)

---

[1] Anaphoric reference was resolved manually in a pre-processing step.

Table 1. Evaluation Metrics

Most system evaluations, to date, have been based on simple answer grading. An alternative way of evaluating answer assessment technology is through precision, recall and the F-measure. Precision, P, indicates how accurate your system is when it classifies a piece of text as having matched a nugget of interest. These nuggets might be correct answer facets, contradictory facets, phrases indicative of a common misconception, etc. However, many systems focus primarily on detecting only correct answer facets. Specifically, precision is the fraction of classifications that are correct out of all of the text fragments the system classified as falling within one of the categories of interest.

$$P = \frac{\text{number of correct nugget classifications by the system}}{\text{number of total nugget classifications by the system}}$$

Recall, R, measures how good the system is at detecting the presence of these nuggets of interests. It is the fraction of classifications that are correct out of all of the text fragments that should have been classified as falling within one of the categories of interest according to the human gold standard annotation.

$$R = \frac{\text{number of correct nugget classifications by the system}}{\text{number of total nuggets in the human reference annotation}}$$

The F-measure provides a means of weighting the precision and recall to provide a single performance metric.

$$F = \frac{(\beta^2 + 1)PR}{\beta^2 P + R}$$

Setting $\beta > 1$, results in precision being more significant than recall – it suggests that when the system classifies a nugget as being of interest, it is important that the system be right in its classification. Setting $\beta < 1$, results in recall being more significant than precision – this suggests that it is more important that the system identify as many of the nuggets of interest as possible, at the cost of classifying some irrelevant text fragments as being of interest. When $\beta = 1$, precision and recall are regarded to be equally important – this is the most common scenario in natural language processing, as it is difficult to quantify the importance of the two factors.

Racer had problems with his back legs.
LegProblems:[Problems_agent].[Dog_Name].racer

Fang always bit Racer and ran away faster than Racer could run.
Bother:[Bother_agent].[Dog_Name].fang
Bother:[Bother_theme].[Dog_Name].racer
Faster:[Run_Away_agent].[Dog_Name].fang
Faster:[Run_Away_theme].[Dog_Name].racer

Fig. 1. Example Phoenix parses of the reference summary

um racer got his leg hurt
LegProblems:[Problems_agent].[Dog_Name].racer

fang kept on teasing racer
Bother:[Bother_agent].[Dog_Name].fang
Bother:[Bother_theme].[Dog_Name].racer

Fig. 2. Example Phoenix parse of a child's summary

The preceding results are for human transcriptions of the children's spoken summaries. We also decoded the speech files with the University of Colorado SONIC speech recognition system (Pellom 2001; Pellom and Hacioglu 2003) and processed the SONIC output. For the same test set, the recall was 83% (30/36) and the precision was 100% (30/30) – speech recognition errors resulted in missing five points that were extracted from the human transcribed summaries, but generated no erroneous points. As expected by such systems, the precision was quite high, while the recall was slightly less so.

A significant disadvantage of these more conventional systems is that they require a considerable investment in labor to cope with a new subject area or even to handle a small change in subject matter coverage. This effort is required to generate new hand-crafted parsers, knowledge-based ontologies, and dialog control mechanisms. In this regard, the use of Latent Semantic Analysis (LSA), a statistical soft computing technique, to assess student's summaries represents an improvement over FSM dialogs, in that the system is more flexible in handling the unconstrained responses of a learner (Landauer and Dumais 1997; Landauer, Foltz and Laham 1998). LSA begins with a term by document matrix, where the cells in the matrix indicate the number of occurrences of the given term in the associated document. This matrix is given a TF-IDF (term frequency-inverse document frequency) weighting to account for the relative importance of a term in the document adjusted for its significance across all documents. LSA then utilizes singular value decomposition and retains only the top k (usually around 300) dimensions to represent the key information in the original matrix. This process effectively smoothes the data and brings out the latent semantics in the original document set – providing connections between terms and documents that were not explicit in the source text. The resulting rank-k approximation of the original matrix is then used to determine the similarity of terms and documents by calculating the dot product or cosine of related vectors.

The Institute of Cognitive Science (ICS) and The Center for Spoken Language Research (CSLR), both at the University of Colorado, Boulder, have worked with the Colorado Literacy Tutor program to develop educational software that helps children learn to read and comprehend text (Cole et al. 2003; Franzke et al. 2005). A significant part of this program is Summary Street®, a tool for improving and training text comprehension through summarization. Summary Street utilizes LSA to grade children's text summaries and provide feedback on the quality of the summary, including completeness, relevance and redundancy. Summary Street's feedback has been shown to improve student scores by on average a letter grade with much more improvement for lower performing students.

AutoTutor (Mathews et al. 2003) is an interactive text-based tutor that utilizes LSA to engage college students in dialogs regarding conceptual physics and introductory computer science. Because AutoTutor is based on soft computing techniques, it is more flexible in handling the unconstrained responses of a learner than are tutors based on FSM dialogs. The AutoTutor architecture requires the lesson planner or system designer to provide the following information:

1. A statement of the problem to be solved, in the form of a question.
2. A set of expectations in an ideal answer, with each expectation being a sentence in natural language of 10-20 words
3. A set of tutor dialog moves that express or elicit from the learner each expectation in #2 (i.e., hints, prompts, and assertions)
4. A set of anticipated bad answers and corrections for those bad answers
5. A set of [subject matter] misconceptions and corrections to those misconceptions
6. A set of basic noun-like concepts about [the subject matter] and their functional synonyms in the specific context of the problem.
7. A summary of the answer or solution
8. A latent semantic analysis (LSA) vector for each expectation, bad answer, and misconception.

(Mathews et al. 2003)

The key advantage of their architecture is that it does not require a syntactic match between the learner's dialog turn and the expectation in the ideal answer; nor does it require key phrase spotting. The LSA component analyzes the semantic similarity between the user's open-ended dialog turn and the system's expectation or reference answer by comparing vector representations derived from a bag of words method – a method which ignores the expressed relations between words. The dialog is then driven by checking the degree to which each expectation has been covered by the sum total of all past user turns for the problem. The answer expectation and the compilation of user turns are each represented as pseudo-documents, weighted averages of the reduced-dimensionality vectors associated with the words they contain. The answer is then assessed by computing the cosine between these vectors representing the two pseudo-documents. If the metric exceeds a threshold, the user's turns are assumed to have the same meaning as the answer expectation.

However, the reality is that, while LSA's evaluations are closely correlated with human evaluations, LSA completely disregards syntax in its analysis and is prone to many related weaknesses. In example (3) above, the student used all the right key words (*sounds vibrate and hit the object and it moves*) and LSA would be satisfied despite the fact that the student has the causal relation reversed. The authors note that LSA does a poor job of detecting misconceptions; it seems likely that this might be due to the relatedness between the bag of words in the misconception's description and the bag of words in the expected answer to the question. LSA also performs very poorly on the short answers that are typical in tutoring settings. This is the reason that they must combine the input from all prior user turns into one cumulative bag of words, rather than compare strictly with the learner's current response. Furthermore, it is not possible with typical LSA-based approaches to perform a detailed assessment of a learner's contribution or to identify the specific reasons that a short answer might not be correct.

Consequently, LSA provides little help in classifying learner contributions and determining the best tutor response or dialog strategy to correct misconceptions.

While the LSA-based approaches are not typically trained for individual questions, they do generally require a fair amount of corpus tuning to ensure adequate coverage of the topic area and the cosine threshold to judge similarity is always tuned to the domain or question. As noted above, AutoTutor also requires "a set of basic noun-like concepts about [the subject matter] and their functional synonyms in the specific context of the problem". Lastly, there is no evidence that LSA is an effective tool for interacting with young children in the K-5 grade range. In preliminary investigations using story summaries written by third and fourth graders, we found that LSA was unable to appropriately assess the quality of these children's summary-length responses. This is further confirmed by LSA's poor performance in grading elementary students' constructed responses to short answer questions in experiments by the creators of AutoTutor (Graesser, personal communication).

Rosé et al. (2003a) and Jordan et al. (2004) have improved on the accuracy of the LSA-based approach to verifying short answers in tutorial dialogs by incorporating a deep syntactic analysis into their evaluation and integrating multiple soft computing technologies. A decision tree is used to learn from the output of a Naïve Bayes classifier and from deep syntactic parse features.[2]

The system classifies each student sentence to determine which, if any, of the set of good answer aspects it matches. This is one of a very few systems for assessing open-ended short answers that provides any finer grained measure of performance than a simple answer grade; they evaluate the system using precision, recall, and F-scores on the task of detecting the good answer aspects in the student responses. They compare their combined technique, CarmelTC, to LSA as a baseline, as well as to the Naïve Bayes classifier without the syntactic parse features, and a classifier based only on the syntactic features without the Naïve Bayes output. The results, as seen in Table 2, show that the combined approach performed much better than LSA or either of the individual systems.

Table 2. Evaluation of CarmelTC

Method	Precision	Recall	F
LSA	93%	54%	0.70
Naïve Bayes	81%	73%	0.77
Symbolic-only	88%	72%	0.79
CarmelTC	90%	80%	0.85

Though CarmelTC does provide slightly more information than a simple grading system, it does not provide the sort of fine-grained assessment of a learner's contribution necessary to understand their mental model and drive high-quality tutoring dialog. Furthermore, much of the parsing in their system is dependent on domain-specific, hand-coded rules, in order to capture the semantics of the domain lexicon and language. While they are building tools to ease this process (Rosé et al. 2003b), there appears to be a long way to go before a lesson planner, with no formal

---

[2] Their deep parse includes information beyond a typical syntactic parse (e.g., mood, tense, negation, etc.).

linguistics or computer science background, would see a positive cost-benefit analysis in utilizing these tools to build lessons, especially on a regular basis. For example, to handle a new question the user must generate first order propositional representations of each good answer aspect and hand annotate example learner answers as to their semantic interpretation within those representations.

Perhaps more important than the time-consuming nature of this process, it defeats one of the ultimate goals of an Intelligent Tutoring System (ITS), namely to be able to interact naturally without being constrained to communication within the realm of a fixed set of question-answer pairs. Finally, the methods described by Rosé and Jordan require that classifiers be trained for each possible proposition associated with the reference answers. This not only requires additional effort (currently on the part of the systems builders) to train the system, but it also requires the collection and grading of a large corpus of learner responses to each question that the tutor might ask; again defeating the ultimate goal of natural dynamic tutoring interactions.

Makatchev et al. (2004) develop an abductive theorem prover with the shared long-term goal of providing more specific, higher quality feedback to learners. Abductive theorem provers provide a softer pattern matching technique than conventional logical theorem provers in that, if a fact does not exist in their knowledge base, they can assume the fact at a cost related to its truth probability. However, their approach is domain dependent, requiring extensive knowledge engineering for each new qualitative physics problem to be solved; their reasoning engine has 105 domain rules that handle seven specific physics problems.

Recall and precision vary by proof cost[3]; at a proof cost of 0.6, recall and precision are approximately .38 & .25, respectively, where at a proof cost of 0.2, they are around .63 & .15, leading to F-measures of around .30 and .24 respectively. In initial analyses, their system did not statistically improve learning outcomes above those achieved by having the students simply read the text. They also indicate that the system performed poorly at correctly identifying misconceptions in the student essays.

Many other ITS researchers are also striving to provide more refined learner feedback (e.g., Aleven et al. 2001; Peters et al. 2004; Pon-Barry et al. 2004; Roll et al. 2005). However, they too are developing very domain-dependent approaches, requiring a significant investment in hand-crafted logic representations, parsers, knowledge-based ontologies, and dialog control mechanisms. Simply put, these domain-dependent techniques will not scale to the task of developing general purpose Intelligent Tutoring Systems and will not enable the long-term goal of effective unconstrained interaction with learners or the pedagogy that requires it.

4   Related Research

4.1 Short-Answer Free-Text Response Scoring

There is a small, but growing, body of research in the area of scoring free-text responses to short answer questions (e.g., Callear et al. 2001; Leacock 2004; Leacock

---

[3] Most systems can achieve higher recall at the cost of lower precision and vice versa, if they classify only the fraction of text fragments that they are the most confident about, they can achieve nearly perfect precision at the cost of much lower recall. With abductive theorem provers this tradeoff is made largely based on the proof cost.

and Chodorow 2003; Mitchell et al. 2002, 2003; Pulman 2005; Sukkarieh 2003, 2005). Most of the systems that have been implemented and tested are based on Information Extraction (IE) techniques (Cowie and Lehnert 1996). They hand-craft a large number of pattern rules, directed at detecting the key aspects of correct answers or common incorrect answers. The results of these approaches range from about 84% accuracy up to nearly 100%. However, the implemented systems are nearly all written by private companies that keep much of the nature of the questions and systems proprietary and the best results seem to frequently be achieved by tuning with the test data. Still, these results provide good evidence that answers can be accurately assessed, at least at the coarse-grained level of assigning a score to high-school and college level students' answers.

Work in this area is typified by c-rater (Leacock and Chodorow 2003; Leacock 2004), which was designed for large-scale tests administered by the Educational Testing Service (ETS). A user examines 100 example student answers and manually extracts the common syntactic variants of the answer (200 or more examples are used for problematic questions that have more syntactic variation or fewer correct answers in the dataset, etc). They build a model for each syntactic variant and specify which aspects of the subject-verb-object structure are required to give credit for the answer. For each of the subject, verb, and object, the model authoring tool provides the user with a list of potential synonyms and they select those synonyms that are appropriate within the given context. The list of synonyms extracted via the fuzzy similarity metrics described in Lin (1998). Lin's similarity metric, which is based on Information Theory, computes the mutual information between words that are connected via syntactic dependencies and then uses this to calculate the similarity between any pair of words based on the ratio of information in the dependencies the two words have in common to the sum of information in all dependencies involving either word. The more dependencies two words have in common, the closer this ratio is to 1.0. Given the list of similar words, the user of c-rater then selects those synonyms that are appropriate within the given context.

During the process of scoring a student's answer, the verb lemma is automatically determined and it must match a lemma in one of the model answers; the system also resolves pronouns and attempts to correct non-word misspellings. Scoring consists of assigning a value of 0 (no credit), 1 (partial credit), or 2 (full credit). In a large-scale reading comprehension exam administered across Indiana (16,625 students), c-rater achieved an 84% accuracy as measured relative to the scores of human judges on a random sample of 100 answers for each of the seven questions that they were able to score using c-rater.

In general, short-answer free-text response scoring systems are designed for large scale assessment tasks, such as those associated with the tests administered by ETS. Therefore, they are not designed with the goal of accommodating dynamically generated, previously unseen questions. Similarly, these systems do not provide feedback regarding the specific aspects of answers that are correct or incorrect; they merely provide a raw score for each question. As with the related work directed specifically at ITSs, these approaches all require in the range of 100-500 example student answers for each planned test question to assist in the creation of IE patterns or to train a machine learning algorithm used within some component of their solution.

4.2 Paraphrasing and Entailment

In recent years, there has been a tremendous increase in interest in the areas of para-phrase acquisition and entailment proof or identification. These technologies have broad applications in numerous areas and great relevance to this work. Paraphrasing is the most common means for a learner to express a correct answer in an alternative form; in fact, Burger and Ferro (2005) note that even in the Pascal Recognizing Tex-tual Entailment (RTE) challenge (Dagan et al. 2005), 94% of the development corpus consisted of paraphrases, rather than true entailments.

The target of a fair amount of the work on paraphrasing is in acquiring paraphrases to be used in information extraction or closely related factoid question answering (Agichtein and Gravano 2000; Duclaye et al. 2002; Ravichandran and Hovy 2002; Shinyama and Sekine 2003; Shinyama et al. 2002; Sudo et al. 2001) and assumes there are several common ways of expressing the same information, (e.g., when and where someone was born). For example, Agichtein and Gravano (2000) use a boot-strapping approach to automatically learn paraphrase patterns from text, given just five seed examples of the desired relation. The technique does not identify general patterns, just those associated with the relation in the seeds. Beginning with the seed examples they extract patterns in the form of a five-tuple <left-context, entity-A, mid-dle-context, entity-B, right-context>, where the contexts are represented in a soft form as weighted vectors that disregard word order. The terms in each context are weighted according to their frequency across the seeds. These patterns are used to find addi-tional entity pairs assumed to have the same relation, which are then used as seed ex-amples to find additional patterns and the process repeats. They achieved almost 80% recall and 85% precision in the task of extracting headquarters' locations for organi-zations. This could be applied to answer assessment, if you can automatically identify the "entities" in an answer using Named Entity (NE) recognizers, then via a similar technique you can determine whether the two encompassing contexts are essentially paraphrases. Since fact-based questions are also somewhat common in testing envi-ronments, these paraphrase extraction techniques could be very useful in tutoring sys-tems. However, for tutor questions requiring deeper reasoning than simple fact recol-lection, these techniques, at minimum, require significant modifications.

Much of the remaining research on paraphrase acquisition presupposes parallel corpora (e.g., Barzilay and McKeown 2001; Pang et al. 2003) or comparable corpora known to cover the same news topics (e.g., Barzilay and Lee 2003). The parallel cor-pora are aligned at the sentence level and the sentence pairs are considered to be para-phrases. From these alignments, Barzilay and McKeown extract patterns for valid lexical or short phrasal paraphrases. They use a machine learning algorithm that is a variant of Co-Training based on (a) context features and (b) lexical and POS features. First, they initialize the seed set of paraphrases to match identical word sequences in aligned sentences, then they iteratively find contexts based on the paraphrases and paraphrases based on the contexts, until no new paraphrases are found or a specified number of iterations passes. They found 9483 paraphrases, of which about 29% were multi-word phrases, and 25 morpho-syntactic rules. Judges found the paraphrases (with context provided) to be valid about 91.6% of the time. Only around 35% of the lexical paraphrases were synonyms; the remaining were 32% hypernyms, 18% sib-lings of a hypernym, 5% from other WordNet relations, and 10% were not related in

WordNet, demonstrating the need for softer assessment versus requiring strict synonyms.

Pang et al. assess syntactic constituency trees to generate Finite State Automata that represent paraphrases. Barzilay and Lee also generate word lattices, but from a single corpus by combining the surface forms of several similarly written sentences about distinct events. These lattices form a database of potential paraphrasing techniques or transformations where the areas of high variability represent the event arguments (e.g., the actors, location, etc.). They then cluster sentences written on the same day in different articles to decide which are paraphrases and use the lattice associate with one sentence to generate paraphrases for another sentence in the same cluster – probabilistically, any path through the lattice, with appropriate entity substitution, is considered a paraphrase. Other techniques exist for lexical paraphrase acquisition, which do not rely on parallel corpora (e.g., Glickman and Dagan 2003). The use of patterns extracted a priori by these systems to verify a paraphrase between the learner's answer and the desired or reference answer in a tutoring environment is unlikely to be of much benefit, since they are not broad coverage patterns, but rather are specific to high frequency news topics. However, these algorithms could be modified to perform online paraphrase recognition.

Research in the area of entailment also has much to offer. Lin and Pantel (2001a, 2001b) extract inference rules from text by looking for dependency parse patterns that share common argument fillers according to pointwise mutual information (Church and Hanks 1989). This work stimulated much of the later work in paraphrasing and entailment. The main idea is to find patterns in text that share similar key content words, where these content words fill slot values at each end of the pattern. The patterns are extracted from paths in a syntactic dependency parse tree and the similarity of patterns is determined by the pointwise mutual information (PMI) computed for the patterns' slots. PMI for a pair of slots indicates whether the sets of words that fill the two slots are more similar than would be expected by chance. This same technique could be applied to determine whether a child's answer to a question is a paraphrase of the reference (desired) answer.

Again the difficulty in directly applying this work to the task of answer assessment is that the inference rules extracted do not tend to have broad coverage. This is evidenced by the fact that several researchers participating in the First Pascal RTE challenge made use of Lin and Pantel's patterns or a variant of their algorithm and still performed quite poorly, barely exceeding chance on any but the easiest task, the comparable documents task, (Braz et al. 2005; Haghighi et al. 2005; Herrera et al. 2005; Raina et al. 2005).

The RTE challenge has brought the issue of textual entailment before a broad community of researchers in a task independent fashion. The challenge requires systems to make simple yes-no judgments as to whether a human reading a text t of one or more full sentences would typically consider a second, hypothesis, text h (usually one shorter sentence) to most likely be true. Fig. 3 shows a typical t-h pair from the RTE challenge. In this example, the entailment decision is *no* – and that is similarly the extent to which training data is annotated. There is no indication of whether some facets of, the potentially quite long, h are addressed in t (as they are in this case) or conversely, which facets are not discussed or are explicitly contradicted.

t: At an international disaster conference in Kobe, Japan, the U.N. humanitarian chief said the United Nations should take the lead in creating a tsunami early-warning system in the Indian Ocean.
h: Nations affected by the Asian tsunami disaster have agreed the UN should begin work on an early warning system in the Indian Ocean.

Fig. 3. Example text-hypothesis pair from the RTE challenge

However, in the third RTE challenge, there is an optional pilot task that begins to address some of these issues. Specifically, they have extended the task by including an Unknown label, where h is neither entailed nor contradicted, and have requested justification for decisions. The form that these justifications will take has been left up to the groups participating, but could conceivably provide some of the information about which specific facets of the hypothesis are entailed, contradicted and unaddressed.

Submitters to the RTE challenge take a variety of approaches including very soft lexical similarity approaches (e.g., Glickman et al. 2005), lexical-syntactic feature similarity (e.g., Nielsen et al. 2006), syntactic/description logic subsumption (Braz et al. 2006), logical inference (e.g., Tatu and Moldovan 2007), and numerous other approaches, most based on machine learning. Many of these systems make use of the lexical similarity metrics discussed earlier in this chapter, or lexical relations described more formally in WordNet, or more formally still in most of the logical inference or abductive reasoning systems. Nearly all of these systems begin with the output of various subcomponents such as part-of-speech taggers, named-entity recognizers, coreference resolution systems, syntactic parsers, shallow semantic (predicate-argument) parsers, and document clustering systems.

The best performing systems in the RTE challenge have been the Hickl et al. (2006, 2007) approaches. In their 2006 entry, they perform a soft lexical alignment and then generate four Boolean semantic role features indicating roughly whether the predicates have the same semantic roles and are aligned similarly. Their soft lexical alignment features include cosine similarity, word co-occurrence statistics, WordNet similarity metrics, NE and POS similarity, and string-based similarity. They consider arguments to probabilistically match if the hypothesis' argument head is lexically aligned with (entailed by) something in the corresponding text's argument. A second set of features includes simpler lexical alignment information based on the longest common substring, the number of unaligned chunks and web-based lexical co-occurrence statistics. A third set of features indicates whether the polarity of the two text fragments is consistent. Their final set of features indicates whether the two text fragments would be grouped together when running a text clustering algorithm on a set of documents retrieved from a query on their keywords. Using a decision tree classifier, Hickl et al. achieved the best results at the second RTE challenge, with an accuracy of 75.4%. Hickl and Bensley (2007) extended this approach at RTE3, extracting the author's discourse commitments from each text fragment and basing their entailment strategy on these. Each sentence from the text was elaborated into a potentially very long list of propositions determined to be true based on the original statement's assertions, presuppositions, and conversational implicatures. These commitments were then used in a system based on their RTE2 submission to achieve an accuracy of 80%.

5   Robust Semantic Analysis in Intelligent Tutoring

5.1   The Necessity of Finer Grained Analysis

Imagine that you are an elementary school science tutor and that rather than having access to the student's full response to your questions, you are simply given the information that their answer was correct or incorrect, a yes or no entailment decision. Assuming the student's answer was not correct, what question do you ask next? What follow up question or action is most likely to lead to better understanding on the part of the child? Clearly, this is a far from ideal scenario, but it is roughly the situation within which many ITSs exist today.

In order to optimize learning gains in the tutoring environment, there are myriad issues the tutor must understand regarding the semantics of the student's response. Here, we focus strictly on drawing inferences regarding the student's understanding of the low-level concepts and relationships or facets of the reference answer. We use the word facet throughout this chapter to generically refer to some part of a text's (or utterance's) meaning. The most common type of answer facet discussed is the meaning associated with a pair of related words and the relation that connects them.

Rather than have a single yes or no entailment decision for the reference answer as a whole, (i.e., does the student understand the reference answer in its entirety or is there some unspecified part of it that we are unsure whether the student understands), we instead break the reference answer down into what we consider to be its lowest level compositional facets. This roughly translates to the set of triples composed of labeled dependencies in a dependency parse of the reference answer.[4] The following illustrates how a simple reference answer (2) is decomposed into the answer facets (2a-d) derived from its dependency parse and (2a'-d') provide a gloss of each facet's meaning. The dependency parse tree is shown in Fig. 4. As can be seen in 2b and 2c, the dependencies are augmented by thematic roles (c.f., Kipper et al. 2000) (e.g., Agent, Theme, Cause, Instrument...) produced by a semantic role labeling system (c.f., Gildea and Jurafsky 2002). The facets also include those semantic role relations that are not derivable from a typical dependency tree. For example, in the sentence "As it freezes the water will expand and crack the glass", water is not a modifier of crack in the dependency tree, but it does play the role of Agent in a shallow semantic parse.

Breaking the reference answer down into low-level facets provides the tutor's dialog manager with a much finer-grained assessment of the student's response, but a simple yes or no entailment at the facet level still lacks semantic expressiveness with regard to the relation between the student's answer and the facet in question. Did the student contradict the facet? Did they express a related concept that indicates a

---

[4] In a dependency parse, the syntactic structure of a sentence is represented as a set of lexical items connected by binary directed modifier relations called dependencies. The goal of most English dependency parsers is to produce a single projective tree structure for each sentence, where each node represents a word in the sentence, each link represents a functional category relation, usually labeled, between a governor (head) and a subordinate (modifier), and each node has a single governor (c.f., Nivre and Kubler, 2006). Each dependency is labeled with a type, (e.g., subject, object, nmod – noun modifier, vmod – verb modifier, sbar – subordinate or relative clause, det – determiner).

(2)   A long string produces a low pitch.

(2a)   NMod(string, long)
(2b)   Agent(produces, string)
(2c)   Product(produces, pitch)
(2d)   NMod(pitch, low)

(2a')   There is a long string.
(2b')   The string is producing something.
(2c')   A pitch is being produced.
(2d')   The pitch is low.

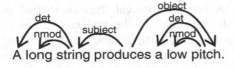

A long string produces a low pitch.

Fig. 4. Dependency parse tree for example reference answer

misconception? Did they leave the facet unaddressed? Can you assume that they understand the facet even though they did not express it, (e.g., it was part of the information given in the question)? It is clear that, in addition to breaking the reference answer into fine-grained facets, it is also necessary to break the annotation (classification) labels into finer levels in order to specify more clearly the relationship between the student's answer and the reference answer aspect. There are many other issues that the system must know to achieve near optimal tutoring, but these two – breaking the reference answer into fine-grained facets and utilizing more expressive annotation labels – are the emphasis of our current work.

5.2 Creating a Training Corpus

Because most text comprehension problems take root in elementary school during the early years of learning to read and comprehend texts, our work focuses on those critical grades, K-6. We acquired data gathered from 3rd-6th grade students utilizing the Full Option Science System (FOSS), a proven system that has been in use across the country for over a decade. Assessment is a major FOSS research focus, of which the Assessing Science Knowledge (ASK) project is a key component.[5] We used 290 free response questions taken from the ASK summative assessments covering their sixteen science teaching and learning modules. These questions had expected responses ranging in length from moderately short verb phrases to a few sentences, could be assessed objectively, and were not overly open ended. We generated a

---

[5] "FOSS is a research-based science program for grades K–8 developed at the Lawrence Hall of Science, University of California at Berkeley with support from the National Science Foundation and published by Delta Education. FOSS is also an ongoing research project dedicated to improving the learning and teaching of science."

Assessing Science Knowledge (ASK) is "designed to define, field test, and validate effective assessment tools and techniques to be used by grade 3–6 classroom teachers to assess, guide, and confirm student learning in science." http://www.lawrencehallofscience.org/foss/

corpus from a random sample of the kids' handwritten responses to these questions. The only special transcription instructions were to fix spelling errors (since these would be irrelevant in a spoken dialog environment), but not grammatical errors (which would still be relevant), and to skip blank answers and non-answers similar in nature to *I don't know* (since these are not particularly interesting from the research perspective).

Ultimately, around 50% of the answers will be set aside for testing the learning algorithms, with the remainder designated for training and development tuning. The test set will consist of all of the questions and answers of some modules – which will be used to test domain independent algorithms, all of the answers for some questions in the remaining modules – which will be used to test question-independent algorithms, and a small fraction of the answers for all of the remaining questions – which will be used to test algorithms designed to handle specific questions. In total including test and training sets, there are nearly 290 questions, 16000 student responses (approximately 40 responses per question for 13 of the modules and 100 responses per question for the remaining three modules) and over 100000 total facet annotations. However, since we have not quite finished annotating the corpus at the time of this writing, the experiments described in this chapter are based on a smaller subset of the data.

The answer assessment annotation described here is intended to be a step toward specifying the detailed semantic understanding of a student's answer that is required for an ITS to interact effectively with a learner. With that goal in mind, annotators were asked to consider and annotate according to what they would want to know about the student's answer if they were the tutor (but a tutor that for some reason could not understand the unstructured text of the student's answer).

Each reference answer (the desired answer as specified by the FOSS/ASK science assessment research team) was decomposed into its constituent facets. Then each student answer was annotated relative to the facets in the corresponding reference answer. The reference answer facets, which are roughly extracted from the relations in a syntactic dependency parse and a shallow semantic parse, are modified slightly to either eliminate most function words or incorporate them into the relation labels (c.f., Lin and Pantel 2001). Example 4 illustrates the decomposition of one of the reference answers into its constituent parts along with their glosses.

(4)   The string is tighter, so the pitch is higher.

(4a)   Is(string, tighter)
(4a') The string is tighter.

(4b)   Is(pitch, higher)
(4b') The pitch is higher.

(4c)   Cause(4b, 4a)
(4c') 4b is caused by 4a

We settled on the eight annotation labels noted in Table 3; see (Nielsen and Ward 2007) for a description of each annotation label and further details on the annotation project.

Table 3. Facet Annotation Labels

Assumed	Facets that are assumed to be understood a priori based on the question
Expressed	Any facet directly expressed or inferred by simple reasoning
Inferred	Facets inferred by pragmatics or nontrivial logical reasoning
Contra-Expr	Facets directly contradicted by negation, antonymous expressions and their paraphrases
Contra-Infr	Facets contradicted by pragmatics or complex reasoning
Self-Contra	Facets that are both contradicted and implied (self contradictions)
Diff-Arg	The core relation is expressed, but it has a different modifier or argument
Unaddressed	Facets that are not addressed at all by the student's answer

There is no compelling reason from the perspective of the automated tutoring system to differentiate between Expressed, Inferred and Assumed facets, since in each case the tutor can assume that the student understands the concepts involved. However, from the systems development perspective there are three primary reasons for differentiating between these facets and similarly between facets that are contradicted by inference versus more explicit expressions. The first reason is that most statistical machine learning systems today cannot hope to detect very many pragmatic inferences (which is the main source of the Inferred and Assumed labels) and including these in the training data sometimes confuses algorithms resulting in worse performance. Having separate labels allows one to remove the more difficult inferences from the training data, thus eliminating this problem. The second rationale is that systems hoping to handle both types of inference might more easily learn to discriminate between these opposing classifications if the classes are distinguished (for algorithms where this is not the case, the classes can easily be combined automatically). Similarly, this allows the possibility of training separate classifiers to handle the different forms of inference. The third reason for separate labels is that it facilitates system evaluation, including the comparison of various techniques and the effect of individual features.

When developing a training corpus for learning algorithms, most annotation efforts, particularly those in natural language processing, are evaluated using the Kappa statistic (see Table 4). We report results under three label groupings: (1) All-Labels, where all labels are left separate, (2) Tutor-Labels, where Expressed, Inferred and Assumed are combined (i.e., the annotator believes the student understands the facet) as are Contra-Expr and Contra-Infr (i.e., the annotator believes the student holds a contradictory view), but the other labels are left as is, and (3) Yes-No, which is a two-way division, Expressed, Inferred and Assumed versus all other labels. Tutor-Labels are the labels that will be used by the system, since it is relatively unimportant to differentiate between the types of inference required in determining that the student understands a reference answer facet (or has contradicted it). We evaluated inter-annotator agreement on a random selection of 50 answers to each of 23 questions in the Water module. Agreement on the Tutor-Labels was 83.8%, with a Kappa statistic of 0.67 corresponding with substantial agreement (Cohen 1960). Inter-annotator agreement was 77.5% on All-Labels and 85.7% on the binary Yes-No decision, each with a Kappa statistic of 0.69, which is in the range of substantial agreement.

5.3 The Answer Assessment Technology

In this section, we describe baseline soft-computing strategies for assessing a student's understanding of individual reference answer facets, present results, and discuss our plans for future work.

Preprocessing and Representation

The features described in the following subsections to be used by a machine learning algorithm are based on document co-occurrence counts. Rather than use the web as our corpus (as did Turney (2001) and Glickman et al. (2005), who generate analogous similarity statistics), we use three publicly available corpora (English Gigaword, The Reuters corpus, and Tipster) totaling 7.4M articles and 2.6B indexed terms. These corpora are all drawn from the news domain, making them less than ideal for assessing

Table 4. Evaluating Annotation Efforts

Most annotation efforts are evaluated by the Kappa statistic, K. This statistic indicates how well humans agree on the labels assigned to the data, (i.e., it is a measure of the inter-annotator agreement or reliability, factoring in chance agreement). Kappa is computed as the ratio of agreement above chance to the maximum possible agreement above chance.

$$\kappa = \frac{P(A) - P(E)}{1 - P(E)}$$

The expected agreement by chance, $P(E)$, is the sum of the chance agreements for the individual labels in L. The chance agreement for a given label l_i is estimated by squaring the fraction of annotations, p_i, that are given the label in question. This effectively estimates the chance that, given an item to be annotated, two annotations will consist of the label in question if both were chosen at chance according to the label's probability.

$$P(E) = \sum_{i=1}^{\|L\|} p_i^2$$

The actual agreement, $P(A)$, is the average proportion of agreement, q_j, over all of the items to be annotated in T. The proportion of agreement, q_j, for a given item t_j is the proportion of annotator-pairs that agreed on that item, which is simply the sum of the number of annotator-pairs that agreed on each of the labels divided by the total number of possible agreeing pairs (the number of combinations of c annotators taken two at a time $= c(c-1)$).

$$P(A) = \frac{1}{\|T\|} \sum_{j=1}^{\|T\|} q_j, \quad q_j = \frac{\sum_{i=1}^{\|L\|} c_{j,i}(c_{j,i} - 1)}{c(c-1)}$$

where $c_{j,i}$ is the number of annotators that assigned label l_i to item t_j. When considering only two annotators, the proportion of agreement on an item, q_j, is either 0 or 1 and $P(A)$ simplifies to the number of items agreed on divided by the total number of items, $\|T\|$.

student's answers to science questions. This will be addressed in the future by indexing relevant information drawn from the web.

The above corpora were indexed and searched using Lucene[6], a publicly available Information Retrieval tool. Two indices were created, the first using their `StandardAnalyzer` and the second adding the `PorterStemFilter`, which replaces words with their lexical stem (e.g., vibrate, vibrates, vibrated, and vibrating are all mapped to the same stem). Each index excludes only three words, {*a, an, the*}. However, when referring to content words in the feature descriptions that follow, Lucene's standard stop-word list was utilized, with the exception of removing the words *no* and *not* from their list.

There are several natural language processing steps that must be performed before we ultimately extract features to train our machine learning classifiers. First, the learner answers must be processed by a classifier that performs sentence segmentation. Then the text of sentences is tokenized, breaking it into the discrete linguistic units (e.g., morpho-syntactic words, numbers, punctuation) required by downstream algorithms. Next, the tokenized sentences are processed by a part-of-speech (POS) classifier. Finally, the POS-tagged sentences are provided as features to a syntactic dependency parser, which in our case is also a classifier model, MaltParser (Nivre et al. 2007), trained on a corpus of hand-annotated dependency parses.

Machine Learning Features

In the following paragraphs, we describe features that loosely assess the overall quality of facet entailment. The features are summarized in Table 5. These features provide an initial performance to compare with the results of more rigorous future work. We use the following example from the Earth Materials science module to clarify some of the features.

(5)   Question: You used three scratch tools in class to test minerals for hardness: your fingernail, a penny, and a paperclip. If a mineral can be scratched by a penny, you can be sure that a (paperclip) will also scratch it. Explain how you know that tool would scratch the mineral.

(6)   Reference Answer: A paperclip is harder than a penny.

(6a)   Is(paperclip, harder)
(6b)   AMod_than(harder, penny)

(7)   Learner Answer: I know because a penny is softer than a paperclip.

Facet-Specific Features: The reference answer facets are comprised primarily of the governing term and modifier derived from semantic and syntactic dependencies, as discussed above. These are shown in (5a-b). Our core features assess the likelihood that these two terms are discussed in the learner's answer. The features come primarily from the lexical entailment calculations in (Turney 2001; Glickman et al. 2005). Here a lexical entailment likelihood is derived from point-wise mutual information, common terms are factored out and maximum likelihood estimates (MLEs) are made

---

[6] http://lucene.apache.org/

based on corpus co-occurrence statistics. For a single content word w from the reference answer r, their methods estimate the probability of lexical entailment as:

$$P\left(Tr_w = 1 \mid l\right) \approx \max_{v \in l} P\left(Tr_w = 1 \mid v\right) \approx \max_{v \in l} \frac{n_{w,v}}{n_v} \qquad (1)$$

where Tr_w is the truth value of w, l is the learner answer, v represents a word in l, n_v is the number of documents in which v occurs, $n_{w,v}$ is the number of documents in which w and v co-occur, and the truth value or entailment of w is assumed to be determined primarily by the single aligned word from l that maximizes this estimate. Turney applied this metric, calling it PMI-IR, to solve the Test of English as a Foreign Language (TOEFL) synonym task. PMI-IR outperformed Latent Semantic Analysis in Turney's experiments, which in turn achieved results that were statistically as good as the performance of college age non-native English speakers on the questions evaluated (Landauer and Dumais 1997).

In example (6), the content words of the reference answer are *paperclip, harder, than,* and *penny*, which align with *paperclip, softer, than,* and *penny* respectively from the learner answer. The exact matching terms are all given co-occurrence probabilities of 1.0. We calculate the word co-occurrence probability of the remaining alignment, *harder* to *softer*, as $n_{harder,softer} / n_{softer} = 457 / 11978 = 0.038$. Since these estimates are subject to significant variance depending on, among other things, n_v – the number of documents in which the entailing word occurs, we expose this count to the classifier so that it can learn how much to trust the estimates. Two sets of these features are generated for each of the governing term and the modifier. The first set is based on the surface morphological form of the words and the second set is based on their stems.

Answer-Global Features: In addition to the facet-specific features described above, we also generate features indicating the extent to which the learner answer addresses all the facets of the reference answer. The rationale for these features is the less the learner answer addresses overall, the less likely it is that they understand any given facet of the reference answer. One technique for doing this, following Glickman et al. (2005), is to estimate the probability of entailment for the reference answer r as a whole as the product of the probabilities of each of its content words w being entailed:

$$P\left(Tr_r = 1 \mid l\right) = \prod_{w \in r} \max_{v \in l} \frac{n_{w,v}}{n_v} \qquad (2)$$

One weakness of this product function is that longer reference answers result in lower entailment probabilities. Therefore, in addition to the product we include features for the average and the geometric mean of the probabilities. We also include a feature for the worst non-zero probability, believing that one very poorly entailed word could affect the validity of the answer as a whole. For example (6), these values are $(1+0.038+1+1)/4 = 0.760$, $(1*0.038*1*1)^{1/4} = 0.442$, and $\min(1,0.038,1,1) = 0.038$.

Again to account for the variance in these estimates, we include a number of features to expose the statistical validity to the classifier. We include n_v – the number of documents in which the entailing word v occurs for the lowest non-zero lexical entailment probability; the more documents v occurs in, the more likely the estimate is to be accurate. We include the largest n_v associated with a lexical entailment probability of zero for similar reasons. The smallest n_v associated with a nonzero probability is included as an indication that the estimate is less reliable. We also provide the classifier with the count of reference answer terms w that do not co-occur with any learner answer terms v, (i.e., the number of words w in r where $n_{w,v}$ is zero for all v in l) and the count of words w that do co-occur with at least one v. For example (6), the outputs for the five features described in this paragraph are 11978 (*harder* had the lowest probability and it aligned with *softer*, which occurred 11978 times in our corpus), -1 (all words had co-occurrences), 11978 (*softer* was the only inexact match), 0 (all 4 content words were aligned), and 4.

Table 5. Core Features (see text for description)

Facet-specific Features
Governing term and modifier lexical entailment prob. (word & stem), (4 ttl ftrs)
Governor and modifier corresponding entailing term frequencies, (4 total features)
Answer-global Features
Product, Ave., Geometric Mean and Worst of the lexical entailment probabilities
Entailing word count associated with the lowest non-zero probability
Largest (smallest) entailing word count with a zero (non-zero) probability
Smallest entailing word count with a non-zero probability
Count of words in r that do and do not co-occur with any words in l

Classification Approach

The feature data was split into a training set and three test sets. The first test set consists of all the data from three of the sixteen science modules (Environment, Human Body and Water), providing what is loosely a domain-independent test set of topics not seen in the training data. The second test set consists of all the student answers associated with 22 questions selected at random from the 233 questions in the remaining 13 modules[7], providing a question-independent test set from within the same topic areas seen in the training data. The third test set was created by randomly assigning all of the facets from approximately 6% of the remaining learner answers to a test set with the remainder comprising the training set.

Following the RTE challenge (Dagan et al. 2005), in this initial investigation, we considered only examples that were given moderately consistent labels by all three annotators. Specifically, we used the adjudicated labels of facets where all annotators provided positive annotations (`Expressed` or `Inferred`), all annotators chose negative annotations (`Contra-Expr`, `Contra-Infr` or `Self-Contra`), the adjudicator and at least one annotator chose `Diff-Arg`, and facets that all annotators labeled `Unaddressed`. We also excluded examples that had a default value of `Assumed`, because these facets require special logic, are expected to be easier to classify, and are not as critical in the tutoring. Since the full dataset annotation was not yet complete at the time of conducting these experiments, this selection resulted in a

---

[7] Though the specific questions selected were random within a module, we, as near is possible, maintained the original proportion of questions for each module.

total of 7273 training examples, 7719 examples in the domain-independent test set, 1239 examples in the question-independent test set and 424 examples in the within domain test-set.

The results for this new task of assessing reference answer facets were obtained by training a Random Forest classifier (Breiman 2001). Several machine learning algorithms from the Weka package (Witten and Frank 2000) were evaluated and Random Forests were selected because they achieved the best results in cross validation on the training data.[8]

5.4 Results

Table 6 shows the classifier's accuracy (percent correctly tagged) in cross validation on the training set as well as on each of our test sets. The row labeled Unseen Answers presents the accuracy when classifying different answers to the same questions that generated the training set answers. Unseen Questions provides the accuracy of classifying answers to questions not used in the training set and Unseen Modules shows the accuracy on the domain-independent test data collected from very different science modules than were used for training. The numbers within parentheses show two simpler baselines, the first being the accuracy of a classifier that always chooses the most frequent label in the training set – Unaddressed, and the second being the accuracy of a classifier that chooses Expressed if the stems of both the governing term and the modifier are present in the learner's answer and votes Unaddressed otherwise. Following the column headings from the corpus annotation, Yes-No presents the accuracy of a two-way classifier that outputs Yes for all facets with positive predictions (Expressed and Inferred) and No for all other facets. This effectively is the accuracy of predicting that the tutor should provide some sort of remediation. The column labeled Tutor Labels outputs five possible predictions, positive (Expressed and Inferred), negative (Contra-Expr and Contra-Infr), Self-Contra, Diff-Arg, and Unaddressed. These are the labels that will drive the type of dialogue provided by the tutor.

In addition to accuracy, we show the confidence weighted score (CWS), which provides an indication of the quality of the classifier's confidence judgments. Loosely speaking, classifiers that are correct more often on their higher confidence judgments score better on CWS. The chance value for CWS is equal to the classification accuracy. It is computed after ranking examples according to the classifier's confidence, from most to least confident. Then CWS is calculated as the average over the system's precision values up to each point in the ranked list:

$$CWS = \frac{1}{n} \sum_{i=1}^{n} \frac{\sum_{j=1}^{i} \delta(\hat{y}_i = y_i)}{i} \tag{3}$$

---

[8] Cross validation consists of splitting the *training* data into N equally sized sets of data, then iteratively holding one set out for evaluation while training on the remaining N-1 sets. In this way, one can evaluate several machine learning algorithms or feature sets without overfitting the test set (i.e., without biasing the system to achieve unrealistically high results on the test data that do not generalize to unseen data).

Table 6. Classifier Accuracy [(most frequent label, stems present) all features] and Confidence Weighted Score (CWS)

	Yes-No Accuracy	Tutor-Labels Accuracy	CWS
Training Set CV	(56.5, 65.2) 81.9%	(54.0, 62.9) 80.3%	87.9%
Unseen Answers	(53.8, 64.4) 80.9%	(50.9, 61.6) 78.8%	85.6%
Unseen Questions	(64.8, 73.3) 69.7%	(62.5, 71.3) 68.4%	78.4%
Unseen Modules	(49.4, 72.0) 76.6%	(45.7, 68.6) 74.6%	83.5%

where n is the number of examples (answer facets) in the test set, i ranges over the examples sorted by decreasing confidence of the classification, and $\delta(z)$ is an indicator function (1 if the z is true and 0 otherwise), so the embedded fraction is the precision through the ith example.

5.5 Discussion and Future Work

This is a new task and new dataset. These early results, based on a very simple lexical similarity approach, are very promising. The results on the Tutor-Labels are 27.9%, 5.9%, and 28.9% over the most frequent label baseline for unseen answers, questions, and modules respectively. Accuracy on unseen answers is 17.2% better than predicting expressed when both of the facet's word stems are present and unaddressed in all other cases, and 6% better on unseen modules. However, this simpler baseline outperforms a classifier trained on the more robust lexical features by 2.9% on our unseen questions test set, showing that there is much room for improvement. Results on the Yes-No Labels, predicting that the tutor should provide some form of remediation, follow a similar trend. The CWS is much higher than chance, indicating that the confidence (class probability estimates) output by the Random Forest can be useful to the dialog manager in deciding how strongly to believe in the classifier's output. For example, if the classification suggests the learner understood a concept, but the confidence is low, the dialog manager could decide to paraphrase the answer as a transition to the next question, rather than assuming the learner definitely understands and simply moving on and rather than asking a confirming question about something the learner probably already expressed. These confidence estimates could be improved further by following the techniques in (Nielsen 2004). These results demonstrate that the task is feasible and we believe with more rigorous feature engineering, accuracy should easily be in a range that allows effective tutoring. Even when the prediction is not correct as long as the tutor acts according to the confidence, no harm and little frustration should result.

The current lexical approach is still prone to the same problems discussed earlier for LSA, (e.g., it cannot interpret the relations between words or distinguish antonyms from synonyms). In future work, we will improve on these baseline results by addressing these weaknesses and extending our feature set as follows. We will include versions of the PMI-IR metric that consider the context of the co-occurring terms and that are intended to differentiate contradictions (e.g., antonyms) from paraphrases and synonyms. We will utilize additional lexical relatedness features such as the Jiang-Conrath distance, the Extended Lesk measure, and latent semantic analysis. The Jiang-Conrath distance measures the distance between words using Information

Content; the distance between two words is the amount of information required to represent their commonality minus the information needed to describe both words (Jiang and Conrath 1997). Budanitsky and Hirst (2006), as well as many other researchers, have found the Jiang-Conrath distance to be the best measure of semantic relatedness they tested in their evaluation framework (they did not test the Extended Lesk measure). The Extended Lesk measure provides a measure of the overlap between the glosses of each word and between the glosses of their various relations such as hyponyms and hypernyms (Banerjee and Pedersen 2003). Banerjee and Pedersen show a slight advantage to this metric in their evaluation framework. These metrics are expected to be good measures of semantic relatedness and, therefore, good alignment metrics, but they are not expected to be useful in distinguishing entailing relations from contradictory relations.

The above metrics are combined to sort the list of learner terms by relatedness to the given reference answer term and to filter out terms assumed to be unrelated due to a poor matching score. First, we simply assess term relatedness, not synonymy or entailment. Most of the metrics discussed in the preceding paragraphs in fact cannot easily recognize the difference between synonyms and antonyms. That is a desirable feature in this case, since we not only want to recognize entailments, but also contradictions. For example, given the reference answer "The pitch rises", if the learner says "The pitch falls", we would like to consider *falls* to be a good alignment with *rises*. At this stage in the processing we want to detect all potentially related terms and then, in a later stage, determine whether the dependency and propositional relationship is one of entailment, contradiction, or neither.

In addition to these co-occurrence features, we will generate collocation features to provide evidence for when two words can be used in similar contexts. These features indicate the extent to which two words tend to co-occur with shared governors, modifiers, and other context words. These co-occurrence and collocation soft semantic relatedness features will be combined with other linguistic cues such as part-of-speech and structural semantic and syntactic dependency information (c.f., Bethard et al. 2007) to determine the relations between terms. We will also implement a paraphrase detection module based on the work described earlier in this chapter. This module will check whether there is evidence in a large corpus to suggest that based on lexical and dependency relations the learner answer is a paraphrase of the reference answer.

When integrating this semantic assessment module in the eventual tutoring system, probabilistic reasoning will be used to decide whether and how to address apparent contradictions, misconceptions and unaddressed issues on the part of the student. Accurate probabilities will allow the dialog manager to decide whether to, for example, assume a misconception with high confidence and take appropriate corrective action or decide, due to low confidence, that it should clarify the learner's understanding on a sub-issue. Even if some dependencies are classified wrong, the tutoring system need only ask for clarification, (i.e., a moderate error rate is not that problematic, especially if the errant classifications have low confidence based on probability estimates).

6 Conclusion

Soft computing is becoming pervasive in Intelligent Tutoring Systems from the initial dialog planning and question generation, to speech recognition and semantic analysis of the student's response, through determining what subsequent tutoring move might optimize learning gains. The two most significant contributions of this work are formally defining and evaluating a learner answer classification scheme, which will help enable more intelligent dialog control, and laying the framework for an answer assessment system that can classify learner responses to previously unseen questions according to this scheme. The classifier used in this paper is a first step toward this more robust system.

The corpora of learner answers described here will be made available for researchers to utilize in improving tutoring systems, educational assessment technologies and entailment systems. This database of annotated answers should stimulate further research in these areas. The corpus provides a shared resource and a standardized annotation scheme allowing researchers to compare work.

All prior work on intelligent tutoring systems has focused on question-specific assessment of answers and even then the understanding of learner responses has generally been restricted to a judgment regarding their correctness or in a small number of cases a classification that specifies which of a predefined set of misconceptions the learner might be exhibiting. The question-independent approach described here enables systems that can easily scale up to new content and learning environments, avoiding the need for lesson planners or technologists to create extensive new rules or classifiers for each new question the system must handle. This is an obligatory first step in creating intelligent tutoring systems that can truly engage children in natural unrestricted dialog, such as is required to perform high quality student directed Socratic tutoring.

This work is an important step toward our goals of significantly increasing the learning gains achieved by intelligent tutoring systems. We hope that the end result will stimulate additional research into the kinds of soft computing and individualized tutoring envisaged below by Bennet:

> With intelligent tutors particularly, student knowledge will be dynamically modeled using cognitive and statistical approaches capable both of guiding instruction on a highly detailed level and of providing a general summary of overall standing. Instruction will be adapted not only to the multiple dimensions that characterize standing in a broad skill area, but to personal interests and background, allowing more meaningful accommodation to diversity than was possible with earlier approaches.
>
> *Bennet R 1998, Reinventing Assessment:*
> *Speculations on the Future of Large-Scale Educational Testing*

References

Agichtein, E., Gravano, L.: Snowball: Extracting relations from large plain-text collections. In: Proc. of the 5th ACM ICDL (2000)

Aleven, V., Popescu, O., Koedinger, K.R.: A tutorial dialogue system with knowledge-based understanding and classification of student explanations. In: IJCAI Workshop on Knowledge and Reasoning in Practical Dialogue Systems (2001)

Anderson, J.R., Corbett, A.T., Koedinger, K., Pelletier, R.: Cognitive tutors: Lessons learned. J of Learning Sciences 4, 167–207 (1995)

Banerjee, S., Pedersen, T.: Extended gloss overlaps as a measure of semantic relatedness. In: IJCAI, pp. 805–810 (2003)

Barzilay, R., Lee, L.: Learning to paraphrase: An unsupervised approach using multiple-sequence alignment. In: Proc. of HLT-NAACL, pp. 16–23 (2003)

Barzilay, R., McKeown, K.: Extracting paraphrases from a parallel corpus. In: Proc. of the ACL/EACL, pp. 50–57 (2001)

Bennett R.: Reinventing assessment: Speculations on the future of large-scale educational testing. Educational Testing Service, on August 16, 2005 (1998) http://www.ets.org/research/pic/bennett.html (Downloaded from August 16, 2005)

Bethard, S., Nielsen, R.D., Martin, J.H., Ward, W., Palmer, M.: Semantic integration in learning from text. In: Machine Reading AAAI Spring Symposium (2007)

Bloom, B.S.: The 2 sigma problem: The search for methods of group instruction as effective as one-on-one tutoring. Educational Researcher 13, 4–16 (1984)

Braz, R.S., Girju, R., Punyakanok, V., Roth, D., Sammons, M.: An inference model for semantic entailment in natural language. In: Proc. of the PASCAL Recognizing Textual Entailment challenge workshop (2005)

Breiman, L.: Random forests. Journal of Machine Learning 45(1), 5–32 (2001)

Budanitsky, A., Hirst, G.: Evaluating WordNet-based measures of lexical semantic relatedness. Computational Linguistics 32(1) (2006)

Burger, J., Ferro, L.: Generating an entailment corpus from news headlines. In: Proc. of the ACL workshop on Empirical Modeling of Semantic Equivalence and Entailment, pp. 49–54 (2005)

Callear, D., Jerrams-Smith, J., Soh, V.: CAA of short non-MCQ answers. In: Proc. of the 5th International CAA conference, Loughborough (2001)

Chi, M.T.H.: Constructing self-explanations and scaffolded explanations in tutoring. Applied Cognitive Psychology 10, 33–49 (1996)

Church, K.W., Hanks, P.: Word association norms, mutual information, and lexicography. In: Proceedings of the 27th ACL, Vancouver BC, pp. 76–83 (1989)

Cohen, J.: A coefficient of agreement for nominal scales. Educational and Psychological Measurement 20, 37–46 (1960)

Cohen, P.A., Kulik, J.A., Kulik, C.L.C.: Educational outcomes of tutoring: A meta-analysis of findings. American Educational Research J 19, 237–248 (1982)

Cole R., VanVuuren S., Pellom B., Hacioglu K., Ma J., Movellan J., Schwartz S., Wade-Stein D., Ward W., Yan J.: Perceptive animated interfaces: First steps toward a new paradigm for human computer interaction (2003)

Cowie, J., Lehnert, W.G.: Information extraction. Communications of the ACM 39(1), 80–91 (1996)

Dagan, I., Glickman, O., Magnini, B.: The PASCAL Recognizing Textual Entailment Challenge. In: Proc. of the PASCAL RTE challenge workshop (2005)

Duclaye, F., Yvon, F., Collin, O.: Using the web as a linguistic resource for learning reformulations automatically. LREC (2002)

Franzke, M., Kintsch, E., Caccamise, D., Johnson, N., Dooley, S.: Summary Street: Computer support for comprehension and writing. Journal of Educational Computing Research 33, 53–80 (2005)

Gildea, D., Jurafsky, D.: Automatic labeling of semantic roles. Computational Linguistics 28(3), 245–288 (2002)

Glickman, O., Dagan, I.: Identifying lexical paraphrases from a single corpus: A case study for verbs. In: Proc. of RANLP (2003)

Glickman, O., Dagan, I., Koppel, M.: Web based probabilistic textual entailment. In: Proc. of the PASCAL RTE challenge workshop (2005)

Graesser, A.C., Hu, X., Susarla, S., Harter, D., Person, N.K., Louwerse, M., Olde, B.: The Tutoring Research Group AutoTutor: An intelligent tutor and conversational tutoring scaffold. In: Proc. of the 10th International Conference of Artificial Intelligence in Education, San Antonio, TX, pp. 47–49 (2001)

Haghighi, A.D., Ng, A.Y., Manning, C.D.: Robust Textual Inference via Graph Matching. In: Proc. HLT-EMNLP (2005)

Herrera, J., Peñas, A., Verdejo, F.: Textual entailment recognition based on dependency analysis and WordNet. In: Proc. of PASCAL Recognizing Textual Entailment challenge workshop (2005)

Hickl, A., Bensley, J.: A discourse commitment-based framework for recognizing textual entailment. In: Proc. of the ACL-PASCAL workshop on textual entailment and paraphrasing (2007)

Hickl, A., Bensley, J., Williams, J., Roberts, K., Rink, B., Shi, Y.: Recognizing textual entailment with LCC's GROUNDHOG system. In: Proc. of the Second PASCAL Recognizing Textual Entailment challenge workshop (2006)

Jiang, J.J., Conrath, D.W.: Semantic similarity based on corpus statistics and lexical taxonomy. In: Proc. of International Conference on Research in Computational Linguistics (ROCLING X), Taiwan, pp. 19–33 (1997)

Jordan, P.W., Makatchev, M., VanLehn, K.: Combining competing language understanding approaches in an intelligent tutoring system. In: Lester, J.C., Vicari, R.M., Paraguacu, F. (eds.) 7th Conference on Intelligent Tutoring Systems, pp. 346–357. Springer, Heidelberg (2004)

Kintsch, W.: Comprehension: A paradigm for cognition. Cambridge University Press, Cambridge (1998)

Kipper, K., Dang, H.T., Palmer, M.: Class-based construction of a verb lexicon. In: AAAI seventeenth National Conference on Artificial Intelligence (2000)

Koedinger, K.R., Anderson, J.R., Hadley, W.H., Mark, M.A.: Intelligent tutoring goes to school in the big city. International J of AI in Education 8, 30–43 (1997)

Landauer, T.K., Dumais, S.T.: A solution to Plato's problem: The latent semantic analysis theory of acquisition, induction, and representation of knowledge. J Psychological Review (1997)

Landauer, T.K., Foltz, P.W., Laham, D.: An introduction to latent semantic analysis. Discourse Processes 25, 259–284 (1998)

Leacock, C.: Scoring free-response automatically: A case study of a large-scale Assessment. Examens 1(3) (2004)

Leacock, C., Chodorow, M.: C-rater: Automated scoring of short-answer questions. Computers and the Humanities 37, 4 (2003)

Lin, D.: Automatic retrieval and clustering of similar words. In: Proc. of the ACL, Montreal, pp. 898–904 (1998)

Lin, D., Pantel, P.: DIRT – Discovery of inference rules from text. In: Proc. of KDD (2001a)

Lin, D., Pantel, P.: Discovery of inference rules for question answering. Natural Language Engineering 7(4), 343–360 (2001b)

MacCartney, B., Grenager, T., de Marneffe, M., Cer, D., Manning, C.: Learning to recognize features of valid textual entailments. In: Proc. of HLT-NAACL (2006)

Madden, N.A., Slavin, R.E.: Effective pullout programs for students at risk. In: Slavin, R.E., Karweit, N.L., Madden, N.A. (eds.) Effective programs for students at risk, Allyn and Bacon, Boston (1989)

Makatchev, M., Jordan, P., VanLehn, K.: Abductive theorem proving for analyzing student explanations and guiding feedback in intelligent tutoring systems. Journal of Automated Reasoning special issue on automated reasoning and theorem proving in education 32(3), 187–226 (2004)

Mathews, E.C., Jackson, G.T., Person, N.K., Graesser, A.C.: Discourse patterns in Why/AutoTutor. In: Proceedings of the 2003 AAAI Spring Symposia on Natural Language Generation, pp. 45–51. AAAI Press, Menlo Park, Palo Alto, CA (2003)

Meichenbaum, D., Biemiller, A.: Nurturing independent learners: Helping students take charge of their learning. Brookline. Brookline, Cambridge (1998)

Mitchell, T., Aldridge, N., Broomhead, P.: Computerized marking of short-answer free-text responses. In: 29th annual conference of the International Association for Educational Assessment, Manchester, UK (2003)

Mitchell, T., Russell, T., Broomhead, P., Aldridge, N.: Towards robust computerized marking of free-text responses. In: Proc. of 6th International Computer Aided Assessment Conference, Loughborough (2002)

Mostow, J., Aist, G., Burkhead, P., Corbett, A., Cuneo, A., Eitelman, S., Huang, C., Junker, B., Sklar, M.B., Tobin, B.: Evaluation of an automated reading tutor that listens: Comparison to human tutoring and classroom instruction. Journal of Educational Computing Research 29(1), 61–117 (2003)

NAEP (2005), http://nces.ed.gov/nationsreportcard/

Nielsen, R.D.: MOB-ESP and other improvements in probability estimation. In: Proceedings of the 20th Conference on Uncertainty in Artificial Intelligence (2004)

Nielsen, R.D., Pradhan, S.: Mixing weak learners in semantic parsing. In: Proc. of EMNLP, Barcelona, Spain (2004)

Nielsen, R.D., Ward, W., Martin, J.H.: Toward dependency path based entailment. In: Proc. of the second PASCAL RTE challenge workshop (2006)

Nielsen, R.D., Ward, W.: A corpus of fine-grained entailment relations. In: Proc. of the ACL workshop on Textual Entailment and Paraphrasing (2007)

Nivre J., Kubler S.: Dependency parsing. Tutorial at COLING-ACL, Sydney (2006)

Nivre, J., Hall, J., Nilsson, J., Chanev, A., Eryigit, G., Kübler, S., Marinov, S., Marsi, E.: Malt-Parser: A language-independent system for data-driven dependency parsing. Natural Language Engineering 13(2), 95–135 (2007)

Pang, B., Knight, K., Marcu, D.: Syntax-based alignment of multiple translations: Extracting paraphrases and generating sentences. In: Proc. HLT/NAACL (2003)

Pellom, B.: SONIC: The University of Colorado continuous speech recognizer. University of Colorado, tech report #TR-CSLR-2001-01, Boulder (2001)

Pellom, B., Hacioglu, K.: Recent improvements in the CU SONIC ASR system for noisy speech: The SPINE task. In: Proc. of IEEE International Conference on Acoustics, Speech, and Signal Processing (ICASSP), Hong Kong (2003)

Peters, S., Bratt, E.O., Clark, B., Pon-Barry, H., Schultz, K.: Intelligent systems for training damage control assistants. In: Proc. of Interservice/Industry Training, Simulation and Education Conference (2004)

Platt, J.: Probabilities for support vector machines. In: Smola, A., Bartlett, P., Scolkopf, B., Schuurmans, D. (eds.) Advances in Large Margin Classifiers. MIT Press, Cambridge (2000)

Pon-Barry, H., Clark, B., Schultz, K., Bratt, E.O., Peters, S.: Contextualizing learning in a reflective conversational tutor. In: Proceedings of the 4th IEEE International Conference on Advanced Learning Technologies (2004)

Pradhan, S., Ward, W., Hacioglu, K., Martin, J.H., Jurafsky, D.: Semantic role labeling using different syntactic views. In: Proceedings of ACL (2005)

Pulman, S.G., Sukkarieh, J.Z.: Automatic short answer marking. In: Proc. of the 2nd Workshop on Building Educational Applications Using NLP, ACL (2005)

Raina, R., Haghighi, A., Cox, C., Finkel, J., Michels, J., Toutanova, K., MacCartney, B., de Marneffe, M.C., Manning, C.D., Ng, A.Y.: Robust textual inference using diverse knowledge sources. In: Proc of the PASCAL RTE challenge workshop (2005)

Ravichandran, D., Hovy, E.: Learning surface text patterns for a question answering system. In: Proc. of the 40th ACL conference, Philadelphia (2002)

Roll, I., Baker, R.S., Aleven, V., McLaren, B.M., Koedinger, K.R.: Modeling students' meta-cognitive errors in two intelligent tutoring systems. In: Ardissono, L., Brna, P., Mitrovic, A. (eds.) User Modeling, pp. 379–388 (2005)

Rosé, C.P., Roque, A., Bhembe, D., VanLehn, K.: A hybrid text classification approach for analysis of student essays. In: Building Educational Applications Using Natural Language Processing, pp. 68–75 (2003a)

Rosé, C.P., Gaydos, A., Hall, B.S., Roque, A., VanLehn, K.: Overcoming the knowledge engineering bottleneck for understanding student language input. In: Proceedings of AI in Education. IOS Press, Amsterdam (2003b)

Shinyama, Y., Sekine, S.: Paraphrase acquisition for information extraction. In: Proc. of the Second International Workshop on Paraphrasing, Sapporo, Japan (2003)

Shinyama, Y., Sekine, S., Sudo, K., Grishman, R.: Automatic paraphrase acquisition from news articles. In: Proc. of HLT, San Diego, CA (2002)

Sudo, K., Sekine, S., Grishman, R.: Automatic pattern acquisition for japanese information extraction. In: Proc. of HLT, San Diego, CA (2001)

Sukkarieh, J.Z., Pulman, S.G., Raikes, N.: Auto-marking: Using computational linguistics to score short, free text responses. In: Proc. of the 29th conference of the International Association for Educational Assessment, Manchester, UK (2003)

Sukkarieh, J.Z., Pulman, S.G.: Information extraction and machine learning: Auto-marking short free text responses to science questions. In: Proc. of AIED (2005)

Sweet, A.P., Snow, C.E. (eds.): Rethinking reading comprehension. Guilford Press, New York (2003)

Tatu, M., Moldovan, D.: COGEX at RTE 3. In: Proc. of the ACL-PASCAL workshop on Textual Entailment and Paraphrasing (2007)

Topping, K., Whitley, M.: Participant evaluation of parent-tutored and peer-tutored projects in reading. Educational Research 32(1), 14–32 (1990)

Turney, P.D.: Mining the web for synonyms: PMI-IR versus LSA on TOEFL. In: Proc. of 12th European Conference on Machine Learning, pp. 491–502 (2001)

Turney, P.D., Littman, M.L., Bigham, J., Shnayder, V.: Combining independent modules to solve multiple-choice synonym and analogy problems. In: Proceedings of RANLP, Borovets, Bulgaria, pp. 482–489 (2003)

VanLehn, K., Lynch, C., Schulze, K., Shapiro, J.A., Shelby, R., Taylor, L., Treacy, D., Weinstein, A., Wintersgill, M.: The Andes physics tutoring system: Five years of evaluations. In: McCalla, G., Looi, C.K. (eds.) Proc. of the 12th International Conference on AI in Education. IOS Press, Amsterdam (2005)

VanLehn, K., Siler, S., Murray, C., Yamauchi, T., Baggett, W.B.: Why do only some events cause learning during human tutoring. In: Cognition and Instruction, vol. 21(3), pp. 209–249. Lawrence Erlbaum Associates (2003)

Voorhees, E.M., Harman, D.K.: Overview of the 6th text retrieval conference (TREC6). In: Proc. of 17th Text Retrieval Conference. NIST (1998)

Ward, W.H.: The Phoenix system: Understanding spontaneous speech. In: Proc. of IEEE ICASSP (1991)

Witten, I.H., Frank, E.: Data mining: Practical machine learning tools with Java implementations. Morgan Kaufmann, San Francisco (2000)

A Decision Making System for the Treatment of Dental Caries

Vijay Kumar Mago[1], Bhanu Prasad[2], Ajay Bhatia[3], and Anjali Mago[4]

[1] Lecturer, Department of Computer Science, DAV College, Jalandhar, India
vijay.mago@gmail.com
[2] Assistant Professor, Department of Computer and Information Sciences,
Florida A&M University, Tallahassee, FL 32307, USA
bhanu.prasad@famu.edu
[3] ICFAI, Hyderabad, India
bhatia_ajay2002@yahoo.co.in
[4] Dental Surgeon, Mago Dental Clinic, Jalandhar, India

Abstract. This paper presents a diagnosis system that helps the dentists to decide the course of treatment for dental caries. The inference mechanism of the system is based on the Bayesian Network (BN) and is designed to decide among various possible treatment plans. The system has been evaluated with the help of 13 different dentists to test its operational effectiveness. The system improves the confidence level of a dentist while deciding the treatment plan. As a result, it improves the effectiveness of the dentist and his/her business. Using this system, patients can also get information regarding the nature of treatment and the associated cost as well.

Keywords: Uncertainty, Bayesian network, Dental caries, Dental treatments, KS test.

1 Introduction

The aim of this research is to help dentists as well as patients determine the treatment of dental caries. During the diagnosis for dental caries, a dentist has to select a treatment plan based on the prevailing sign symptoms of the patient. Whenever a dentist faces uncertainty in diagnosing, he[1] usually prefers to apply a palliative treatment. In this way, due to the lack of confidence, he may not be able to provide the required treatment or he may need to consult another dentist or so. The system presented in this paper is intended to help the dentists in diagnosing and deciding the effective treatment plan for patients having dental caries. As a result, we hope the system will improve the quality of service of the dentist and save his time and efforts and hence improve his business.

In addition, the system will help the patients to know about the treatment and the associated cost. As a result, the patient need not visit a dentist for getting this information. He just needs to fill-in the sign-symptoms at the system's website and the system would suggest the required treatment and the associated cost. The system is named CaDiP. A list of acronyms used in this paper is provided in the appendix.

Bayesian Network (BN) [1], which is based on robust mathematical foundation, is one of the most appropriate techniques for dealing with uncertainties. This paper

---

[1] Masculine pronouns are used, in a generic sense, throughout the paper.

B. Prasad (Ed.): Soft Computing Applications in Business, STUDFUZZ 230, pp. 231–242, 2008.
springerlink.com

presents the decision making process based on BN for diagnosing and suggesting treatment for dental caries.

1.2 Problem Definition

Dental caries is a habitual disease that was also found during Bronze, Iron and Medieval Ages. World's largest population, mainly school going children are under clutches of this disease. Based on a recent study, an estimated 90% of schoolchildren worldwide and also most adults have experienced caries [2]. The study also indicated that this disease is more prevalent in Asian and Latin American countries and least in African countries. An estimated 29-59% of adults over the age of 50 experience caries [3]. Among children in the United States and Europe, 60-80% of cases of dental caries occur in 20% of the population [4]. In conclusion, dental caries is a major problem for millions of people. As a result, the diagnose and treatment for dental caries involves lot of money.

It is very hard for dentist to diagnosis caries by using traditional methods. A minor variation in sign-symptoms can change the course of treatment and hence the cost of the treatment can change drastically. This paper studies the sign-symptoms of the dental caries and suggests a treatment plan based on BN.

1.3 Overview

CaDiP, the BN based diagnosis system, described in Section 3 is yet to be made online for dentists and for public use. The user of the system can either be a dentist or a patient.

A user uses the CaDiP's interface, on the client-side, to provide sign-symptoms as the input. On the server side, CaDiP calculates the probabilities of the related treatments and chooses among the best. The results are passed back to the user. The functional diagram is presented in Fig. 1.

CaDiP is very useful for dentists. For instance, the dentist can observe the sign-symptoms present in a patient suffering from dental caries. These symptoms are,

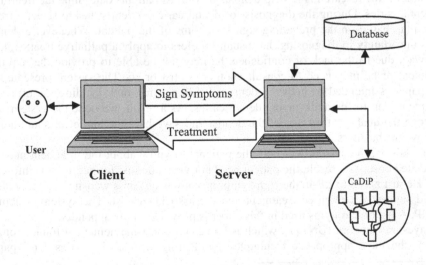

Fig. 1. Functional diagram of CaDiP

'presence of cavities' and 'pain on percussion'. Based on the nature of the cavity (i.e., whether it is deep, shallow, low, etc.) and if the pain is felt on percussion, a conclusion about the treatment plan can be reached. This information is sufficient for diagnosing the caries. But, the same information in the presence of a few more symptoms make it multifaceted problem that is faced by the dentists. So there is a need to handle such an uncertainty.

2 Related Work

The usefulness of Bayes's theorem has been accepted in medical domain long time back [5-7]. It suitable to apply the theorem for medical domains because the information needed in the decision making process is probabilistic in nature [8]. The Bayes's theorem is good for delivering accurate results when specific manifestations have high frequency and have high specificity [9]. Various types of experimentation is undergoing to utilize BNs in current scenarios also. For example, an intelligent pediatric agent deciding the super specialist using BN is provided by Mago et al [10]. Discovering the temporal-state transition patterns during Hemodialysis has been discussed by Lin et al. [11]. There are some limitations for the Bayesian based probabilistic systems. The main limitation is that the realistic prior probabilities need to be obtained. This limitation can be handled by involving the domain experts for deciding the probabilities or by utilizing the statistical data, as being done by Blinowska el al. [12] for diagnosing the hypertension.

Nikovski has constructed a BN for medical diagnosis from incomplete and partially correct statistics [13]. He demonstrated the correct uses of the statistical data with Conditional Probability Tables (CPTs) for medical diagnosis.

Problem-Based Learning (PBL), as an alternative to the traditional didactic medical education, has been introduced by Suebnukarn and Haddawy [14]. Traditional pedagogy is subject-based because the students are told what they need to know. In this, a course outline is provided and a required textbook may also be specified. The lectures (of the instructors) and the students' problem solving will follow later on. In contrast, in the PBL paradigm, a problem situation is posed first. In this, the students must identify what they need to know to solve the problem. They learn it and then use the new knowledge to solve a given problem.

The above discussion indicates that BNs can work well in the medical diagnosis systems for health care domain. In this paper, we design and evaluate the BN for diagnosis of dental caries and for the selection of appropriate treatment.

In the subsequent sections, we introduce the fundamentals of BN. The basics of BN, specific to our problem, are also discussed. Finally, the design of probabilistic network with the help of GeNIe [15] is presented. A web-based Graphical User Interface (GUI) system is designed for the convenience of the users.

3 System Design

Probabilistic decision making system can be made using the extended relational data model, as discussed by Wong et al. [16]. This paper presents the unified approach to design a database and a probabilistic reasoning system. The optimization querying system for probabilistic inference has been discussed by Bravo and Ramakrishnan

[17]. But, to the best of the authors' knowledge, a reliable database for dental caries does not exist. CaDiP is intended to represent the knowledge of a dentist in the form of a BN. In this section, we describe the design of BN and briefly discuss the sign symptoms and treatments for which BN has been constructed.

3.1 Dental Caries

Treatment of dental caries spread over various plans. These treatments depend on the degree of extent of the painful conditions and also on the presence or absence of these painful conditions. The dependencies increase the complexity of deciding a treatment plan. To handle such an uncertainty, a BN is designed for 12 different sign-symptoms and 6 related treatments, which are explained in Tables 1 and 2 respectively.

Table 1. Sign-Symptoms and its description

Sign-Symptoms	Description
Cavity	A hole developed in tooth. It is caused by bacteria that colonize the tooth surface and, under certain conditions, produce sufficient acids to demineralize the enamel covering of the tooth crown or the cementum covering the root, and then the underlying dentin.
Pain	An unpleasant sensory and emotional experience associated with actual or potential tissue damage.
Pulp Exposure	The result of pathological changes in the hard tissue of a tooth caused by carious lesions, mechanical factors, or trauma, which renders the pulp susceptible to bacterial invasion from the external environment.
Pain on Percussion	An abnormal pain felt by hitting with percussion instrument on the effected area of tooth.
Partial Denture	A partial denture is a removable dental appliance that replaces multiple missing teeth. It can be attached to the teeth with clasps (clasp or conventional partial) or it can be attached to the teeth with crowns with precision attachments (hidden clasps).
Food Lodgment	Debris of food which logs in tooth due to cavity.
Fistula	It is due to destruction of intervening tissue, between the two sites and is a major component of a periapical abscess. Inflamed pus forms an abscess causing a pressure increase in the surrounding tooth area.
Swelling (Gingivitis)	Gingivitis is the inflammation of the gums, and often includes redness, swelling, bleeding, exudation, and sometimes pain. Gingivitis can be chronic or acute, but is usually a chronic condition.
High Filling	Presence of high dental filling material used to artificially restore the function, integrity, and morphology of missing tooth surface.
Sensitivity	Tooth sensitivity is hypersensitivity or root sensitivity. If hot, cold, sweet or very acidic foods and drinks, makes your teeth or a tooth sensitive or painful then you have sensitive teeth

Table 2. Treatments and its description

Treatment	Description of treatment
Relieve High Points	Check for the high points with the articulating paper and remove if any. Put on Symptomatic medication, if required.
Root Canal Treatment (RCT)	To cure the infection and save the tooth, drill into the pulp chamber and removes the infected pulp by scraping it out of the root canals. Once this is done, fill the cavity with an inert material and seals up the opening.
Palliative Treatment	Put a patient on antibiotics and analgesics/anti-inflammatory.
Direct Pulp Capping	When a small amount of pulp becomes exposed during removal of decay or following a traumatic injury, medication is placed directly on the exposed but healthy pulp to prevent further damage.
Indirect Pulp Capping	When the decay has come very close to the pulp but does not reach it, most of the infected parts of the tooth are removed and a protective dressing is placed over the slight amount of remaining decay. This prevents exposing the pulp and stimulates healing. A filling is then placed in the tooth.
Restoration	The tooth may be restored with a composite filling material if it is a front tooth and the cavity is small, but back teeth in many cases will need a crown.

3.2 Inference Mechanism

Bayesian networks provide a flexible structure for modeling and evaluating the uncertainty. The network consists of nodes representing the random variables. The nodes are connected by arcs that quantify a causal relationship between the nodes. Bayesian networks have proven useful for a variety of monitoring and predictive purposes.

3.2.1 General Bayesian Inference

Given a hypothesis H and an evidence e, Bayes' theorem may be stated as follows:

$$P(H|e) = \frac{P(e|H).P(H)}{P(e)}$$

Here, P(H|e) is the conditional probability of hypothesis H being true given the evidence e, also known as the posterior probability or posterior. P(H) is the probability of the hypothesis H being true or the prior probability. P(e|H) is the probability of the evidence e occurring, given that the hypothesis H is true. This is also known as the likelihood that evidence e will materialize if hypothesis H is true.

3.2.2 CaDiP's Inference Mechanism

Inference is the process of updating probabilities of outcomes based upon the relationships in the model and the evidence known about the symptoms at hand. In a Bayesian model, the end user applies the evidence about the recent events or observations. This information is applied to the model by "instantiating" a variable to a state that is consistent with the observation. Then the mathematical mechanics are performed to update the probabilities of all the other variables that are connected to the variable representing the new evidence.

The probabilistic inference of the system is governed by the law of total probability with Bayes' theorem as a base. That is,

$$\Pr(Treatment_i) = \sum_n \Pr(Treatment_i \mid Symptom_n) . \Pr(Symptom_n)$$

$$\Pr(RCT) = \Pr(RCT \mid Pulp\_Exp) . \Pr(Pulp\_Exp) +$$
$$\Pr(RCT \mid Pain) . \Pr(Pain) + ... + \Pr(RCT \mid Sensitivity) . \Pr(Sensitivity)$$

3.3 Bayesian Network of CaDiP

Conditional probabilities can be obtained either from extensive interviews with the dentists or capture them from database using an algorithm such as 'maximum likelihood estimator' (MLE) [18]. As mentioned in Section 3, since there are no such reliable databases available, we decided to obtain the estimated values from a dentist and at a latter stage, the behavior of CaDiP to be verified.

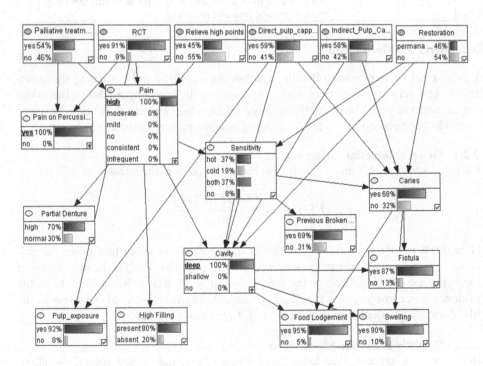

Fig. 2. Bayesian Network for Diagnosing Treatment for Dental Caries

Fig 2 represents the BN of dental caries in the form of a Directed Acyclic Graph (DAG) [19]. Nodes at the top level signify treatments plans, for instance 'Restoration' whereas the nodes at the intermediate level or at the bottom represent the sign-symptoms such as 'Sensitivity' and 'Swelling'. The dependency between the sign symptoms and the treatment plans is represented by using the edges. In this figure, an illustration is shown for conducting Root Canal Treatment (RCT) with 91% probability due to the presence of deep cavity, pain and the pain on percussion. This figure reveals an illustration with the pain being high, cavity being deep and pain on percussion being true. CaPiD suggests conducting RCT on the patient undergoing the diagnostic procedure.

4   Implementation

CaDiP's algorithm is provided below:

```
function GetRecommendedTreatment is
     input:

             network N,
             symptom S_i,
             treatment T_i
     output:
             treatment T_i

     for each Selected(S_i) as evidence in N do
          SetEvidence(S_i)

     while (there exists a treatment T_i
             which is effected by above operation)
          do:
                  if(GetMaxValueOfTraetment(T_i))
                      return T_i
                  end if
     end while
end function
```

The BN has been designed using GeNIe while ASP .net and C# [21] have been used for server-side programming. For the purpose of user management and fulfilling the security constraints, the MS SQL Server 2005 [21] has been used.

Using the system shown in Fig 3, a user can supply the sign-symptoms. The decision making inference engine, based on BN, suggests the appropriate treatment plan as shown in Fig 4.

An example test case is provided. A user supplied the following sign-symptoms, as shown in Fig 3:

- Sensitivity to Cold,
- Partial Denture as High, and
- Swelling as present.

The inference mechanism, after setting these inputs as evidence, concludes that 'Palliative Treatment Plan' is recommended with 'very high probability'.

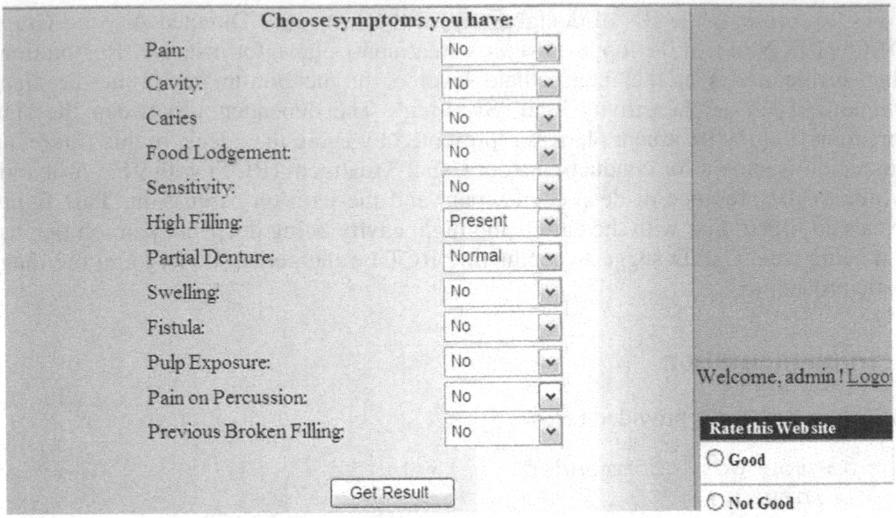

Fig. 3. Sign symptoms provided by the user

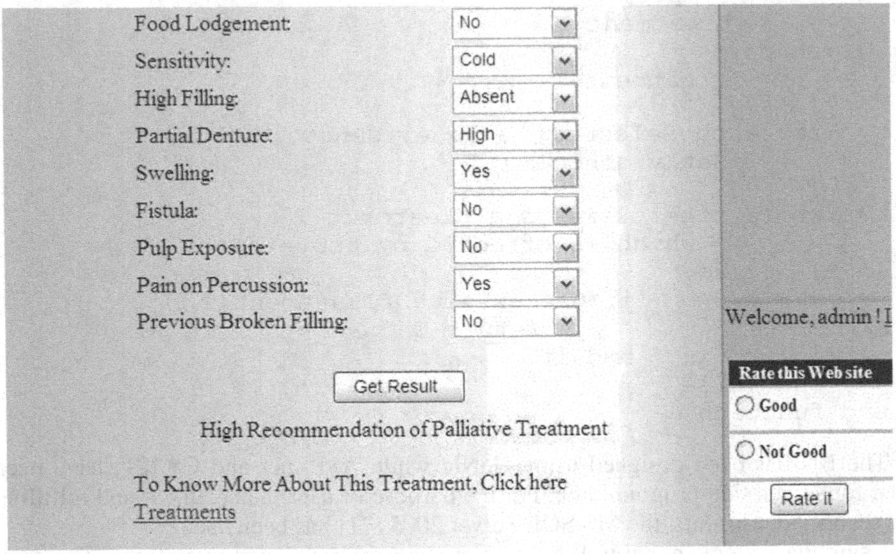

Fig. 4. Recommendation of treatment with probability

5   Evaluation

Thirteen different dentists were involved in evaluating the results produced by CaDiP. These dentists were supplied a questionnaire that contains 10 distinct test cases. A partial list of the questionnaire is provided in Table 4. The dentists were asked to mark the checkbox meticulously in each row. Each row contains the sign-symptom(s) and 6 probable treatment plans. An example test case is mentioned in Section 4.

In order to determine if the two distinct samples (one produced by CaDiP and the other one assembled from the dentists) are significantly different or not, we applied the Kolmogorov-Smirnov (KS) test [22]. The advantage of this test is that it is a non-parametric method and it makes no assumptions about the underlying distributions of the two observed data being tested.

Table 3. Result of the KS test

Two-sample Kolmogorov-Smirnov test :	
D	0.400
p-value	0.087
alpha	0.05

The result of the test is summarized in Table 3 and Figure 5. One test case was erroneously skipped by a dentist; hence we received 129 valid results. The mean of the observation produced by the dentists is 12.9 and the standard deviation is 1.197. The result of the test suggests that the BN is producing 91.33% accurate result. The significance level assumed in the test is 5%.

The aim of the test is to verify the behavior of the inference mechanism of CaDiP. That is, to verify the conditional probabilities those were earlier filled with the consultation of a dentist. The result suggests that the CPTs were quite accurately designed and as a result, CaDiP can be made available for general public. At the same time, CaDiP can also be used by the dentists as it is a reliable decision making system.

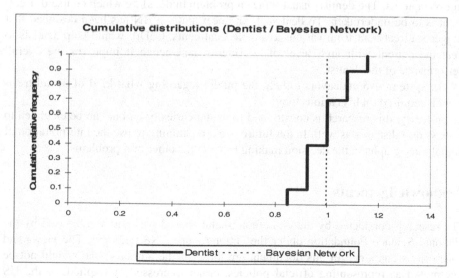

Fig. 5. Comparison of outputs using Kolmogorov-Smirnov Test

Table 4. Partial list of test cases distributed to dentists

Sign Symptoms	Treatment Plans					
	Palliative Treatment	RCT	Relieve High Point	Direct Pulp Capping	Indirect Pulp Capping	Restoration
Cavity (deep), Sensitivity (Cold)	☐	☐	☐	☐	☐	☐
Food Lodgment (yes), Swelling (yes), Caries (yes)	☐	☐	☐	☐	☐	☐
Pain (moderate), Sensitivity (hot)	☐	☐	☐	☐	☐	☐
Fistula (yes), Swelling (yes)	☐	☐	☐	☐	☐	☐
Partial Denture (high)	☐	☐	☐	☐	☐	☐

6 Conclusions

The aim of this research is to assist dentists in deciding the treatment plan for dental caries. There are many treatment plans that can be carried out as per the prevalent sign-symptoms. The dentists usually face a problem in deciding which course of treatment is to be undertaken. To deal such an uncertainty, a BN has been designed that suggests a treatment from various available treatments. CaDiP would help dentists to treat the patient with high level of confidence and hence it improves the overall performance of the dentist.

The system also endeavors to help the public regarding what kind of dental treatment is required on his carious tooth.

Currently, this research is constrained to dental caries only, but can be extended to address oral diseases as well. In the future, we are planning to use the interpretation of digital radiographs in the decision making process for other oral problems.

Acknowledgements

The research conducted by the co-author Bhanu Prasad was partly supported by the National Science Foundation under the Grant Award CNS-0424556. The views and conclusions contained in this document are those of the author(s) and should not be interpreted as representing official policies, either expressed or implied, of the US government or any of the sponsoring organizations.

References

1. McCabe, B.: Belief networks for engineering applications. International Journal of Technology Management 21(3–4), 257–270 (2001)
2. The World Oral Health Report. Continuous improvement of oral health in the 21st century – the approach of the WHO Global Oral Health Program (2003)
3. Jamison, D.T., Breman, J.G., Measham, A.R., Alleyne, G., Claeson, M., Evans, D.B., Jha, P., Mills, A., Musgrove, P.: Disease Control Priorities Project in De-veloping Countries, 2nd edn., The World Bank Group (2006)
4. Touger-Decker, R., van Loveren, C.: Sugars and dental caries. The American Journal of Clinical Nutrition, 881–892 (2003)
5. Zheng, B., Chang, Y.H., Wang, X.H., W.F.: Good Comparison of artificial neural network and Bayesian belief network in a computer-assisted diagnosis scheme for mammography. In: International Joint Conference on Neural Networks, vol. 6, pp. 4181–4185. IEEE press, Washington DC (1999)
6. Kahn, C.E., Laur, J.J., Carrera, G.F.: A Bayesian network for diagnosis of primary bone tumors. Journal of Digital Imaging 14(2), 56–57 (2001)
7. Antal, P., Verrelst, H., Timmerman, D., Moreau, Y., Van Huffel, S., De Moor, B., Vergote, I.: Bayesian networks in ovarian cancer diagnosis: potentials and limitations. In: IEEE Symposium on Computer based Medical Systems, pp. 103–108. IEEE press, Washington DC (2000)
8. Russell, S.J., Norvig, P.: Artificial Intelligence -A modern Approach. 2nd Indian Reprint, Pearson Education, India (2002)
9. Peng, C., Xiao, S., Nie, Z., Wong, Z.: Applying Bayes's Theorem in Medical Expert Systems. IEEE Engineering in Medicine and Biology 15(3), 76–79 (1996)
10. Mago, V.K., Devi, M.S., Mehta, R.: Decision Making System: Agent Diag-nosing Child Care Diseases. In: Multi-agent systems and Applications V, pp. 316–318. Springer, Germany (2007)
11. Lin, F., Chiu, C., Wu, S.: Using Bayesian Networks for Discovering Tem-poral-State Transition patterns in Hemodialysis. In: 35th Annual Hawaii International Conference on System Sciences, pp. 1995–2002. IEEE press, Washington DC, USA (2003)
12. Blinowska, A., Chatellier, G., Bernier, J., Lavril, M.: Bayesian Statistics as Applied to Hypertension Diagnosis. IEEE Transactions on Biomedical Engineering 38(7), 699–706 (1991)
13. Nikovski, D.N.: Constructing Bayesian Network for Medical Diagnosis from Incomplete and Partially Correct Statistics. IEEE Transactions on Knowledge and Data Engineering 12(4), 509–516 (2000)
14. Siriwan, S., Peter, H.: A Bayesian approach to generating tutorial hints in a collaborative medical problem-based learning system. Artificial intelligence in Medicine 38, 5–24 (2006)
15. http://genie.sis.pitt.edu, Accessed on (February 13, 2008)
16. Wong, S.K.M., Butz, C.J., Xiang, Y.: A Method for Implementing a Probabilistic Model as a Relational Database. In: 11th Conference on Uncertainty in Artificial Intelligence, pp. 556–564. Morgan Kaufmann Publishers, San Mateo, USA (1955)
17. Bravo, H.C., Ramakrishnan, R.: Optimizing mpf queries: Decision support and probabilistic inference. In: The 2007 ACM SIGMOD international conference on Management of data, pp. 701–712. ACM, New York, USA (2007)
18. Armitage, P., Matthews, J.N.S., Berry, G.: Statistical Methods in Medical Research. Blackwell Science, USA (2001)

19. Richardson, T.: A discovery algorithm for directed cyclic graphs. In: Proceedings of the 12th Conference of Uncertainty in AI, Portland, OR, pp. 454–461. Morgan Kaufmann, San Francisco (1996)
20. Hart, C., Kauffman, J., Sussman, D., Ullman, C.: Beginning ASP.NET 2.0. Wiley Publishing, Indianapolis, USA (2006)
21. Schneider, R.D.: Microsoft SQL Server 2005 Express Edition For Dummies. Wiley Publishing, New Jersey, USA (2006)
22. Laha, R., Chakravarti, I.M., Roy, J.: Handbook of Methods of Applied Statistics. John-Wiley and Sons, New York, USA (1967)

Appendix

List of Acronyms used in this Paper

BN	Bayesian Network
CaDiP	Caries Diagnostic System based on Probabilistic Network
CPT	Conditional Probability Table
DAG	Directed Acyclic Graph
GUI	Graphical User Interface
KS	Kolmogorov Smirnov
MLE	Maximum Likelihood Estimator
PBL	Problem-Based Learning
RCT	Root Canal Treatment

CBR Based Engine for Business Internal Control

M.L. Borrajo[1], E.S. Corchado[3], M.A. Pellicer[3], and J.M. Corchado[2]

[1] Dept. Informática, University of Vigo, Escuela Superior de Ingeniería Informática, Edificio Politécnico, Campus Universitario As Lagoas s/n, 32004, Ourense, Spain
[2] Dept. de Ingeniería Civil, University of Burgos, Esc. Politécnica Superior, Edificio C, C/ Francisco de Vitoria, Burgos, Spain
[3] Dept. de Informática y Automática, University of Salamanca, Plaza de la Merced s/n, 37008 Salamanca, Spain

Abstract. The complexity of current organization systems and the increase in importance of the realization of internal controls in firms makes the construction of models that automate and facilitate the work of the auditors crucial. A tool for the decision support process has been developed based on a multi-cbr system that incorporates two case-based reasoning systems and automates the business control process. The objective of the system is to facilitate the process of internal audit in small and medium firms. The multi-cbr system analyses the data that characterises each one of the activities carried out by the firm, then determines the state of each activity, calculates the associated risk, detects the erroneous processes, and generates recommendations to improve these processes. Therefore, the system is a useful tool for the internal auditor in order to make decisions based on the obtained risk. Each one of the case-based reasoning systems that integrates the multi-agent system uses a different problem solving method in each step of the reasoning cycle: a Maximum Likelihood Hebbian learning-based method that automates the organization of cases and the retrieval phase, an Radial Based Function neural network and a multi-criterion discreet method during the reuse phase and a rule based system for recommendation generation. The multi-cbr system has been tested in 14 small and medium size companies during the last 26 months in the textile sector, which are located in the northwest of Spain. The achieved results have been very satisfactory.

1 Introduction

Nowadays, the organization systems employed in enterprises are increasing in complexity. In the last years, also the number of regulatory norms has increased considerably. As a consequence of this, the need for periodic internal audits has arisen. As a matter of fact the evaluation and the estimation of the evolution of this type of systems are generally complicated, because these systems are characterized by their great dynamism. It is necessary to create models that facilitate the analysis work in changing environments such as the financial environment.

The processes carried out inside a company can be divided into functional areas (Corchado et al. 2004). Each one of these areas is denominated a "Function". A Function is a group of coordinated and related activities which are necessary to reach the objectives of the company and are carried out in a systematic and reiterated way (Mas and Ramió 1997). Purchases, Cash Management, Sales, Information Technology, Fixed Assets Management, Compliance to Legal Norms and Human Resources are the functions that are usually carried out in a company.

B. Prasad (Ed.): Soft Computing Applications in Business, STUDFUZZ 230, pp. 243–260, 2008.
springerlink.com
© Springer-Verlag Berlin Heidelberg 2008

Further on, each one of these functions is divided into a series of activities, which aim to achieve different objectives. For example, the function Information Technology is divided into the following activities: Computer Plan Development, Study of Systems, Installation of Systems, Treatment of Information Flows and Security Management.

Each activity is composed of a number of tasks. For example, the activity Computer Plan Development, belonging to the function Information Technology, can be divided in the following tasks:

1. Definition of the required investment in technology in the short and medium time of period.
2. Coordination of the technology investment plan and the development plan of the company.
3. Periodic evaluation of the established priorities on the technology investment plan to identify their relevance.
4. Definition of a working group focused in the identification and control of the information technology policy.
5. Definition of a communication protocol in both directions: bottom-up and top-down, to involve the company's employees in the maintenance strategic plan.

Control procedures are also to be established in the tasks to ensure that the established objectives are achieved.

Therefore, an adaptive system has been developed. The system possesses the flexibility of functioning in different ways and to evolve, depending on the environment state where it operates. The developed multi-cbr system is composed of two fundamental agents (Borrajo 2003):

- ISA CBR (Identification of the State of the Activity): the objectives are to identify the state or situation of each one of activities of the company and to calculate the risk associated with this state.
- GR CBR (Generation of Recommendations), the goal of which is to generate recommendations to reduce the number of inconsistent processes in the company .

Both agents are implemented using a case-based reasoning (CBR) system (Aamodt and Plaza 1994; Kolodner 1993; Lenz et al. 1998; Watson 1997). The CBR system integrated within each agent uses different problem solving techniques, but shares the same case memory (Hunt and Miles 1994; Medsker 1995).

The rest of this article is structured as follows: at first the description of the concept of internal control (IC) and its importance in the current company, followed by the presentation of the basic concepts that characterize the case based reasoning. It will be shown how this methodology facilitates the construction of intelligence hybrid systems. Finally, the developed system is discussed and the achieved results are presented.

2 Internal Control

Small to medium enterprises require an internal control mechanism in order to monitor their modus operandi and to analyse if they achieve their goals. Such mechanisms are constructed around a series of organizational policies and specific procedures dedicated to giving reasonable guarantees to their executive bodies. This group of policies and procedures is named "controls", and they all conform to the company's internal control structure. The establishment of objectives (which is not a function of the Internal Control) is a previous condition for control risk evaluation, which is the main goal of the Internal Control.

In general terms, the administration of a firm has three large categories of objectives when designing an internal control structure (AICPA 1996):

1. Reliability of financial information
2. Efficiency and effectiveness of operations
3. Fulfilment of the applicable rules and regulations

The internal auditor must monitor the internal controls directly and recommend improvements on them. Therefore, all the activities carried out inside the organization can be included, potentially, within the internal auditors' remit. Essentially, the activities of the auditor related to IC can be summarized as follows:

- To be familiar with and possess the appropriate documentation related to the different components of the system that could affect financial aspects.
- To assess the quality of internal controls in order to facilitate the planning of the audit process with the aim of obtaining necessary indicators.
- To assess internal controls in order to estimate the level of errors and reach a decision on the final opinion to be issued in the memorandum on the system under consideration.

A lot of changes have happened in firms because of current technological advances. These changes involved considerable modifications in the area of auditing, basically characterized by the following features (Sánchez 1995):

- Progressive increase in the number and level of complexity of audit rules and procedures.
- Changes in the norms of professional ethics, which demand greater control and quality in auditing.
- Greater competitiveness between auditing firms, consequently resulting in lower fees; the offer of new services to clients (e.g. financial or computing assessment ...).
- Development of new types of auditing (e.g. operative management auditing, computer auditing, environmental auditing ...).

Together these circumstances have made the audit profession increasingly competitive. Consequently, the need has arisen for new techniques and tools, which can be provided by information technology and artificial intelligence. The aim is to achieve more relevant and more suitable information in order to help auditors make decisions faster and thereby increase the efficiency and quality of auditing. The following section presents a system developed to facilitate the auditing process.

3 Case-Based Reasoning for Internal Control

Rule-based systems (RBS) have traditionally been used with the purpose of delimiting the audit decision-making tasks (Denna et al. 1991). However, Messier and Hansen

(Messier and Hansen 1988) found many situations in which auditors resolved problems by referring to previous situations. This contrasts with the very nature of RBS systems, since they have very little capacity for extracting information from past experience and present problems in order to adapt to changes in the environment.

In contrast, case based reasoning (CBR) systems are able to relate past experiences or cases to current observations, solving new problems through the memorization and adaptation of previously tested solutions. This is an effective way of learning, similar to the general structure of human thought. CBR systems are especially suitable when the rules that define a knowledge system are difficult to obtain, or the number and complexity of the rules is too large to create an expert system. Moreover, CBR systems have the capacity to update their memory dynamically, based on new information (new cases), as well as, improving the resolution of problems (Riesbeck and Schank 1989).

However, dealing with problems like the one presented in this document, standard techniques of monitoring and prediction cannot be applied due to the complexity of the problem, the existence of certain preliminary knowledge, the great dynamism of the system, etc. In this type of systems, it is necessary to use models that combine the advantages of several mechanisms of problem solving, able to resolve concrete parts of the general problem and attend the other ones.

In this sense, a CBR adaptive system has been developed. The system possesses the flexibility to function in different ways and to evolve, depending on the environment in which it operates. A case based reasoning system solves a given problem by means of the adaptation of previous solutions to similar problems (Aamodt and Plaza 1994). The CBR memory stores a certain number of cases. A case includes a problem and the solution to this problem. The solution of a new problem is obtained recovering similar cases stored in the CBR memory.

A CBR is a dynamic system in which new problems are added continuously to its memory, the similar problems are eliminated and gradually new ones are created by combining several existing ones. This methodology is based on the fact that humans use the knowledge learned in previous experiences to solve present problems.

CBR systems record past problem solving experiences and by means of indexing algorithms, retrieve previously stored problems with their solutions (cases) and match and adapt them to a given situation. This means that the set of cases stored in the memory of CBR systems represents the knowledge concerning the domain of the CBR. As discussed below, this knowledge is updated constantly.

A typical CBR system is composed of four sequential steps which are recalled every time a problem needs to be solved (Aamodt and Plaza 1994; Kolodner 1993; Watson 1997) (see Fig. 1):

1. *Retrieve* the most relevant case(s).
2. *Reuse* the case(s) in order to solve the problem.
3. *Revise* the proposed solution if necessary.
4. *Retain* the new solution as a part of a new case.

Like other mechanisms of problem solving, the objective of a CBR is to find the solution for a certain problem. A CBR is a system of incremental learning, because each time a problem is solved, a new experience is retained, thereby making it available for future reuse.

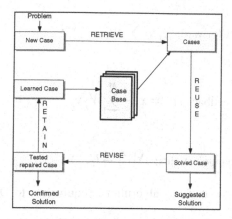

Fig. 1. The CBR cycle

CBR systems have proven to be an effective method for problem solving in multiple domains, for example, prediction, diagnosis, control and planning (Lenz et al. 1998). This technology has been successfully used in several disciplines: law, medicine, diagnosis systems and so on (Watson 1997).

Case based reasoning can be used by itself, or as part of another conventional or intelligent system (Medsker 1995). Although there are many successful applications based on CBR methods alone, CBR systems can be improved by combining them with other technologies (Hunt and Miles 1994). Their suitability for integration with other technologies, creating a global hybrid reasoning system, stems from the fact that CBR systems are very flexible. Therefore, they are capable of absorbing the beneficial properties of other technologies.

4 Maximum Likelihood Hebbian Learning Based Method

The use of Maximum Likelihood Hebbian Learning Based Method has been derived from the work of (Corchado et al. 2002; Fyfe and Corchado 2002a, 2002b; Fyfe and MacDonald 2001), etc. in the field of pattern recognition as an extension of Principal Component Analysis (PCA) (Oja 1989, Oja et al. 1992).

4.1 Principal Component Analysis (PCA)

Principal Component Analysis (PCA) is a standard statistical technique for compressing data; it can be shown to give the best linear compression of the data in terms of least mean square error. There are several artificial neural networks which have been shown to perform PCA e.g. (Oja 1989, Oja et al. 1992). We will apply a negative feedback implementation (Fyfe and Baddeley 1995).

The basic PCA network is described by equations (1)-(3). Let us have an N-dimensional input vector at time t, $x(t)$, and an M-dimensional output vector, y, with W_{ij} being the weight linking input j to output i. η is a learning rate. Then the activation passing and learning is described by

$$\text{Feedforward: } y_i = \sum_{j=1}^{N} W_{ij} x_j , \forall i \tag{1}$$

$$\text{Feedback: } e_j = x_j - \sum_{i=1}^{M} W_{ij} y_i \tag{2}$$

$$\text{Change weights: } \Delta W_{ij} = \eta e_j y_i \tag{3}$$

We can readily show that this algorithm is equivalent to Oja's Subspace Algorithm (Oja 1989):

$$\Delta W_{ij} = \eta e_j y_i = \eta (x_j - \sum_k W_{kj} y_k) y_i \tag{4}$$

and so this network not only causes convergence of the weights but causes the weights to converge to span the subspace of the Principal Components of the input data. We might ask why we should be interested in the negative feedback formulation rather than the formulation (4) in which the weight change directly uses negative feedback. The answer is that the explicit formation of residuals (2) allows us to consider probability density functions of the residuals in a way which would not be brought to mind if we use (4).

Exploratory Projection Pursuit (EPP) is a more recent statistical method aimed at solving the difficult problem of identifying structure in high dimensional data. It does this by projecting the data onto a low dimensional subspace in which we search for its structure by eye. However, not all projections will reveal the data's structure equally well. We therefore define an index that measures how "interesting" a given projection is, and then represent the data in terms of projections that maximise that index.

The first step in our exploratory projection pursuit is to define which indices represent interesting directions. Now "interesting" structure is usually defined with respect to the fact that most projections of high-dimensional data onto arbitrary lines through most multi-dimensional data give almost Gaussian distributions (Diaconis and Freedman 1984). Therefore, if we wish to identify "interesting" features in data, we should look for those directions onto which the data-projections are as far from the Gaussian as possible.

It was shown in (Karhunen and Joutsensalo 1994) that the use of a (non-linear) function creates an algorithm to find those values of W which maximise that function whose derivative is f() under the constraint that W is an orthonormal matrix. This was applied in (Fyfe and Baddeley 1995) to the above network in the context of the network performing an Exploratory Projection Pursuit. Thus if we wish to find a direction which maximises the kurtosis of the distribution which is measured by s4, we will use a function $f(s) \approx s3$ in the algorithm. If we wish to find that direction with maximum skewness, we use a function $f(s) \approx s2$ in the algorithm.

4.2 ε-Insensitive Hebbian Learning

It has been shown (Xu 1993) that the nonlinear PCA rule

$$\Delta W_{ij} = \eta \left(x_j f(y_i) - f(y_i) \sum_k W_{kj} f(y_k) \right) \tag{5}$$

can be derived as an approximation to the best non-linear compression of the data. Thus we may start with a cost function

$$J(W) = 1^T E\left\{ (x - Wf(W^T x))^2 \right\} \tag{6}$$

which we minimise to get the rule (5). (Lai et al. 2000) used the residual in the linear version of (6) to define a cost function of the residual

$$J = f_1(e) = f_1(x - Wy) \tag{7}$$

where $f_1 = \|.\|^2$ is the (squared) Euclidean norm in the standard linear or nonlinear PCA rule. With this choice of $f_1(\ )$, the cost function is minimised with respect to any set of samples from the data set on the assumption that the residuals are chosen independently and identically distributed from a standard Gaussian distribution. We may show that the minimisation of J is equivalent to minimising the negative log probability of the residual, e. , if e is Gaussian.

$$\text{Let } p(e) = \frac{1}{Z} \exp(-e^2) \tag{8}$$

Then we can denote a general cost function associated with this network as

$$J = -\log p(e) = (e)^2 + K \tag{9}$$

where K is a constant. Therefore performing gradient descent on J we have

$$\Delta W \propto -\frac{\partial J}{\partial W} = -\frac{\partial J}{\partial e} \frac{\partial e}{\partial W} \approx y(2e)^T \tag{10}$$

where we have discarded a less important term. See (Karhunen and Joutsensalo 1994) for details.

In general (Smola and Scholkopf 1998), the minimisation of such a cost function may be thought to make the probability of the residuals greater dependent on the probability density function (pdf) of the residuals. Thus if the probability density function of the residuals is known, this knowledge could be used to determine the optimal cost function. Fyfe and MacDonald (Fyfe and MacDonald 2001) investigated this with the (one dimensional) function:

$$p(e) = \frac{1}{2 + \varepsilon} \exp\left(-|e|_\varepsilon\right) \tag{11}$$

$$\text{where } |e|_\varepsilon = \begin{cases} o & \forall |e| < \varepsilon \\ |e| - \varepsilon & otherwise \end{cases} \qquad (12)$$

with ε being a small scalar ≥ 0.

Fyfe and MacDonald (Fyfe and MacDonald 2001) described this in terms of noise in the data set. However, we feel that it is more appropriate to state that, with this model of the pdf of the residual, the optimal $f_1(\ )$ function is the ε-insensitive cost function:

$$f_1(\mathbf{e}) = |\mathbf{e}|_\varepsilon \qquad (13)$$

In the case of the negative feedback network, the learning rule is

$$\Delta W \propto -\frac{\partial J}{\partial W} = -\frac{\partial f_1(\mathbf{e})}{\partial \mathbf{e}} \frac{\partial \mathbf{e}}{\partial W} \qquad (14)$$

which gives:

$$\Delta W_{ij} = \begin{cases} o & if |e_j| < \varepsilon \\ otherwise & \eta y(sign(e)) \end{cases} \qquad (15)$$

The difference with the common Hebb learning rule is that the sign of the residual is used instead of the value of the residual. Because this learning rule is insensitive to the magnitude of the input vectors x, the rule is less sensitive to outliers than the usual rule based on mean squared error. This change from viewing the difference after feedback as simply a residual rather than an error, permits us to consider a family of cost functions, each member of which is optimal for a particular probability density function associated with the residual.

4.3 Applying Maximum Likelihood Hebbian Learning

The Maximum Likelihood Hebbian Learning algorithm is now constructed on the bases of the previously presented concepts as outlined here. Now the ε-insensitive learning rule is clearly only one of a possible family of learning rules which are suggested by the family of exponential distributions. This family was called an exponential family in (Hyvärinen et al. 2002) though statisticians use this term for a somewhat different family. Let the residual after feedback have probability density function

$$p(\mathbf{e}) = \frac{1}{Z} \exp(-|\mathbf{e}|^p) \qquad (16)$$

Then we can denote a general cost function associated with this network as

$$J = E(-\log p(\mathbf{e})) = E(|\mathbf{e}|^p + K) \qquad (17)$$

where K is a constant independent of W and the expectation is taken over the input data set. Therefore performing gradient descent on J we have

$$\Delta W \propto -\frac{\partial J}{\partial W}\,|_{W(t-1)} = -\frac{\partial J}{\partial e}\frac{\partial e}{\partial W}\,|_{W(t-1)} \approx E\{\mathbf{y}(p\,|\,\mathbf{e}\,|^{p-1}\,sign(\mathbf{e}))^{T}\,|_{W(t-1)}\} \tag{18}$$

where T denotes the transpose of a vector and the operation of taking powers of the norm of e is on an element wise basis as it is derived from a derivative of a scalar with respect to a vector.

Computing the mean of a function of a data set (or even the sample averages) can be tedious, and we also wish to cater for the situation in which samples keep arriving as we investigate the data set and so we derive an online learning algorithm. If the conditions of stochastic approximation (Kashyap et al. 1994) are achieved, we may approximate this with a difference equation. The function to be approximated is clearly sufficiently smooth and the learning rate can be made to satisfy $\eta_k \geq 0, \sum_k \eta_k = \infty, \sum_k \eta_k^2 < \infty$ and so we have the rule:

$$\Delta W_{ij} = \eta.y_i.sign(e_j)|\,e_j\,|^{p-1} \tag{19}$$

We would expect that for leptokurtotic residuals (more kurtotic than a Gaussian distribution), values of p<2 would be appropriate, while for platykurtotic residuals (less kurtotic than a Gaussian), values of p>2 would be appropriate. Researchers from the community investigating Independent Component Analysis (Hyvärinen 2001; Hyvärinen et al. 2002) have shown that it is less important to get exactly the correct distribution when searching for a specific source than it is to get an approximately correct distribution i.e. all supergaussian signals can be retrieved using a generic leptokurtotic distribution and all subgaussian signals can be retrieved using a generic platykutotic distribution. Our experiments will tend to support this to some extent but we often find that accuracy and speed of convergence are improved when we are accurate in our choice of p. Therefore the network operation is:

$$\text{Feedforward: } y_i = \sum_{j=1}^{N} W_{ij}x_j, \forall_i \tag{20}$$

$$\text{Feedback: } e_j = x_j - \sum_{i=1}^{M} W_{ij}y_i \tag{21}$$

$$\text{Weights change: } \Delta W_{ij} = \eta.y_i.sign(e_j)|\,e_j\,|^{p-1} \tag{22}$$

Fyfe and MacDonald (Fyfe and MacDonald 2001) described their rule as performing a type of PCA, but this is not strictly true since only the original (Oja) ordinary Hebbian rule actually performs PCA. It might be more appropriate to link this family of learning rules to Principal Factor Analysis since PFA makes an assumption about the noise in a data set and then removes the assumed noise from the covariance structure of the data before performing a PCA. We are doing something similar here in that we are basing our PCA-type rule on the assumed distribution of the residual. By maximising the likelihood of the residual with respect to the actual distribution, we are matching the learning rule to the probability density function of the residual.

More importantly, we may also link the method to the standard statistical method of Exploratory Projection Pursuit: now the nature and quantification of the interestingness is in terms of how likely the residuals are under a particular model of the probability density function of the residuals. In the results reported later, we also sphere the data before applying the learning method to the sphered data and show that with this method we may also find interesting structure in the data.

4.4 Sphering of the Data g

Because a Gaussian distribution with mean a and variance x is not less interesting than a Gaussian distribution with mean b and variance y - indeed this second order structure can obscure higher order and more interesting structure - we remove such information from the data. This is known as "sphering". That is, the raw data is translated till its mean is zero, projected onto the principal component directions and multiplied by the inverse of the square root of its eigenvalue to give data which has mean zero and is of unit variance in all directions. So for input data X we find the covariance matrix.

$$\Sigma = \left\langle \left(X - \langle X \rangle\right)\left(X - \langle X \rangle\right)^T \right\rangle = UDU^T \qquad (23)$$

Where U is the eigenvector matrix, D the diagonal matrix of eigenvalues, T denotes the transpose of the matrix and the angled brackets indicate the ensemble average. New samples, drawn from the distribution are transformed to the principal component axes to give y where

$$y_i = \frac{1}{\sqrt{D_i}} \sum_{j=1}^{n} U_{ij}\left(X_i - \langle X_i \rangle\right), \; for \, 1 \le i \le m \qquad (24)$$

Where n is the dimensionality of the input space and m is the dimensionality of the sphered data.

5 Phases of the Proposed System

This section describes the developed intelligent system in detail. The objective of the system is to facilitate the internal control process in small to medium enterprises in the textile sector. This system, after analyzing the relative data to the activities that are developed in the firm, is able to: carry out an estimation of the state of each activity and calculate the associated risk, detect the erroneous processes and generate recommendations to improve these processes.

In this way, the system helps the internal auditor to make decisions being based on the risk associated to the current state of activities in the firm.

The cycle of operations of the developed case based neuro-symbolic system is based on the classic life cycle of a CBR system (Aamodt and Plaza 1994; Watson and Marir 1994). This cycle is executed twice, since the system bases its operation on two

CBR subsystems (subsystem ISA-Identification of the State of the Activity and subsystem GR-Generation of Recommendations). Both subsystems share the same case base (Table 1 shows the attributes of a case) and a case represent the "shape" of a given activity developed in the company. Each subsystem is used as a framework of hybridization of other technologies.

For each activity to analyse, the system uses the data for this activity, introduced by the firm's internal auditor, to construct the problem case. For each task making up the activity analyzed, the problem case is composed of the value of the realization state for that task, and its level of importance within the activity (according to the internal auditor).

The subsystem ISA identifies the state or situation of each of the firm's activities and calculates the risk associated with it. In the retrieval step, the system retrieves K cases – the most similar ones to the problem case. This is done with the Maximum Likelihood Hebbian Learning proposed method. Applying equations 20 to 22 to the

Table 1. Case structure

	IDENTIFICATION	DESCRIPCTION
	Case number	Unique identification: positive integer number.
PROBLEM	Input vector	Information about the tasks (n sub-vectors) that compose an industrial activity: $(GI_1,V_1),(GI_2,V_2),...,(GI_n,V_n))$ for n tasks. Each task sub-vector has the following structure (GI_i,V_i): <ul><li>GI_i: importance rate for this task inside the activity. Can only take one of the following values: - VHI (Very high importance). - HI (High Importance). - AI (Average Importance) - LI (Low Importance) - VLI (Very low importance)</li><li>V_i: Value of the realization state of a given task. Positive integer number (between 1 and 10).</li></ul>
	Function number	Unique identification number for each function
	Activity number	Unique identification number for each activity
	Reliability	Percentage of probability of success. It represents the obtained percentage of success using the case like a reference to generate recommendations.
SOLUTION	Activity State	Degree of perfection of the development of the activity, expressed by percentage.

case-base, the MLHL algorithm groups the cases in clusters automatically. The proposed indexing mechanism classifies the cases/instances automatically, clustering together those of similar structure. One of the great advantages of this technique is its unsupervised method so we do not need any information about the data before hand. When a new case is presented to the CBR system, it is identified as a belonging to a particular type, also by applying equations 20 to 22 to it. This mechanism may be used as a universal retrieval and indexing mechanism to be applied to any problem similar to the ones presented here.

Maximum Likelihood Hebbian Learning techniques are used because of its database size and the need to group the most similar cases together in order to help retrieve the cases that most resemble the given problem.

Maximum Likelihood Hebbian Learning techniques are especially interesting for nonlinear or ill-defined problems, making it possible to treat tasks involved in the processing of massive quantities of redundant or imprecise information. It allows the available data to be grouped into clusters with fuzzy boundaries, expressing uncertain knowledge.

The following step, the re-use phase, aims to obtain an initial estimation of the state of the activity analysed using a RBF networks (Corchado et al. 2000; Fdez-Riverola and Corchado 2004; Fritzke, 1994). As at the previous stage, the number of attributes of the problem case depends on the activity analyzed. Therefore it is necessary to establish an RBF network system, one for each of the activities to be analysed.

The retrieved K cases are used by the RBF network as a training group that allows it to adapt its configuration to the new problem encountered before generating the initial estimation.

The RBF network is characterized by its ability to adapt and to learn rapidly, as well as to generalize (especially in interpolation tasks). Specifically, within this system the network acts as a mechanism capable of absorbing knowledge about a certain number of cases and generalizing from them. During this process the RBF network interpolates and carries out predictions without forgetting part of those already carried out. The system's memory acts as a permanent memory capable of maintaining many cases or experiences while the RBF network acts as a short term memory, able to recognize recently learnt patterns and to generalize from them.

The objective of the revision phase is to confirm or refute the initial solution proposed by the RBF network, thereby obtaining a final solution and calculating the control risk. In view of the initial estimation or solution generated by the RBF network, the internal auditor will be responsible for deciding if the solution is accepted. For this it is based on the knowledge he/she retains, specifically, knowledge about the company with which he/she is working. If he/she considers that the estimation given is valid, the system will store it as the final solution and in the following phase of the CBR cycle, a new case will be stored in the case base consisting of the problem case and the final solution. The system will assign the case an initial reliability of 100%. On the other hand, if the internal auditor considers the solution given by the system to be invalid, he will give his own solution, which the system will store as the final one. This together with the problem case will form the new case to be stored in the case base in the following phase. This new case will be given a reliability of 30%. This value has been defined taking into account the opinion of various auditors in terms of the weighting that should be assigned to the personal opinion of the internal auditor.

From the final solution: state of activity, the system calculates the control risk associated with the activity. Every activity developed in the business sector has a risk associated with it that can influence negatively the good operation of the firm. In other words, the control risk of an activity measures the impact that the current state of the activity has on the business process as a whole. In this study, the level of risk is valued at three levels: low, medium and high. The calculation of the level of control risk associated with an activity is based on the current state of the activity and its level of importance. This latter value was obtained after analysing data obtained from a series of questionnaires (98 in total) carried out by auditors throughout Spain. In these questionnaires the auditors were asked to rate subjects from 1-10 according to the importance or weighting of each activity in terms of the function that it belonged to. The higher the importance of the activity, the greater its weighting within the business control system. The level of control risk was then calculated from the level of importance given to the activity by the auditors and the final solution obtained after the revision phase. For this purpose, if-then rules are employed.

If the internal auditor in the analyzed company thinks, that the initial estimation is coherent, the problem, together with his solution (the estimation proposed by the system) is stored in the case base, as it is a new case, that means a new piece of knowledge. The addition of a new case to the case base causes the redistribution of the clusters. From this estimation, also the level of risk inherent to the state of the activity is calculated.

Maximum Likelihood Hebbian Learning technique contained within the prototypes related to the activity corresponding to the new case is reorganised in order to respond to the appearance of this new case, modifying its internal structure and adapting itself to the new knowledge available.

The RBF network uses the new case to carry out a complete learning cycle, updating the position of its centres and modifying the value of the weightings that connect the hidden layer with the output layer.

In the subsystem GR the recommendations are generated to improve the state of the analyzed activity. In order to recommend changes in the execution of the processes in the firm, the subsystem compares the initial estimation to the cases belonged to the cluster used in the initial phase of subsystem ISA, that reflect a better situation of this activity in the firm. Therefore, in the retrieve phase, those cases with solution or state of the activityhigher (in an interval between 15% and 20%) than the solution generated as final estimation in the subsystem ISA are obtained. In the reuse phase, the multi-criteria decision-making method Electre (Barba-Romero and Pomeral 1997) is employed, obtaining the most favourable case among all retrieved cases, depending on the importance degree of tasks.

A RBS takes the obtained case and compares it to the initial case problem, generating recommendations related to the found erroneous processes.

By means of the estimations and generated recommendations, the internal auditor can make a report with the inconsistent processes and the steps that must be followed to improve the firm. After correcting the detected errors, the firm is evaluated again. Experience of the experts show, that three months are considered enough to evolve the company toward a more favourable state. If verified, that the erroneous processes and the level of risk have diminished, the retain phase is carried out modifying the case used to generate the recommendations. The reliability (percentage of successful estimations obtained with this case) of this case is increased. In contrast, when the firm happens not to have evolved to a better state, the reliability of the case is decreased.

6 Results

The hybrid system developed has been tested over 26 months in 14 small to medium companies (9 medium-sized and 5 small) in the textile sector, located in the northwest of Spain. The data employed to generate prototype cases, in order to construct the system's case bases, have obtained after performing 98 surveys with auditors from Spain, as well as 34 surveys from experts within different functional areas within the sector.

In order to test this system, various complete operation cycles were carried out. In other words, for a given company, each one of its activities was evaluated, obtaining a level of risk and generating recommendations. These recommendations were communicated to the company's internal auditor and he was given a limit of three months in order to elaborate and apply an action plan based on those recommendations. The action plan's objective was to reduce the number of inconsistent processes within the company. After three months, a new analysis of the company was made and the results obtained were compared with those of the previous three months. This process was repeated every three months.

Results obtained demonstrate that the application of the recommendations generated by the system causes a positive evolution in firms. This evolution is reflected in the reduction of erroneous processes. The indicator used to determine the positive evolution of the companies was the state of each of the activities analysed. If, after analysing one of the company's activities, it is proven that the state of the activity (valued between 1 and 100) has increased over the state obtained in the previous three month period, it can be said that the erroneous processes have been reduced within the same activity. If this improvement is produced in the majority of activities (above all in those of most relevance within the company), the company has improved its situation.

In order to reflect as reliably as possible the suitability of the system for resolving the problem, the results from the analysis of the 14 companies were compared with those of 5 companies in which the recommendations generated by the system have not been applied. In these five companies, the activities were analysed from the beginning of the three month period until the end, using the ISA (Identification of the State of the Activity). The recommendations generated by the second subsystem were not presented to the firm managers (and consequently, the recommendations were not applied).

In order to analyse the results obtained, it is necessary to consider that some of the recommendations implied costs that the companies were not able to afford or that some of them involved a long term implementation. Therefore, companies are considered to have followed the recommendations if they applied more than 70% of them.

The results obtained were as follows:

1. Of the 14 companies analysed, in which the recommendations generated by the system were applied, the results were (see Fig. 2):

- In 9 companies, the number of inconsistent processes was reduced, improving the state of activities by an average of 14.3%.
- In 4 of these companies, no improvement was detected in the state of activities. In other words, the application of the recommendations generated by the system did not have any effect on the activities of the company.
- In 1 company, the inconsistent processes increased, in other words, the application of recommendations generated by the system, prejudiced the positive evolution of the company. Once the situation in this company had been analysed, it was concluded, that there was a high level of disorganisation without a clearly defined set of objectives. This means that any attempt to change the business organisation will actually lead to a worse situation.

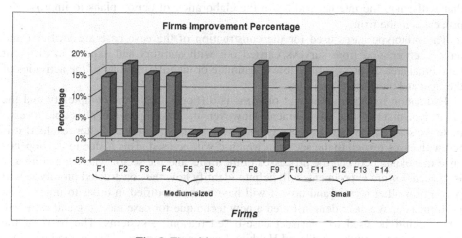

Fig. 2. Firms' improvement percentage

2. On the other hand, for the 5 companies in which the recommendations generated by the system were not applied, the results were as follows: 4 of them improved their results, though reaching an average productivity that was 6.8% below the same measurement of other companies that used the system. The fifth company analysed ceased operations before the end of the first year of evaluation.

In general, it could be said that these results demonstrate the suitability of the techniques used for their integration in the developed intelligent control system. The best results occurred in the companies of smaller size. This is due to the fact that these firms have a greater facility to adapt and adopt the changes suggested by the system's recommendations.

7 Conclusions

This article presents a neuro-symbolic system that uses two CBR systems employed as a basis for hybridization of a multicriteria decision-making method, a radial based function neural network, and a maximum likelihood hebbian learning-based method.

Therefore, the developed model combines the complementary properties of the connectionist methods with the symbolic methods of Artificial Intelligence.

The used reasoning model can be applied in situations that fulfil the following conditions:

- Each problem can be represented in the form of a vector of quantified values.
- The case base should be representative of the totality of the spectrum of the problem.
- Cases must be updated periodically.
- Enough cases should exist to train the net.

The system is able to carry out estimations about the state of the activities of the firm and their associated risk. Furthermore, the system generates recommendations that will guide the internal auditor in the elaboration of action plans to improve the processes in the firm.

The prototype cases used for the construction of the case base are artificial and have been created from surveys carried out with auditors and experts in different functional areas. The system is able to estimate or identify the state of the activities of the firm and their associated risk.

Estimation in the environment of firms is difficult due to the complexity and the great dynamism of this environment. However, the developed model is able to estimate the state of the firm with precision. The system will produce better results if it is fed with cases related to the sector in which it will be used. This is due to the dependence that exists between the processes in the firms and the sector where the company is located. Future experiments will help to identify how the constructed prototype will perform in other sectors and how it will have to be modified in order to improve its performance. We have demonstrated a new technique for case indexing and retrieval, which could be used to construct case-based reasoning systems. The basis of the method is a Maximum Likelihood Hebbian Learning algorithm. This method provides us a very robust model for indexing the data and retrieving instances without any need of information about the structure of the data set.

References

Aamodt, A., Plaza, E.: Case-Based Reasoning: Foundational Issues, Methodological Variations, and System Approaches. AICOM 7(1) (1994)

American Institute of Certified Public Accountants (AICPA), Statements on Auditing Standards No. 78 (SAS No. 78). Consideraciones de la Estructura del Control Interno en una Auditoría de Estados Financieros (Amendment to SAS núm. 55), The Auditing Standards Executive Committee, New York (1996)

Barba-Romero, S., Pomeral, J.: Decisiones Multicriterio. Fundamentos teóricos y utilización práctica. Colección de Economía. Servicio de publicaciones Universidad de Alcalá (1997)

Borrajo, L.: Sistema híbrido inteligente aplicado a la auditoría de los sistemas internos. Ph.D. thesis. Universidade de Vigo (Spain) (2003) ISBN: 84-8158-274-3.

Corchado, E., MacDonald, D., Fyfe, C.: Optimal Projections of High Dimensional Data. In: ICDM 2002 the 2002 IEEE International Conference on Data Mining, Maebashi TERRSA, Maebashi City, Japan, IEEE Computer Society, Los Alamitos (2002)

Corchado, J.M., Borrajo, L., Pellicer, M.A., Yáñez, J.C.: Neuro-symbolic System for Business Internal Control. In: Perner, P. (ed.) ICDM 2004. LNCS (LNAI), vol. 3275, pp. 302–9743. Springer, Heidelberg (2004)

Corchado, J.M., Díaz, F., Borrajo, L., Fdez-Riverola, F.: Redes Neuronales Artificiales: Un enfoque práctico. Departamento de publicaciones de la Universidad de Vigo (2000)

Denna, E.L., Hansen, J.V., Meservy, R.: Development and application of expert systems in audit services. Transactions on Knowledge and Data Engineering (1991)

Diaconis, P., Freedman, D.: Asymptotics of Graphical Projections. The Annals of Statistics 12(3), 793–815 (1984)

Fdez-Riverola, F., Corchado, J.M.: FSfRT: Forecasting System for Red Tides. Applied Intelligence. Special Issue on Soft Computing in Case-Based Reasoning 21(3), 251–264 (2004)

Fritzke, B.: Fast Learning with Incremental RBF Networks. Neural Processing Letters 1(1), 2–5 (1994)

Fyfe, C., Baddeley, R.: Non-linear data structure extraction using simple Hebbian networks. Biological Cybernetics 72(6), 533–541 (1995)

Fyfe, C., Corchado, E.: Maximum Likelihood Hebbian Rules. In: 10th European Symposium on Artificial Neural Networks, ESANN 2002, Bruges, April 24-25-26 (2002a)

Fyfe, C., Corchado, E.: A New Neural Implementation of Exploratory Projection Pursuit. In: Yin, H., Allinson, N.M., Freeman, R., Keane, J.A., Hubbard, S. (eds.) IDEAL 2002. LNCS, vol. 2412, Springer, Heidelberg (2002)

Fyfe, C., MacDonald, D.: ε-Insensitive Hebbian learning. Neuro Computing (2001)

Hunt, J., Miles, R.: Hybrid case-based reasoning. The Knowledge Engineering Review 9(4), 383–397 (1994)

Hyvärinen, A.: Complexity Pursuit: Separating interesting components from time series. Neural Computation 13, 883–898 (2001)

Hyvärinen, A., Karhunen, J., Oja, E.: Independent Component Analysis. Wiley, Chichester (2002)

Karhunen, J., Joutsensalo, J.: Representation and Separation of Signals Using Non-linear PCA Type Learning. Neural Networks 7, 113–127 (1994)

Kashyap, R.L., Blaydon, C.C., Fu, K.S.: Stochastic Approximation. In: Mendel, J.M. (ed.) A Prelude to Neural Networks: Adaptive and Learning Systems. Prentice Hall, Englewood Cliffs (1994)

Kolodner, J.: Case-Based Reasoning, San Mateo, CA. Morgan Kaufmann, San Francisco (1993)

Lai, P.L., Charles, D., Fyfe, C.: Seeking Independence using Biologically Inspired Artificial Neural Networks, in Developments. In: Girolami, M.A. (ed.) Artificial Neural Network Theory: Independent Component Analysis and Blind Source Separation. Springer, Heidelberg (2000)

Lenz, M., Bartsch-Spörl, B., Burkhard, D., Wees, S. (eds.): Case-based Reasoning Technology: From Fundations to Applications. LNCS (LNAI), vol. 1400. Springer, Heidelberg (1998)

Mas, J., Ramió, C.: La Auditoría Operativa en la Práctica. Ed. Marcombo, Barcelona (1997)

Medsker, L.R.: Hybrid Intelligent Systems. Kluwer Academic Publishers, Dordrecht (1995)

Messier, W.F., Hansen, J.V.: Inducing rules for expert systems development: An example using default and bankruptcy data. Management Science 34(12), 1403–1415 (1988)

Oja, E.: Neural Networks, Principal Components and Subspaces. International Journal of Neural Systems 1, 61–68 (1989)

Oja, E., Ogawa, H., Wangviwattana, J.: Principal Components Analysis by Homogeneous Neural Networks, part 1, The Weighted Subspace Criterion. IEICE Transaction on Information and Systems, E75D, 366–375 (1992)

Riesbeck, C.K., Schank, R.C.: Inside case-based reasoning. Lawrence Erlbaum Associates, Hillsdale, NJ (1989)

Sánchez, A.: Los Sistemas Expertos en la Contabilidad. Biblioteca Electrónica de Contabilidad 1(2) (1995)

Smola, A.J., Scholkopf, B.: A Tutorial on Support Vector Regression. Technical Report NC2-TR-1998-030, NeuroCOLT2 Technical Report Series (1998)

Watson, I.: Applying Case-Based Reasoning: Techniques for Enterprise Systems. Morgan Kaufmann, San Francisco (1997)

Watson, I., Marir, F.: Case-Based Reasoning: A Review. The Knowledge Engineering Review 9(4), 355–381 (1994)

Xu, L.: Least Mean Square Error Reconstruction for Self-Organizing Nets. Neural Networks 6, 627–648 (1993)

An EMD-Based Neural Network Ensemble Learning Model for World Crude Oil Spot Price Forecasting

Lean Yu[1,2], Shouyang Wang[1], and Keung Lai[2]

[1] Institute of Systems Science, Academy of Mathematics and Systems Science,
Chinese Academy of Sciences, Beijing 100080, P.R. China
[2] Department of Management Sciences, City University of Hong Kong, Tat Chee Avenue,
Kowloon, Hong Kong

Abstract. In this study, an empirical mode decomposition (EMD) based neural network ensemble learning model is proposed for world crude oil spot price modeling and forecasting. For this purpose, the original crude oil spot price series were first decomposed into a finite and often small number of intrinsic mode functions (IMFs). Then the three-layer feed-forward neural network (FNN) model was used to model each extracted IMFs so that the tendencies of these IMFs can be accurately predicted. Finally, the prediction results of each IMFs are combined with an adaptive linear neural network (ALNN) to formulate a ensemble output for the original oil series. For verification, two main crude oil price series, West Texas Intermediate (WTI) crude oil spot price and Brent crude oil spot price are used to test the effectiveness of this proposed neural network ensemble methodology.

Keywords: Empirical mode decomposition, ensemble learning, feed-forward neural network, adaptive linear neural network, crude oil price prediction.

1 Introduction

Crude Oil plays an increasingly significant role in the world economy since nearly two-thirds of the world's energy consumption comes from crude oil and natural gas (Alvarez-Ramirez et al., 2003). Worldwide consumption of crude oil exceeds $500 billion — roughly 10% of US GDP and crude oil is also the world's largest and most actively traded commodity, accounting for about 10% of total world trade (Verleger, 1993). Crude oil is traded internationally among many different players — oil producing nations, oil companies, individual refineries, oil importing nations and oil speculators. The crude oil price is basically determined by its supply and demand, and is strongly influenced by many irregular past/present/future events like weather, stock levels, GDP growth, political aspects and people's expectations. Furthermore, since it takes considerable time to ship crude oil from one country to another, oil prices vary in different parts of the world. These facts lead to a strongly fluctuating and interacting market and the fundamental mechanism governing the complex dynamics is not understood. As Maurice (1994) reported, the oil market is the most volatile of all the markets except Nasdaq. In addition, as sharp oil price movements are likely to disturb aggregate economic activity. From the different perspectives, the impact of crude oil price fluctuation on national economic performance reflects two-fold. On one hand,

B. Prasad (Ed.): Soft Computing Applications in Business, STUDFUZZ 230, pp. 261–271, 2008.
springerlink.com   © Springer-Verlag Berlin Heidelberg 2008

sharp increases of crude oil prices adversely affect the economic growth and inflation in oil importing economies. On the other hand, crude oil price falls, like the one in 1998, create serious budgetary problems for oil exporting countries (Abosedra and Baghestani, 2004). Therefore, volatile oil prices are of considerable interest to many institutions, business practitioners as well as academic researchers. As such, crude oil prices forecasting is a very important topic, albeit an extremely hard one due to its intrinsic difficulty and high volatilities (Wang et al., 2005). Because the crude oil spot price formulation is a complex process, which is affected by many factors, it is very challenging for crude oil price prediction.

When it comes to crude oil price forecasting, most of the studies focus only on oil price volatility analysis (see e.g. Watkins and Plourde, 1994; Alvarez-Ramirez et al., 2003) and oil price determination within the supply and demand framework (see e.g. Hagen, 1994; Stevens, 1995). This is why there are only a few studies about crude oil spot price forecasting, including quantitative and qualitative methods. Among the qualitative methods, Nelson et al. (1994) used the Delphi method to predict oil prices for the California Energy Commission. Abramson and Finizza (1991) used belief networks, a class of knowledge-based models, to forecast crude oil prices. For the quantitative methods, Huntington (1994) used a sophisticated econometric model to forecast crude oil prices in the 1980s. Abramson and Finizza (1995) utilized a probabilistic model for predicting oil prices and Barone-Adesi et al. (1998) suggested a semi-parametric approach for oil price forecasting. Morana (2001) adopted the same approach to predict short-term oil prices and reported some performance improvement. Recently, in order to incorporate the effect of irregular events on crude oil price, a novel methodology — TEI@I methodology — is constructed for crude oil interval forecasting. The empirical results obtained showed the superiority of the integration methodology (Wang et al., 2005). Similarly, Xie et al. (2006) applied a support vector machine (SVM) model to crude oil spot price forecasting and obtained some promising results.

Motivated by TEI@I methodology, this chapter attempts to applied "divide-and-conquer" principle to construct a novel crude oil spot price forecasting methodology. In this chapter, the basic idea of "divide-and-conquer" principle can be understood as "decomposition-and-ensemble". The main aim of decomposition is to simplify the forecasting task while the goal of ensemble is to formulate a total consensus of original data. Inspired by this basic idea, a novel empirical mode decomposition (EMD) (Huang et al., 1998) based neural network ensemble methodology is proposed for crude oil spot price modeling and forecasting. In this proposed methodology, we simplify the difficult forecasting task by decomposing the original crude oil price series into some separate components and predict each component separately. Detailed speaking, the original crude oil spot price series were first decomposed into a finite and often small number of intrinsic mode functions (IMFs). From the methodological point of view, the EMD technique is very suitable for decomposing nonlinear and nonstationary time series, which has been reported to work better in describing the local time scale instantaneous frequencies than the wavelet transform and fast Fourier transform (FFT). After these simple IMF components are adaptively extracted via EMD from a nonstationary time series, each IMF component is modeled by a separated three-layer feed-forward neural network (FNN) model (Hornik et al., 1989; White, 1990) so that the tendencies of these IMF components can be accurately predicted. Finally, the prediction results of each IMF components are combined with an adaptive linear neural network (ALNN) (Hagan et al., 1996) to output the forecasting result of the original oil series.

The main motivation of this chapter is to propose an EMD-based neural network ensemble model for crude oil spot price prediction and compare its performance with some existing forecasting techniques. The rest of this chapter is organized as follows. Section 5.2 describes the formulation process of the proposed EMD-based neural network ensemble model in detail. For illustration and verification purposes, two main crude oil price series, West Texas Intermediate (WTI) crude oil spot price and Brent crude oil spot price are used to testing the effectiveness of the proposed methodology and the results are reported in Section 5.3. In Section 5.4, some concluding remarks are drawn.

2 Methodology Formulation Process

In this section, the overall process of formulating the EMD-based neural network ensemble methodology is presented. First of all, the EMD technique is briefly reviewed. Then the EMD-based neural network ensemble model is proposed. Finally, the overall steps of the EMD-based neural network ensemble model are summarized.

2.1 Empirical Mode Decomposition (EMD)

The empirical mode decomposition (EMD) method first proposed by Huang et al. (1998) is a form of adaptive time series decomposition technique using Hilbert transform for nonlinear and nonstationary time series data. The basic principle of EMD is to decompose a time series into a sum of oscillatory functions, namely intrinsic mode functions (IMFs). In the EMD, the IMFs must satisfy the following two prerequisites:

1. In the whole data series, the number of extrema (sum of maxima and minima) and the number of the zero crossings must be equal or differ at most by one, and
2. The mean value of the envelope defined by local maxima and the envelope defined by local minima must be zero at all points.

With these two requirements, the meaningful IMFs can be well defined. Otherwise, if one blindly applied the technique to any data series, the EMD may result in a few meaningless harmonics (Huang et al., 1999). Usually, an IMF represents a simple oscillatory mode compared with the simple harmonic function. With the definition, any data series $x(t)$ ($t = 1, 2, ..., n$) can be decomposed according to the following sifting procedure.

1. Identify all the local extrema including local maxima and minima of $x(t)$.
2. Connect all local extrema by a cubic spline line to generate its upper and lower envelopes $x_{up}(t)$ and $x_{low}(t)$.
3. Compute the point-by-point envelope mean $m(t)$ from upper and lower envelopes, i.e.,

$$m(t) = \left(x_{up}(t) + x_{low}(t)\right)/2 \tag{1}$$

4. Extract the details using the following equations:

$$d(t) = x(t) - m(t) \tag{2}$$

5. Check the properties of $d(t)$: (i) if $d(t)$ meets the above two requirements, an IMC is derived and replace $x(t)$ with the residual $r(t) = x(t) - d(t)$; (ii) if $d(t)$ is not an IMC, replace $x(t)$ with $d(t)$.
6. Repeat Step (1) – (5) until the residual satisfies the following stopping condition:

$$\sum\nolimits_{t=1}^{T} \frac{[d_j(t) - d_{j+1}(t)]^2}{d_j^2(t)} < SC \tag{3}$$

where $d_j(t)$ is the sifting result in the jth iteration, and SC is the stopping condition. Typically, it is usually set between 0.2 and 0.3.

The EMD extracts the next IMC by applying the above procedure to the residual term $r_1(t) = x(t) - c_1(t)$, where $c_1(t)$ denotes the first IMC. The decomposition process can be repeated until the last residue $r_n(t)$ only has at most one local extremum or becomes a monotonic function from which no more IMCs can be extracted.

At the end of this sifting procedure, the data series $x(t)$ can be expressed by

$$x(t) = \sum\nolimits_{j=1}^{n} c_j(t) + r_n(t) \tag{4}$$

where n is the number of IMCs, $r_n(t)$ is the final residue, which is the main trend of $x(t)$, and $c_j(t)$ ($j = 1, 2, \ldots, n$) are the IMCs, which are nearly orthogonal to each other, and all have nearly zero means. Thus, one can achieve a decomposition of the data series into n-empirical modes and a residue. The frequency components contained in each frequency band are different and they change with the variation of data series $x(t)$, while $r_n(t)$ represents the central tendency of data series $x(t)$.

Relative to traditional Fourier analysis, the EMD has several distinct advantages. First of all, it is relatively easy to understand and use. Second, the fluctuations within a time series are automatically and adaptively selected from the time series and it is robust in the presence of nonlinear and nonstationary data. Third, it lets the data speak of themselves. EMD can adaptively decompose a time series into a set of independent IMCs and a residual component. The IMCs and residual component displaying linear and nonlinear behavior only depend on the nature of the time series being studied.

2.2 Overall Process of the EMD-Based Neural Network Ensemble Methodology

Supposed there is a time series $x(t)$, $t = 1, 2, \ldots, N$, one would like to make the l-step ahead prediction, i.e., $x(t+l)$. For example, $l = 1$ means single-step ahead prediction and $l = 30$ represents 30-step ahead prediction. Depending on the previous techniques and methods, an EMD-based neural net ensemble methodology can be formulated, as illustrated in Fig. 1.

As can be seen from Fig. 1, the proposed methodology is generally composed of the following three steps:

1. The original time series $x(t)$, $t = 1, 2, ..., N$ is decomposed into n IMFs, $c_j(t)$, $j = 1, 2, ..., n$, and a residual term $r_n(t)$ via EMD.
2. For each extracted IMFs and residue, the three-layer feed-forward neural network (FNN) model is used to model the decomposed components and make the corresponding prediction. Through many practical experiments, we find that the high-frequency IMFs (e.g., c_1 and c_2) often require more hidden neurons while the low-frequency IMFs (e.g., the last IMF c_n and the residue) only need less hidden nodes.
3. The prediction results of all IMFs and residue produced by FNN are combined to generate an aggregated output using an adaptive linear neural network (ALNN), which can be seen as the prediction results of the original time series.

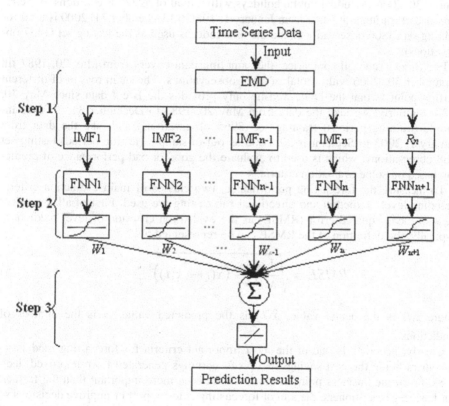

Fig. 1. Overall process of EMD-based neural network ensemble methodology

To summarize, the proposed methodology is actually an "EMD-FNN-ALNN" model. In order to test the effectiveness of the proposed EMD-based neural network ensemble methodology, two main crude oil price series, West Texas Intermediate (WTI) crude oil spot price and Brent crude oil spot price, are used in the next section.

3 Experiments

3.1 Research Data and Evaluation Criteria

There are a great number of indicators of crude oil prices. In this chapter, two main crude oil price series, West Texas Intermediate (WTI) crude oil spot price and Brent crude oil spot price are chosen as the experiment sample data since the two crude oil prices are most famous benchmark prices used widely as the basis of many crude oil price formulas. The two crude oil price data used in this chapter are daily data and are obtained from the energy information administration website of Department of Energy (DOE) of US (http://www.eia.doe.gov/).

For WTI crude oil spot price, we take the daily data from January 1, 1986 to September 30, 2006 excluding public holidays with a total of 5237 observations. For convenience of training, the data from January 1, 1986 till December 31, 2000 is used for training set (3800 observations) and the remainder is used as the testing set (1437 observations).

For Brent crude oil spot price, the sampling data covers from May 20, 1987 till September 30, 2006 with a total of 4933 observations. The main reason of different starting point is that the DOE website only provides the Brent data since May 20, 1987. Similarly, we take the data from May 20, 1987 to December 31, 2002 as in-sample (training periods) training set (3965 observations) and take the data from January 1, 2003 to September 30, 2006 as out-of-sample (testing period) testing set (968 observations) which is used to evaluate the good or bad performance of prediction based on some evaluation criteria.

To measure the forecasting performance, two classes of main evaluation criteria including level prediction and directional forecasting are used. First of all, we select the root mean squared error (RMSE) as the evaluation criterion of level prediction. Typically, the definition of the RMSE can be represented by

$$RMSE = \sqrt{\frac{1}{N} \sum_{t=1}^{N} \left(\hat{x}(t) - x(t) \right)^2} \qquad (5)$$

where $x(t)$ is the actual value, $\hat{x}(t)$ is the predicted value, N is the number of predictions.

Clearly, accuracy is one of the most important criteria for forecasting models — the others being the cost savings and profit earnings generated from improved decisions. From the business point of view, the latter is more important than the former. For business practitioners, the aim of forecasting is to support or improve decisions so as to make more money. Thus profits or returns are more important than conventional fit measurements. But in the crude oil price forecasting, improved decisions often depend on correct forecasting directions or turning points between the actual and predicted values, $x(t)$ and $\hat{x}(t)$, respectively, in the testing set with respect to directional change of crude oil price movement (expressed in percentages). The ability to forecast movement direction or turning points can be measured by a directional statistic (D_{stat}) (Yu et al., 2005), which can be expressed as

$$D_{stat} = \frac{1}{N} \sum_{t=1}^{N} a_t \times 100\%$$ (6)

where $a_t = 1$ if $(x_{t+1} - x_t)(\hat{x}_{t+1} - x_t) \geq 0$, and $a_t = 0$ otherwise.

In order to compare the forecasting capability of the proposed EMD-based neural network ensemble methodology with other forecasting approaches, the auto-regressive integrated moving average (ARIMA) model and individual feed-forward neural network (FNN) model are used as the benchmark models. In addition, three ensemble model variants, EMD-ARIMA-ALNN model, EMD-BPNN-Averaging model and EMD-ARIMA-Averaging model, are also used to predict the crude oil price for comparison purpose.

3.2 Experimental Results

In this chapter, all ARIMA models are implemented via the Eviews software package, which is produced by Quantitative Micro Software Corporation. The individual FNN model and the EMD-based neural network ensemble model are built using the Matlab software package, which is produced by Mathworks Laboratory Corporation. The EMD-ARIMA-ALNN model uses ARIMA model to predict the each IMFs extracted by EMD and applied ALNN for combination. EMD-FNN-Averaging model applies the FNN to predict all IMFs and the predicted results are integrated using a simple averaging method. While EMD-ARIMA-Averaging model utilizes ARIMA method to predict the IMFs and then the prediction results are combined with a simple averaging strategy. Therefore, the single ARIMA model, the single FNN model, the EMD-ARIMA-ALNN ensemble model, EMD-FNN-Averaging model, the EMD-ARIMA- Averaging ensemble model, and the proposed EMD-FNN-ALNN ensemble model are used to predict the two main crude oil prices for comparison. For space reasons the computational processes are omitted but can be obtained from the authors upon request.

For prediction results, Tables 1 to 2 show the forecasting performance of different models from different perspectives. From the tables, we can generally see that the forecasting results are very promising for all crude oil prices under study either where the measurement of forecasting performance is goodness of fit such as *RMSE* (refer to Table 1) or where the forecasting performance criterion is D_{stat} (refer to Table 2).

Subsequently, the forecasting performance comparisons of different models for the two main crude oil prices via *RMSE* and D_{stat} are reported in Tables 1 to 2 respectively. Generally speaking, these tables provide comparisons of *RMSE* and D_{stat} between these different methods, indicating that the prediction performance of the EMD-based neural network ensemble forecasting model is better than those of other models including these empirical analyses for the two main crude oil prices.

Focusing on the *RMSE* indicator, the EMD-based neural network ensemble model performs the best in all cases, followed by EMD-FNN-Averaging model, the EMD-ARIMA-ALNN model, individual BPNN model, EMD-ARIMA-Averaging model and the individual ARIMA model. To summarize, the EMD-based neural network ensemble model outperforms the other different models presented here in terms of *RMSE*. Interestingly, the *RMSEs* of the individual BPNN model for two crude oil prices are better than those of the EMD-ARIMA-ALNN and EMD-ARIMA - Averaging as well as the individual GLAR models. The possible reasons reflect

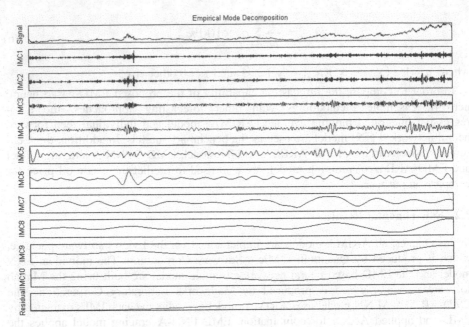

Fig. 2. The decomposition of WTI crude oil spot price

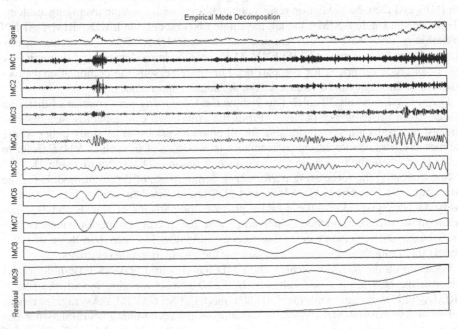

Fig. 3. The decomposition of Brent crude oil spot price

Table 1. The *RMSE* comparisons for different methods

Methodology	WTI		Brent	
	RMSE	Rank	*RMSE*	Rank
EMD-FNN-ALNN	0.273	1	0.225	1
EMD-FNN-Averaging	0.509	2	0.457	2
EMD-ARIMA-ALNN	0.975	4	0.872	4
EMD-ARIMA-Averaging	1.769	5	1.392	5
Individual FNN	0.841	3	0.743	3
Individual ARIMA	2.035	6	1.768	6

two-fold. On one hand, the BPNN is a class of nonlinear forecasting model which can capture nonlinear patterns hidden in the crude oil prices. On the other hand, the crude oil market is a high-volatility market and the crude oil price often show nonlinear and nonstationary patterns, while the EMD-ARIAM-Averaging model and the individual ARIMA model are a class of linear model, which are not suitable for predicting crude oil price with high volatility and irregularity.

However, the low *RMSE* does not necessarily mean that there is a high hit rate of forecasting direction for crude oil price movement direction prediction. Thus, the D_{stat} comparison is necessary for business practitioners. Focusing on D_{stat} of Table 2, we find the EMD-based neural network ensemble model also performs much better than the other models according to the rank; furthermore, from the business practitioners' point of view, D_{stat} is more important than *RMSE* because the former is an important decision criterion. With reference to Table 2, the differences between the different models are very significant. For example, for the WTI test case, the D_{stat} for the individual ARIMA model is 52.47%, for the individual FNN model it is 69.03%, and for the EMD-FNN-Averaging model D_{stat} is only 77.45%; while for the EMD-based neural network ensemble forecasting model, D_{stat} reaches 86.99%. For the other three variant ensemble models, the rank of forecasting accuracy is always in the middle for any of the test crude oil prices. The main cause of this phenomenon is that the bad performance of the individual models and ensemble strategy (i.e., simple averaging) has an important effect on the holistic forecast efficiency. Similarly, the individual FNN can model nonlinear time series such as crude oil price series well, and the D_{stat}

Table 2. The D_{stat} (%) comparisons for different methods

Methodology	WTI		Brent	
	D_{stat} (%)	Rank	D_{stat} (%)	Rank
EMD-FNN-ALNN	86.99	1	87.81	1
EMD-FNN-Averaging	77.45	2	79.44	2
EMD-ARIMA-ALNN	62.98	4	68.91	4
EMD-ARIMA-Averaging	60.13	5	62.29	5
Individual FNN	69.03	3	75.41	3
Individual ARIMA	52.47	6	57.75	6

rank is also in the third position for any of the test crude oil prices. In the same way, we also notice that the D_{stat} ranking for ARIMA is the last. The main reason is that the high noise, nonlinearity and complex factors are contained in crude oil price series, and unfortunately the ARIMA is a class of linear model.

From the experiments presented in this chapter we can draw the following conclusions.

1. The experimental results show that the EMD-based neural network ensemble forecasting model is superior to the individual ARIMA model, the individual FNN model, the EMD-ARIMA-Averaging and EMD-ARIMA-ALNN model as well as the EMD-FNN-Averaging model for the test cases of two main crude oil prices in terms of the measurement of level prediction measurement (*RMSE*) and directional change statistics (D_{stat}), as can be seen from Tables 1 and 2.
2. The proposed EMD-based neural network ensemble forecasts are able to improve forecasting accuracy significantly — in other words, the performance of the EMD-based neural network ensemble forecasting model is better than those of other forecasting models in terms of *NMSE* and D_{stat}. This leads to the third conclusion.
3. The EMD-based neural network ensemble model can be used as an alternative solution to world crude oil spot price forecasting.

4 Conclusions

This chapter proposes using an EMD-based neural network ensemble forecasting model to predict crude oil spot prices. In terms of the empirical results, we find that across different forecasting models for the test cases of two main crude oil prices — WTI crude oil spot price and Brent crude oil spot price — on the basis of different criteria, the EMD-based neural network ensemble model performs the best. In the all test cases, the RMSE is the lowest and the Dstat is the highest, indicating that the EMD-based neural network ensemble forecasting model can be used as a viable alternative solution for crude oil prediction for investment managers and business practitioners.

Acknowledgements

This work describe here is partially supported by the grants from the National Natural Science Foundation of China (NSFC No. 70601029, 70221001), the Chinese Academy of Sciences (CAS No. 3547600), the Academy of Mathematics and Systems Sciences (AMSS No. 3543500) of CAS, and the Strategic Research Grant of City University of Hong Kong (SRG No. 7001806).

References

Abosedra, S., Baghestani, H.: On the predictive accuracy of crude oil future prices. Energy Policy 32, 1389–1393 (2004)
Abramson, B., Finizza, A.: Using belief networks to forecast oil prices. International Journal of Forecasting 7(3), 299–315 (1991)

Abramson, B., Finizza, A.: Probabilistic forecasts from probabilistic models: a case study in the oil market. International Journal of Forecasting 11(1), 63–72 (1995)

Alvarez-Ramirez, J., Soriano, A., Cisneros, M., Suarez, R.: Symmetry/anti-symmetry phase transitions in crude oil markets. Physica A 322, 583–596 (2003)

Barone-Adesi, G., Bourgoin, F., Giannopoulos, K.: Don't look back. Risk August 8, 100–103 (1998)

Hagan, M.T., Demuth, H.B., Beale, M.H.: Neural Network Design. PWS Publishing Company, Boston (1996)

Hagen, R.: How is the international price of a particular crude determined? OPEC Review 18(1), 145–158 (1994)

Hornik, K., Stinchcombe, M., White, H.: Multilayer feedforward networks are universal approximators. Neural Networks 2, 359–366 (1989)

Huang, N.E., Shen, Z., Long, S.R.: A new view of nonlinear water waves: The Hilbert spectrum. Annual Review of Fluid Mechanics 31, 417–457 (1999)

Huang, N.E., Shen, Z., Long, S.R., Wu, M.C., Shih, H.H., Zheng, Q., Yen, N.C., Tung, C.C., Liu, H.H.: The empirical mode decomposition and the Hilbert spectrum for nonlinear and nonstationary time series analysis. Proceedings of the Royal Society A: Mathematical, Physical & Engineering Sciences 454, 903–995 (1998)

Huntington, H.G.: Oil price forecasting in the 1980s: What went wrong? The Energy Journal 15(2), 1–22 (1994)

Maurice, J.: Summary of the oil price. Research Report, La Documentation Francaise (1994), http://www.agiweb.org/gap/legis106/oil_price.html

Morana, C.: A semiparametric approach to short-term oil price forecasting. Energy Economics 23(3), 325–338 (2001)

Nelson, Y., Stoner, S., Gemis, G., Nix, H.D.: Results of Delphi VIII survey of oil price forecasts. Energy report, California Energy Commission (1994)

Stevens, P.: The determination of oil prices 1945-1995. Energy Policy 23(10), 861–870 (1995)

Verleger, P.K.: Adjusting to volatile energy prices. Working paper, Institute for International Economics, Washington DC (1993)

Wang, S.Y., Yu, L., Lai, K.K.: Crude oil price forecasting with TEI@I methodology. Journal of Systems Sciences and Complexity 18(2), 145–166 (2005)

Watkins, G.C., Plourde, A.: How volatile are crude oil prices? OPEC Review 18(4), 220–245 (1994)

White, H.: Connectionist nonparametric regression: Multilayer feedforward networks can learn arbitrary mappings. Neural Networks 3, 535–549 (1990)

Xie, W., Yu, L., Xu, S.Y., Wang, S.Y.: A new method for crude oil price forecasting based on support vector machines. In: Alexandrov, V.N., van Albada, G.D., Sloot, P.M.A., Dongarra, J. (eds.) ICCS 2006. LNCS, vol. 3994, pp. 441–451. Springer, Heidelberg (2006)

Yu, L., Wang, S.Y., Lai, K.K.: A novel nonlinear ensemble forecasting model incorporating GLAR and ANN for foreign exchange rates. Computers and Operations Research 32(10), 2523–2541 (2005)

Structured Hidden Markov Models: A General Tool for Modeling Agent Behaviors

Ugo Galassi, Attilio Giordana, and Lorenza Saitta

Dipartimento di Informatica, Università del Piemonte Orientale
Via Bellini 25/G, Alessandria, Italy
{ugo.galassi,attilio.giordana,lorenza.saitta}@unipmn.it

Abstract. Structured Hidden Markov Model (S-HMM) is a variant of Hierarchical Hidden Markov Model that shows interesting capabilities of extracting knowledge from symbolic sequences. In fact, the S-HMM structure provides an abstraction mechanism allowing a high level symbolic description of the knowledge embedded in S-HMM to be easily obtained. The paper provides a theoretical analysis of the complexity of the matching and training algorithms on S-HMMs. More specifically, it is shown that the Baum-Welch algorithm benefits from the so called *locality* property, which allows specific components to be modified and retrained, without doing so for the full model. Moreover, a variant of the Baum-Welch algorithm is proposed, which allows a model to be biased towards specific regularities in the training sequences, an interesting feature in a knowledge extraction task. Several methods for incrementally constructing complex S-HMMs are also discussed, and examples of application to non trivial tasks of profiling are presented.

Keywords: Hidden Markov Model, Keystroking dynamics, User authentication.

1 Introduction

Modeling an agent behavior is a task that is becoming of primary issue in computer networks. In fact, critical applications such as intrusion detection [13, 22, 7], network monitoring [23, 14], and environment surveillance [5, 20] are using the approaches in which modeling the processes and users is a key factor.

However, to automatically infer a model for an agent behavior is not a trivial task. On the one hand, the features that the model should consider, and the ones that should be ignored, are often problem specific. On the other hand, current formal modeling tools become rapidly intractable when the complexity of the patterns to be captured increases. Then, only simple technologies can be used in practice.

Problems of modeling an agent behavior fall into two classes. The first class contains problems where a stationary, long-term behavior has to be modeled, for instance, the daily usage of a set of resources [13]. In this case, building a behavior model can be frequently reduced to collecting statistics related to some set of global variables, and the particular sequences in which actions are performed is not important.

The second class contains more intriguing problems, aimed at explicitly capturing transitory behaviors involving events duration and ordering; for instance, how long it takes for an agent to go from A to B, or what are the actions made by the agent in A and B. Modeling such behavioral aspects requires modeling time and action sequences, a problem for which still no general solution exists.

B. Prasad (Ed.): Soft Computing Applications in Business, STUDFUZZ 230, pp. 273–292, 2008.
springerlink.com © Springer-Verlag Berlin Heidelberg 2008

We focus on this second class of problems, and we propose a variant of Hidden Markov Models (HMM) [19], which we call *Structured Hidden Markov Models* (S-HMM), as a general modeling tool. In fact, an S-HMM has sufficient expressiveness to cope with most practical problems, still with acceptable computational complexity.

In the following we provide a brief introduction to S-HMMs, referring the reader to other papers for the formal analysis of their properties [9, 10, 28]. Then we will investigate the problem of modeling time duration and transitories using specific HMMs. Finally we will describe applications to keystroking dynamics, where modeling durations was the most critical issue.

2 Structured HMMs

The basic assumption underlying an S-HMM (see Bouchaffra and Tan [2]) is that a sequence $O = \{o_1, o_2, o_3, ..., o_T\}$ of observations could be segmented into a set of sub-sequences $O_1, O_2, ..., O_R$, each one generated by a sub-process with only weak interactions with its neighbors. This assumption is realistic in many practical applications, such as, for instance, speech recognition [18, 19], and DNA analysis [6]. S-HMMs aim exactly at modeling such kind of processes, and, hence, they are represented as directed graphs, structured into sub-graphs (*blocks*), each one modeling a specific kind of sub-sequences.

Informally, a block consists of a set of states, only two of which (the *initial* and the *end* state) can be connected to other blocks. Then, an S-HMM can be seen as a forward acyclic graph with blocks as nodes. As an S-HMM is itself a block, a nesting mechanism is immediate to define.

2.1 Structure of a Block

In this section, a formal definition of S-HMMs is provided. Following Rabiner [19], O will denote a sequence of observations $\{o_1, o_2, o_3, ..., o_T\}$, where every observation o_t is a symbol v_k chosen from an alphabet V. An HMM is a stochastic automaton characterized by a set of states Q, an alphabet V, and a triple $\lambda = \langle A, B, \pi \rangle$, being:

- A: $Q \times Q \to [0, 1]$ a probability distribution, a_{ij}, governing the transition from state q_i to state q_j,
- B: $Q \times V \to [0, 1]$ a probability distribution, $b_i(v_k)$, governing the emission of symbols in each state q_i,
- $\pi : Q \to [0, 1]$ a distribution assigning to each state q_i in Q the probability of being the start state.

A state q_i will be said a *silent* state if $\forall\, v_k \in V : b_i(v_k) = 0$, *i.e.*, q_i does not emit any observable symbol. When entering a silent state, the time counter must not be incremented.

Definition 1. In an S-HMM, a *basic block* is a 4-tuple $\lambda = \langle A, B, I, E \rangle$, where I, E $\in$ Q are silent states such that:

$\pi(I) = 1, \forall\, q_i \in Q : a_{iI} = 0$, and $\forall\, q_i \in Q : a_{Ei} = 0$.

In other words, I and E are the input and the output states, respectively. Therefore, a composite block can be defined by connecting, through a forward transition network, the input and output states of a set of blocks.

Definition 2. Given an ordered set of blocks $\Lambda = \{\lambda_i \mid 1 \leq i \leq R\}$, a *composite block* is a 4-tuple $\lambda = \langle A_I, A_E, I, E \rangle$, where:

- $A_I : E \times I \to [0,1]$, $A_E : I \times E \to [0,1]$ are probability distributions governing the transitions from the output states **E** to the input states **I**, and from the input states **I** to the output states **E** of the component blocks in Λ, respectively.
- For all pairs $\langle E_i, I_j \rangle$ the transition probability $a_{E_i I_j} = 0$ if $j \leq i$.
- $I = I_1$ and $E = E_R$ are the input and output states of the composite block, respectively.

According to Definition 2 the components of a composite block can be either basic blocks or, in turn, composite blocks. In other words, composite blocks can be arbitrarily nested. Moreover, we will keep the notation S-HMM to designate non-basic blocks only.

As a special case, a block can degenerate to the *null block*, which consists of the start and end states only, connected by an edge with probability $a_{IE} = 1$. A *null block* is useful in order to provide dummy input, output states (I, E), when no one of the component blocks is suited to this purpose.

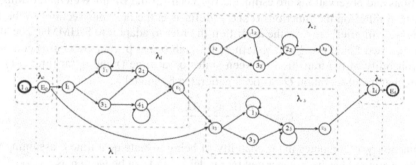

Fig. 1. Example of a Structured Hidden Markov Model, composed of three interconnected blocks, plus two *null blocks*, λ_0 and λ_4, providing the start and end states. Distribution A is non-null only for explicitly represented arcs.

An example of S-HMM structured into three blocks $\lambda_1, \lambda_2, \lambda_3$, and two *null blocks* λ_0, λ_4, providing the start and the end states, is described in Figure 1.

2.2 Estimating Probabilities in S-HMM

Well-known algorithms are used to solve typical problems arising in using HMMs; in particular, the *Forward-Backward* algorithm computes the probability $P(O \mid \lambda)$ that a model λ generates the observation O. The *Viterbi* algorithm computes the most likely sequence of λ's states that generates O. Finally; the *Baum-Welch* algorithm allows the parameters of λ (*i.e.*, the matrices A and B and the vector π) to be estimated from sequences.

The mentioned algorithms are based on two functions, α, β, plus two auxiliary functions, γ, ξ, which are defined on α, β. The latter have an intuitive meaning. Function $\alpha_t(i)$ evaluates, for every time t $(1 \leq t \leq T)$ and every state q_i $(1 \leq i \leq N)$, the joint probability of being in state q_i *and* observing the symbols from o_1 to o_t. Funtion $\beta_t(i)$ is complementary to $\alpha_t(i)$, and computes, for every time t $(1 \leq t \leq T)$ and every state q_i $(1 \leq i \leq N)$, the conditional probability of observing the symbols o_{t+1}, o_{t+2}, ... , o_T, given that q_i is the state at time t. Then, $\alpha_T(E) = \beta_1(I) = P(O \mid \lambda)$.

All these functions are defined on basic HMMs, but we have extended them to the case of nested S-HMMs [9], and provided algorithms that calculate them incrementally. In an S-HMM each block interacts with others only through the *initial* and *end* states. Then, the computation of α and β inside one block does not affect the one in other blocks, and, hence, parameters inside each block can be estimated independently of the other ones, reducing the standard $O(T\ N^2)$ complexity to $O(T\ \Sigma_h\ N_h^2)$, where N_h is the number of states inside block λ_h and $N = \Sigma_h\ N_h$.

3 S-HMMs Are Locally Trainable

The classical algorithm for estimating the probability distributions governing state transitions and observations are estimated by means of the Baum-Welch algorithm [1, 19], which relies on the functions α and β defined in the previous section. In the following we will briefly review the algorithm in order to adapt it to S-HMMs. The algorithm uses two functions, ξ and γ, defined through α and β. Function $\xi_t(i,j)$ computes the probability of the transition between states q_i (at time t) and q_j (at time t+1), assuming that the observation O has been generated by model λ:

$$\xi_t(i,j) = \frac{\alpha_t(i)\ a_{ij}\ b_j(O_{t+1})\ \beta_{t+1}(j)}{P(O \mid \lambda)} \qquad (1)$$

Function $\gamma_t(i)$ computes the probability of being in state q_i at time t, assuming that the observation O has been generated by model λ, and can be written as :

$$\gamma_t(i) = \frac{\alpha_t(i)\ \beta_t(i)}{P(O \mid \lambda)} \qquad (2)$$

The sum of $\xi_t(i,j)$ over t estimates the number of times transition $q_i \rightarrow q_j$ occurs when λ generates the sequence O. In an analogous way, by summing $\gamma_t(i)$ over t, an estimate of the number of times state q_i has been visited is obtained. Then a_{ij} can be re-estimated (a-posteriori, after seeing O) as the ratio of the sum over time of $\xi_t(i,j)$ and $\gamma_t(i)$:

$$\overline{a}_{ij} = \frac{\sum_{t=1}^{T-1} \alpha_t(i)\ a_{ij}\ b_j(O_{t+1})\ \beta_{t+1}(j)}{\sum_{t=1}^{T-1} \alpha_t(i)\ \beta_t(i)} \qquad (3)$$

With a similar reasoning it is possible to obtain an a-posteriori estimate of the probability of observing $o = v_k$ when the model is in state q_j. The estimate is provided by the ratio between the number of times state q_j has been visited and symbol v_k has been observed, and the total number of times q_j has been visited:

$$\bar{b}_j(k) = \frac{\sum_{t=1, o_t = v_k}^{T-1} \alpha_t(j)\, \beta_t(j)}{\sum_{t=1}^{T-1} \alpha_t(j)\, \beta_t(j)} \tag{4}$$

Inside the basic block λ_k, it clearly appears that equations (3) and (4) are immediately applicable. Then the Baum-Welch algorithm can be used without any change to learn the probability distributions inside basic blocks.

On the contrary, equation (3) must be modified in order to adapt it to re-estimate transition probabilities between output and input states of the blocks, which are silent states. As there is no emission, α and β propagate through transitions without time change; then, equation (3) must be modified as follows:

$$\bar{a}_{E_i I_j} = \frac{\sum_{t=1}^{T-1} \alpha_t(E_i)\, a_{E_i I_j}\, \beta_t(I_j)}{\sum_{t=1}^{T-1} \alpha_t(E_i)\, \beta_t(I_j)} \tag{5}$$

It is worth noticing that functions α, β depend upon the states in other blocks only through the value of $\alpha_t(I_k)$ and $\beta_t(E_k)$, respectively. This means that, in block λ_k, given the vectors $\alpha_1(I_k)$, $\alpha_2(I_k)$, ... $\alpha_T(I_k)$ and $\beta_1(E_k)$, $\beta_2(E_k)$, ... $\beta_T(E_k)$, the Baum-Welch algorithm can be iterated inside a block without the need of computing again α and β in the external blocks. We will call this a *locality property*. The practical implication of the locality property is that a block can be modified and trained without any impact on the other components of an S-HMM.

4 Modeling Motifs and Duration

The basic assumption underlying profiling is that an agent periodically performs (usually short) sequences of actions, typical of the task it executes, interleaved with phases where the activity cannot be modeled, because it is non-repetitive. By analogy with the DNA sequences in molecular biology, we will call *motifs* such a kind of characteristic sequences of actions. In fact, under the previous assumptions, the problem of agent profiling presents a strong analogy to the problem of discovering and characterizing coding subsequences in a DNA chromosome.

A model for interpreting a sequence must be a *global* model, able to identify both interesting patterns that occur with significant regularity, and *gaps*, *i.e.*, regions where no regularities are found. Generating a global model of the sequence is important, because it allows inter-dependencies among motifs to be detected. Nevertheless, a global model must account for the distribution of the observations on the entire sequence, and hence it could become intractable. We tamed this problem by introducing special basic blocks, designed to keep low the complexity of modeling the irrelevant parts of sequences. In the following we address the following issues:

(a) how to construct basic blocks modeling motifs;
(b) how to construct models of gaps between motifs;
(c) how to extract (insert) knowledge in readable form from (to) an S-HMM;
(d) how to find the interpretation of a sequence.

4.1 Modeling Motifs

A motif is a subsequence occurring frequently in a reference sequence set. Motif occurrences may be different from one another, provided that an assigned equivalence relation is satisfied. In the specific case, the equivalence relation is encoded through a basic block of an S-HMM.

Many proposals exist for HMM architectures oriented to capture specific patterns. Here, we will consider the Profile HMM (PHMM), a model developed in Bioinformatics [6], which well fits the needs of providing a readable interpretation of a sequence. The basic assumption underlying PHMM is that the different instances of a motif originate from a canonical form, but they are subject to insertion, deletion and substitution errors.

As described in Figure 2, a PHMM has a left-to-right structure with a very restricted number of arcs. Moreover, it makes use of *typed* states: *Match* states, where the observation corresponds to the expectation, *Delete* states (silent states) modeling deletion errors, and *Insert* states, modeling insertion errors supposedly due to random noise. According to this assumption, the distribution of the observations in all insert states is the same, and it can be estimated just once.

After training, the canonical form can be easily extracted from a PHMM by collecting the maximum likelihood observation sequence from the match states.

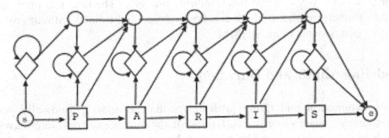

Fig. 2. Example of Profile Hidden Markov Model. Circles denote states with *no-observable* emission, rectangles denote *match states*, and diamonds denote *insert states*.

4.2 Modeling Duration and Gaps

The problem of modeling durations arises when the time span covered by an observation or the interval length between two observations is important. In the HMM framework, this problem has been principally faced in Speech Recognition and in Bioinformatics. However, the problem setting is slightly different in the two fields, and consequently the dominant approach tends to be different. In speech recognition, the input is a continuous signal, which, after several steps of signal processing, is segmented into variable length intervals, each one labeled with a symbol. Then, the obtained symbolic sequence is fed into a set of HMMs, which accomplish the recognition of long range structures, such as syllables or words, requiring thus to deal with interval durations [15]. In Bioinformatics the major application for HMMs is the analysis of DNA strands [6]. Here, the input sequence is a string of equal length symbols. The need of modeling duration comes from the presence of gaps, *i.e.*, substrings where no coding information is present. The gap duration is often a critical cue in order to interpret the entire sequence.

The approach first developed in Speech Recognition is to use Hidden Semi-Markov Models (HSMM), which are HMMs augmented with probability distributions over the state permanence [5, 12, 17, 26, 24].

An alternative approach is what has been called *Expanded HMM* [12]. Every state, where it is required to model duration, is expanded into a network of states. In this way, the duration of the permanence in the original state is modeled by a sequence of transitions through the new state network in which the observation remain constant. The advantage of this method is that the Markovian nature of the HMM is preserved. Nevertheless, the complexity increases according to the number of new states generated by expansion.

A similar solution is found in Bioinformatics for modeling long gaps. In this case, the granularity of the sequence is given, and so there is no need of expansion. However, the resulting model of the gap duration is similar to the one mentioned above.

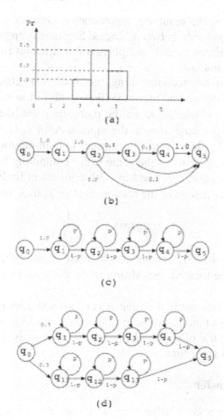

Fig. 3. Possible HMM architectures for modeling duration

A Profile HMM [6] naturally models the duration of observations according to the expansion technique, but it is only able to model short gaps inside a motif, attributed to random noise controlled by a Poisson statistics. Nevertheless, single insertion states do not correctly model long gaps occurring in between two motifs. The most appropriate probability distribution for this kind of gaps may vary from case to case, but it is never the exponential decay defined by an insert state with a self-loop.

Several HMM topologies, proposed for modeling gaps, are reported in Figure 3. The architecture in Figure 3(b) can be used to model any duration distribution over a finite and discrete interval. However, the drawback of this model is the potentially large number of parameters to estimate. In all cases, the observation is supposed to be produced by random noise. The model in Figure 3(c) exhibits an Erlang's distribution, when the Forward-Backward algorithm is used. Unfortunately, the distribution of the most likely duration computed by Viterbi algorithm still follows an exponentially low. Therefore, this model, which is more compact with respect to the previous one, is not useful for the task of segmenting and tagging sequences by means of Viterbi algorithm. However, sequence segmentation and interpretation is not a primary task in the class of problems we are considering in this paper.

4.3 Sequence Segmentation

In the S-HMM framework, sequence segmentation consists in identifying the most likely locations of boundaries between blocks. Segmentation provides a probabilistic interpretation of a sequence model, and plays a fundamental role when an S-HMM is used for knowledge extraction.

Two methods exist for accomplishing this task. The classical one is based on Viterbi algorithm in order to find the most likely path in the state space of the model. Then, for every pair of blocks λ_r, λ_s on the path, the most likely time instant for the transition from the output state of λ_r to the input state of λ_s is chosen as the boundary between the two blocks. In this way a unique, non-ambiguous segmentation is obtained, with a complexity which is the same as for computing functions α, and β.

The second method, also described in [19], consists in finding the maximum likelihood estimation of the instant of the transition from λ_r to λ_s by computing:

$$\tau_{rs} = \arg\max_t \left(\frac{\xi_t(E_r, I_s)}{\gamma_t(E_r)} \right) \tag{6}$$

Computing boundaries by means of (6) requires a complexity O(T) for every boundary that has to be located, in addition to the complexity of computing the α and β values.

The advantage of this method is that it can provide alternative segmentations by considering also blocks that do not lie on the maximum likelihood path. Moreover, it is compatible with the use of gap models of the type described in Figure 3(c), because it does not use Viterbi algorithm.

4.4 Knowledge Transfer

When a tool is used in a knowledge extraction task, two important features are desirable: (a) the extracted knowledge should be readable for a human user; (b) a human user should be able to elicit chunks of knowledge, which the tool will exploit during the extraction process.

The basic HMM does not have such properties, whereas task oriented HMMs, such as Profile HMM, may provide such properties to a limited extent. On the contrary, the S-HMM structure naturally supports high level logical descriptions.

An example of how an S-HMM can be described in symbolic form is provided in Figure 4. Basic blocks and composite blocks must be described in different ways. Basic blocks are either HMMs (modeling subsequences), or gap models. In both cases, a precise description of the underlying automaton will be complex, without providing readable information to the user. Instead, an approximate description, characterizing at an abstract level the knowledge captured by a block, is more useful. For instance, blocks corresponding to regularities like *motifs* can be characterized by providing the maximum likelihood sequence (MLS) as the nominal form of the sequence, and the

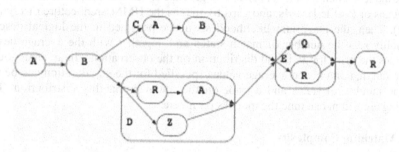

(b) **Basic block description:**

motif(x)∧MLS(x, "ctgaac")∧AvDev(x, 0.15) → A(x)
motif(x)∧MLS(x, "cctctaaa")∧AvDev(x, 0.15) → R(x)
motif(x)∧MLS(x, "tatacgc")∧AvDev(x, 0.15) → Q(x)
gap(x)∧AvDr(x, 11.3)∧MnDr(x, 8)∧MxDr(x, 14) → B(x)
gap(x)∧AvDr(x, 15.6∧ MnDr(x, 12)∧MxDr(x, 19) → Z(x)

(c) **Block structure logical description:**

A(x) ∧ B(y) ∧ follow(x, y) → C([x,y])
R(x) ∧ A(y) ∧ follow(x, y) → D([x,y])
Z(x) → D(x), Q(x) → E(x), R(x) → E(x)
B(x) ∧ C(y) ∧ follow(x, y) → G([x,y]
B(x) ∧ D(y) ∧ follow(x, y) → G([x,y]
A(x) ∧ G(y) ∧ E(z) ∧ R(w) ∧ follow(x, y) ∧
    ∧ follow(y, z) ∧ follow(z, w) → MySEQ([x,y,z,w)]

(d) **Block structure as a regular expression:**

A (B (AB | (RA | Z)))(Q|R) R

Fig. 4. Structured HMMs are easy to translate into an approximate logic description

average deviation (AvDv) from the nominal form. Instead, gaps can be characterized by supplying the average duration (AvDr), and the minimum (MnDr) and maximum (MxDr) duration.

On the contrary, the model's composite structure is easy to describe by means of a logic language. For instance, Figure 4(c) provides an example of translation into Horn clauses, whereas Figure 4(d) provides an example of translation into regular expressions. In both cases, richer representations can be obtained by annotating the expressions with numerical attributes.

By using a logical description language, or regular expressions, a user can also provide the specification of an S-HMM structure, or part of it, which will be completed and trained by a learning algorithm. Logical formulas as in Figure 4(c) can be immediately translated into the structure of composite blocks. Nevertheless, also an approximate specification for basic blocks, as described in Figure 4, can be mapped to block models when the model scheme is given. For instance, suppose to model motifs by means of Profile HMMs, and gaps by means the HMM architecture of Figure 3(b) or (c). Then, the maximum likelihood sequence provided in the logical description implicitly sets the number of match states and, together with the average deviation, provides the prior for an initial distribution on the observations. In an analogous way, minimum, maximum and average values specified for the gap duration can be used to set the number of states and a prior on the initial probability distribution. Then, a training algorithm can tune the model's parameters.

4.5 Matching Complexity

In the previous section the upper bound on the complexity of computing functions α and β has been reported. However, adopting the basic block structure suggested in the previous sections for modeling motifs and gaps, this complexity becomes quasi-linear in the number of states. Considering Profile HMMs (see Figure 2), it is immediate to verify that computing α and β has complexity $O(3TN)$, being N the number of states. On the contrary, the complexity decreases to $O(2TN)$ for both gap models in Figure 3(a), and 3(b). Then, the only nonlinear term can be due to the matrix interconnecting the blocks. An experimental evaluation is reported in Figure 5, using a set of S-HMMs of different size (from 160 to 920 states) and sequences of different length (from 633 to 2033 symbols). The number of basic blocks ranges from 6 to 23. Figure 5 reports the Cpu time obtained on a PowerBook G4 for the evaluation of $P(O|\lambda)$ using function α_T. It appears that very complex models can be reasonably used to mine quite long sequences.

5 S-HMM Construction and Training

In classification tasks, the goal is to maximize the probability $P(O|\lambda)$ when O belongs to the target class, while $P(O|\lambda)$ remains as low as possible when O does not belong to the target class. Letting the Baum-Welch algorithm tune the S-HMM parameters is the simplest solution.

On the contrary, when the goal is knowledge extraction, the strategy for constructing and training the model must satisfy different requirements. In fact, it is important

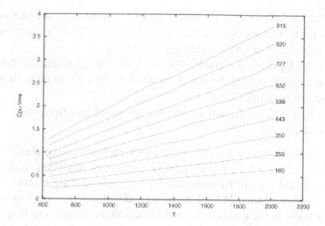

Fig. 5. Complexity for a sequence interpretation task: (a) CPU time versus string length. Different curves correspond to different number of states.

to preserve the structure of interesting patterns. This is not implicit in the classic training strategy, because the Baum-Welch algorithm simply maximizes the probability $P(O|\lambda)$, and is not guaranteed to preserve specific structures in the patterns.

In Section 3 we have discussed the locality property and how the Baum-Welch algorithm must take into account the S-HMM structure in order to treat in a different way the silent states interconnecting the blocks. In a similar way, the locality property and the S-HMM structure can be exploited in order to bias the training phase preserving the knowledge already embedded in the model, and increasing the probability of discovering specific kinds of regularities.

In the following we will first outline an algorithm scheme for constructing an S-HMM with a nested block structure. The description of a fully implemented algorithm can be found in [10]. Then, we will discuss how single blocks can be constructed and trained.

5.1 Discovery Algorithm

Let LS be a set of training sequences for which an S-HMM must be constructed. Suppose that a model λ, possibly built up in previous learning steps, already exists. Initially, λ can be provided as a kind of a-priori knowledge, or it can just represent a single gap, interpreting all the sequences in LS as noise. In this last case, λ can be made more specific by replacing it with a composed block containing some motifs possibly interleaved with new, shorter gaps.

Let us assume that $P(O|\lambda)$ has been already evaluated for all sequences $O \in LS$ by computing functions $\alpha_t(I_k)$ and $\beta_t(E_k)$ ($1 \le t \le T$) for each one of the blocks λ_k constituting λ. The basic step for extending λ encompasses the following actions:

1. Find a not yet analyzed basic block λ_k encoding a gap.
2. For each observation $O \in LS$ determine the most likely location of the boundaries $t_I(k)$ and $t_E(k)$ of λ_k, using relation (6), and insert the observed subsequences from $t_I(k)$ to $t_E(k)$ into a temporary learning set LS_k.

3. Search for regularities (motifs) occurring in the sequences accumulated in LS_k. If no new motifs are discovered, then exit, otherwise build a composite block λ'_k containing the new motifs and replace λ_k with λ'_k.
4. Locally train λ'_k using the Baum-Welch algorithm, and validate the final model according to some test. If the test is successful, then continue with the next step, otherwise restore model λ_k and exit.
5. Compute α_t and β_t for I_k, and E_k and for all initial and final states of the blocks internal to λ'_k.

The above procedure is iterated until no gap block remains unexplored.

The problem of describing regularities in sequences has been faced in Bioinformatics and in data mining. In Bioinformatics, well assessed algorithms, based on edit distance and string alignment [11, 6], have been developed to discover local similarities in DNA strands. In the data mining field, algorithms for mining frequent item sets have been adapted to extract *frequent subsequences* (see, for instance [21, 27, 25]). The current version of our algorithm (called EDY) integrates the two approaches and uses a simplified version of the algorithm GSP [21], combined with an implementation of FASTA [16], in order to discover frequent motif chains. However, very accurate algorithms for mining motifs in sequences are not really necessary in our case. In fact, even a rough hypothesis about the structure of a candidate motif is sufficient to construct a first version of a PHMM, which, tuned by means of the Baum-Welch algorithm, will converge to an accurate model. It is also worth noticing that the approach we suggest is not the one usually followed in Bioinformatics, were PHMMs are constructed from sequence multi-alignment without going through the complete training process. In this case the accuracy of the algorithm for motif discovering has a much greater impact. However, the cost for training the PHMM is saved.

5.2 Locally Training Nested Blocks

In this section we will briefly discuss the Baum-Welch training in basic blocks encoded as PHMMs or gaps, as well as in composite blocks, showing how it is possible to preserve some basic characteristics of the knowledge they encode. As described in Section 4, both gap models and PHMMs handle observations in a non standard way with respect to basic HMMs. A PHMM uses *insertion states*, where the observation is assumed to be just noise. Even further, in gap models all non silent states are supposed to produce an observation which is irrelevant from the point of view of the model, which aims at capturing only durations. Training this kind of models would definitely change their initial purpose (modeling noise), because the probability distribution in matrix B will adapt to the observations, which are actually irrelevant, interfering thus with the global distribution the block is supposed to encode. A simple solution to this problem consists in avoiding training the distribution B, preserving the initial distribution.

For structured blocks, a similar problem exists. In fact, given the model structure, (*i.e.*, the number of states) a PHMM can adapt to any distribution of sequences having length in the same range. Then, the training process can change the role of basic blocks, as well as of composite blocks. The problem arises when a structured block is constructed starting from frequent subsequences *mined* in a learning set LS. In this case, each block encodes a piece of knowledge (given by an expert or learned) that should be maintained when blocks are interconnected.

Suppose we want to connect two models λ_1 and λ_2, describing two sets of sequences S_1 and S_2, respectively, drawn from two distributions D_1 and D_2. Connecting them in series does not generate problems, owing to the locality property. However, if we connect the models in parallel (by letting them share the same input and output states, namely I and E), continuing training may lead the two original distributions to be "forgotten" in favor of a new, intermediate one. If D_1 and D_2 were close, that would be OK, but the blurring may happen also when D_1 and D_2 are quite different. In order to avoid this problem, we use the following procedure:

- For each sequence O, compute $p_1 = P(O| \lambda_1)$ and $p_2 = P(O| \lambda_2)$
- If $p_1 \geq p_2$, then apply Baum-Welch on λ_1 only, leaving λ_2 unchanged. Do the opposite if $p_2 > p_1$
- Apply Baum-Welch to estimate the transition probability from I to λ_1 and to λ_2 (they sum up to 1)

The above algorithm enforces the original diversity between λ_1 and λ_2, and converges to a single distribution only when the two models are close. The degree of closeness can be stated in terms of a distance between the estimates of D_1 and D_2.

6 Modeling Keystroking Dynamics of a Human Agent

The problem of modeling the behavior of an agent may have a very different setting, depending on the kind of agent and on the type of features available. Here, we will focus on the problem of modeling the behavior of a human agent interacting with a computer system through a keyboard. In other words, the goal is to construct a model capturing the dynamics of a user typing on a keyboard.

This task has been widely investigated in the past, in order to develop biometric authentication methods (see, for instance [8, 3]), which led to patented solutions [4]. Our purpose is not to provide a new challenging solution to this task, but simply to show how easy it is to generate user models based on S-HMMs, which are performing quite well.

Two case studies have been investigated, characterized by two different targets. The first one addresses the problem of building a discriminant profile of a user, during the activity of typing a free text. This kind of task plays an important role when the goal is to build a monitoring system, which checks on typical behaviors of a group of agents, and sets off an alarm when someone of them is not behaving according to its profile.

The second case study aims at producing an authentication procedure based on keystroke dynamics during the login phase, which can be used in order to make more difficult to break into a system in case of password theft.

Not withstanding the deep difference in the targets, the two experiments share the problem setting, i.e., the input information to match against the behavior model, and the structure of the model.

6.1 Input Information

In both case studies, the input data are sequences of triples $\langle c, t_p, t_r \rangle$, where c denotes the ASCII code of the stroked key, t_p the time at which the user begun to press the key, and t_r the time at which the key has been released. The input data are collected by a key-logger transparent to the user.

As the S-HMMs used for building the behavior models rely on the state expansion method for modeling temporal durations, the input data are transformed into symbolic strings, according to the following method: Each keystroke is transformed into a string containing a number of repetitions of the typed character proportional to the duration of the stroke. In a similar way, delay durations have been represented as repetitions of a dummy symbol (".") not belonging to the set of symbols occurring in the typed text. The transformation preserves the original temporal information up to 10 ms accuracy.

```
SSSSSSSS...............AAAAIIIII.................TTTTT..........TTTTTTTT......AAAAAAAAAAAAAAAA
SSSSSSS............AAAAIIIIII...............TTTTT..........TTTTTT...AAAAAAAAAAAAA
SSSSSSSSSS................AAAAAIIIIII.................TTTT..........TTTTTT....AAAAAAAAAAAAAAAA
SSSSSSSSSSS...........IIIIIIAAAAAAAA.................TTTTT..........TTTTTTT..AAAAAAAAAAAA
SSSSSSSSSDDDD.......IIIIIAAAAAAAAAA..............TTTTTT..........TTTTTTT...AAAAAAAAAAAAAAAA
SSSSSSSSSSS..........IIIIIAAAAAAA.................TTTTT..........TTTTTTT.....AAAAAAAAAAAAAAA
SSSSSSSSSS...........AAAAAAIIIIII.................TTTT..........TTTTTTTT....AAAAAAAAAAAAAAA
SSSSSSS........IIIIIAAAAAAAAA...............TTTTT..........TTTTTTT.....AAAAAAAAAAAAAAA
IIIIIIAAAAAAAAAA...............TTTTT..........TTTTTT....AAAAAAAAAAAAAA
SSSSSSSSSS............AAAAIIIIII.................TTTT..........TTTTTT...AAAAAAAAAAAA
SSSSSSSSSSSSSSSSS......IIIIIIAAAAAAAAAAAAA..............TTTTTT..........TTTTTTTTT.AAAAAAAAAAAA
SS...........IIIIIIAAAAAAAAAA...............TTTTT..........TTTTTTT..AAAAAAAAAAAAAAAA
SSSSSS.........IIIIIAAAAAAAAA.................TTTTT..........TTTTTTTT..AAAAAAAAAAAAAAAAA
SSSSSSDDD.......IIIIIIIAAAAAAAAAAAA..............TTTTTT..........TTTTTTTT.AAAAAAAAAAAAAAAA
IIIIIIAAAAAAAAAAAA..............TTTT..........TTTTTTTT...AAAAAAAAAAAAAAAA
SSSSSSSSSS...................IIIIIAAAAAAAAAAAA............TTTTT.........TTTTTTT....AAAAAAAAAAAA
SSSSSSSSSSSSSSS...........AAAAIIIIII.................TTTTT..........TTTTTTT.....AAAAAAAAAAAAAA
SSSSSS...........IIIIIAAAAAAAAA...............TTTTT..........TTTTTTTTT..AAAAAAAAAAAAAAA
SSSSSSSSSSSSSSSS........IIIIIIAAAAAAAAAAAAA...........TTTTT..........TTTTTTTT....AAAAAAAAAAAAAAAA
```

Fig. 6. Example of string set obtained by expansion of a word. Typical typing errors are evident: The sequence "A", "I" is frequently inverted, and key S and D are pressed simultaneously. The correct keyword is "SAITTA".

6.2 Modeling User Behavior

Concerning the user profiling case study, we notice that a *good* user profile should be as much as possible independent from the specific text the user is typing, and, at the same time, should capture its characteristic biometric features. In order to meet these requirements, we constructed a user profile based on specific keywords (denoted as K), such as conjunctive particles, articles, prepositions, auxiliary verbs, and word terminations frequently occurring in any text of a given language. In the specific case, nine Ks, from three to four consecutive stroke long, have been selected. That means that the adherence of a user to its profile is checked only considering the dynamics of the selected keywords as long as they occur during the typing activity. Then, the user profile is reduced to be a collection of nine S-HMM, one for each keyword.

Concerning the user authentication case study, the problem is naturally posed as the one of constructing an S-HMM for each one of the words in the login phase. However, for both cases some more considerations are necessary. During an editing activity errors due to different factors frequently happen. In many cases, the way in which errors occur is strongly related to the user personality and should be included in the user profile itself. As an example, a very frequent error is the exchange of the order of two keys in the text, when the user types two hands and doesn't have a good synchronization. Another similar situation is when two adjacent keys are simultaneously stroked because of an imperfect perception of the keyboard layout. Some examples of this kind of typing error are evident in Figure 6. Most mistakes of this kind are self-corrected by the word processors today available, and so do not leave a permanent trace in the final text. Nevertheless, capturing this kind of misbehavior in the user profile greatly improves the robustness of the model.

For the above reason, we made the a-priori choice of modeling substrings corresponding to keystroke expansion by means of basic blocks encoding PHMMs, in order to automatically account for most typing errors. Gaps between consecutive keystrokes are simply modeled by a gap model of the kind described in Figure 3(b).

6.3 Model Construction

For every selected word K the corresponding S-HMM, modeling the keystroking dynamics of a specific user, has been constructed using the algorithm EDY, which has been described in other papers [10, 9]. More specifically, EDY starts with a database LS of learning sequences, and automatically builds an S-HMM λ_{LS} modeling all sequences in LS. In the present case LS contains the strings encoding the keystroking dynamics for a specific word observed when a specific user was typing (see Figure 6 for an example of string dataset collected for one of the key-word).

A description of EDY and a general evaluation of its performances are outside the scope of this paper. In the following section we will report the performances obtained on the two case studies described above.

7 User Profiling

Ten users collaborated to the experiment by typing a corpus of 2280 words while a key-logger was recording the duration of every stroke and the delay between two consecutive strokes. Each one of the chosen Ks occurs in the corpus from 50 to 80 times.

Then, for every user, nine datasets have been constructed, each one collecting the temporal sequences of a specific keyword K.

Let D_{ij} denote the dataset containing the sequences corresponding to keyword K_i typed by the user u_j. Every D_{ij} has been partitioned into a learning set LS_{ij}, containing 25 instances, and a test set TS_{ij}, containing the remaining ones. From every LS_{ij} an S-HMM λ_{ij} has been constructed using the algorithm EDY [10]. EDY was instructed to use PHMM for modeling motifs and the model scheme in Figure 3(b) for modeling gaps.

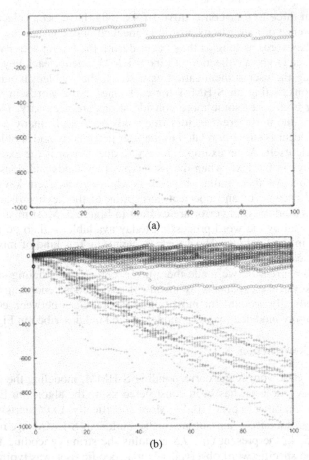

Fig. 7. Time evolution, of the logarithm of $P_{est}(u_j)$ (a) for a single user profile, and (b) for all user profiles. Circles describe $P_{est}(u_j)$ when the performer was u_j. Crosses correspond to $P_{est}(u_j)$ when the performer was another user.

Then, EDY modeled the strings corresponding to keystrokes as motifs, and the ones corresponding to delays as gaps. It is worth noticing that the mistyping rate was relevant, being key overlapping or key inversion the most frequent mistypes. Consequently, the strings obtained by expansion frequently contained mixtures of two overlapped keys. The S-HMM proved to be an excellent tool for modeling such features, which are highly user specific. Then, the set $U_j = \{ \lambda_{ij} \mid 1 \leq i \leq 9 \}$ of the nine models constructed for every users u_j constitute her profile.

The profile performances have been tested by simulating the editing activity of users typing a text written in the same language as the corpus used to collect the data. The simulation process consisted in generating series of sequences extracted with replacement from the datasets TS_{ij}, which has been compared to the user profile under test. More precisely, the evaluation procedure was as in the following. Let $P_{est}(u_j)$ be the probability estimated by means of profile U_j that the observed performance belongs to user u_j. The procedure for testing profile U_j against user u_k ($1 \leq k \leq 10$) is as in the following:

Let initially $P_{est}(u_j) = 1$
repeat
1. Select a sequence O_{ij} from the set TS_{ij} according to the probability distribution of K_i in the corpus.

2. Evaluate the probability $P(O_{ij}|\lambda_{ij})$ that model λ_{ij} has generated O_{ij}.
3. Set $\overline{P}(u_j) = P(u_j)\dfrac{P(O_{ij}|\lambda_{ij})}{\overline{P}_{ij}}$.

end

In the above procedure, $\overline{P}_{ij}$ denotes the average probability that model λ_{ij} generates a sequence belonging to its domain, estimated from the corresponding dataset LS_{ij}.

Figure 7(a) reports the evolution of $P_{est}(u_j)$ versus the number of keywords progressively found in a text, when $k = j$ and when $k \neq j$. It appears that when the performer corresponds to the profile, $P_{est}(u_j)$ remains close to 1, whereas it decreases exponentially when the performer does not correspond to the model. Figure 7(b) summarizes in a single diagram the results for all pairing $\langle u_k, U_j \rangle$.

8 User Authentication

As it was impractical to run the experiment using real passwords, it has been supposed that username and password coincide with the pair (name N, surname S) of the user. Under this assumption the conjecture that people develop specific skills for typing words strictly related the their own person still holds.

The experiment followed the same protocol described previously. The same group of users typed a number of times their own name and surname, and, then, the name and surname of the other users.

For every user two S-HMMs, namely λ_{N_j}, λ_{S_j}, have been constructed, for the name and surname, respectively. The learning sets contained 30 sequences. Figure 8 reports the results of the evaluation of the two models learned for one of the users. For every sequence pair $\langle N_j, S_j \rangle$ the probabilities $P(N_j) = P(O_i | \lambda_{N_j})$ and $P(S_j) = P(O_i | \lambda_{S_j})$ have been evaluated and represented with a point in the plane $\langle P(N_j), P(S_j) \rangle$.

It is evident that the sequences typed by the user u_j and the ones typed by the other users are separated by a wide margin: in this case, a simple linear discriminator provides a perfect separation. Using a testing set containing 150 negative examples (provided by 9 users different from u_j) and 100 positive examples (provided by the user u_j) for evaluating each one of the models, a discrimination rate of 99% has been obtained. This evaluation does not take into account a small percentage (5%) of positive sequences, which have been rejected because containing abnormal mistakes (for instance, several "backspaces" when the user tried to go back and correct errors he/she noticed).

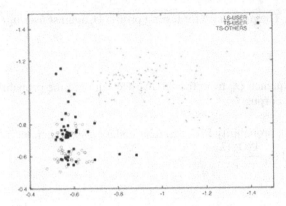

Fig. 8. Results produced by the pair of models learned for the name and the surname of a user. The x-axis reports the probability assigned to a sequence by the *name* model. The y-axis reports the probability assigned by the *surname* model. Circles denote sequences belonging to the learning set. Black squares denote sequences belonging to the positive testing set, and '+' denotes sequences typed by other users.

The S-HMM modeling the name of one of the users is reported in Figure 9. The structure of the model accounts for different typing modalities typical of the user, both concerning the gap duration and the stroke duration. In some cases the model also learned typical mistakes made by the user, such as key inversion or key overlapping. This kind of mistakes happens when the user strike keys with both hands at the same time. The global number of states of the model in Figure 9 is 513. In general, the size of the models constructed for the name or the surname of the considered users ranges from 200 to 600 states, while the response time remains below one second (real-time).

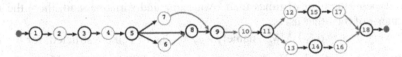

Fig. 9. Example of the S-HMM learned for the name of a user. Circles represent basic blocks encoding models of strokes (thick line) and of gaps between one stroke and another (thin line).

9 Conclusions

In this paper, S-HMMs have been proposed as a general tool for modeling the behavior of an agent, intended either as a human user or as a computer process. The most important property of S-HMM is its ability to model durations by means of the state expansion technique. This is made possible by the low computational cost inherent to S-HMMs, which allows matching in quasi-linear time very complex models.

The ability of S-HMM to accomplish the task of modeling behaviors has been demonstrated on two non-trivial case studies of user profiling and of user authentication. The obtained results look very promising, taking also into account the very little effort required to develop the applications.

References

[1] Baum, L.E., Petrie, T., Soules, G., Weiss, N.: A maximisation techniques occurring in the statistical analysis of probabilistic functions of markov chains. The Annals of Mathematical Statistics 41(1), 164–171 (1970)

[2] Bouchaffra, D., Tan, J.: Structural Hidden Markov models using a relation of equivalence: Application to automotive designs. Data Mining and Knowledge Discovery 12, 79–96 (2006)

[3] Brown, M., Rogers, S.J.: User identification via keystroke characteristics of typed names using neural networks. International Journal of Man-Machine Studies 39, 999–1014 (1993)

[4] Brown, M.E., Rogers, S.J.: Method and apparatus for verification of a computer user's identification, based on keystroke characteristics. Patent n. 5,557,686, U.S. Patent and Trademark Office, Washington, DC (1996)

[5] Hai Bui, H., Venkatesh, S., West, G.A.W.: Tracking and surveillance in wide-area spatial environments using the abstract Hidden Markov model. IJPRAI 15(1), 177–195 (2001)

[6] Durbin, R., Eddy, S., Krogh, A., Mitchison, G.: Biological sequence analysis. Cambridge University Press, Cambridge (1998)

[7] Fawcett, T., Provost, F.: Adaptive fraud detection. Data Mining and Knowledge Discovery Journal 1, 291–316 (1997)

[8] Furnell, S., Orrissey, J.M., Sanders, P., Stockel, C.: Applications of keystroke analysis for improved login security and continuous user authentication. In: Proceedings of the Information and System Security Conference, pp. 283–294 (1996)

[9] Galassi, U., Botta, M., Giordana, A.: Hierarchical Hidden Markov models for user/process profile learning. Fundamenta Informaticae 78, 1–19 (2007)

[10] Galassi, U., Giordana, A., Saitta, L.: Incremental construction of structured Hidden Markov models. In: Proceedings IJCAI-2007 (Hyderabad, India), pp. 2222–2227 (2007)

[11] Gussfield, D.: Algorithms on Strings, Trees, and Sequences. Cambridge University Press, Cambridge (1997)

[12] Bonafonte, A.: Duration modeling with expanded HMM applied to speech recognition

[13] Lee, W., Stolfo, S.J.: Data mining approaches for intrusion detection. In: Proceedings of the Seventh USENIX Security Symposium (San Antonio, TX) (1998)

[14] Lee, W., Fan, W., Miller, M., Stolfo, S.J., Zadok, E.: Toward cost-sensitive modeling for intrusion detection and response. Journal of Computer Security 10, 5–22 (2002)

[15] Levinson, S.E.: Continuous variable duration Hidden Markov models for automatic speech recognition. Computer Speech and Language 1, 29–45 (1986)

[16] Lipman, D.J., Pearson, W.R.: Rapid and sensitive protein similarity searches. Science 227, 1435–1476 (1985)

[17] Pylkknen, J., Kurimo, M.: Using phone durations in finnish large vocabulary continuous speech recognition (2004)

[18] Rabiner, L., Juang, B.: Fundamentals of Speech Recognition. Prentice Hall, Englewood Cliffs (1993)

[19] Rabiner, L.R.: A tutorial on Hidden Markov models and selected applications in speech recognition. Proceedings of IEEE 77(2), 257–286 (1989)

[20] Schulz, D., Burgard, W., Fox, D., Cremens, A.: People tracking with mobile robots using sample-based joint probabilistic data association filters (2003)

[21] Srikant, R., Agrawal, R.: Mining sequential patterns: Generalizations and performance improvements. In: Apers, P.M.G., Bouzeghoub, M., Gardarin, G. (eds.) EDBT 1996. LNCS, vol. 1057, pp. 3–17. Springer, Heidelberg (1996)

[22] Stolfo, S.J., Fan, W., Lee, W., Prodromidis, A., Chan, P.: Cost-based modeling for fraud and intrusion detection: Results from the jam project. In: Proceedings of the 2000 DARPA Information Survivability Conference and Exposition (DISCEX 2000) (2000)

[23] Stolfo, S.J., Li, W., Hershkop, S., Wang, K., Hu, C., Nimeskern, O.: Detecting viral propagations using email behavior profiles. In: ACM Transactions on Internet Technology (TOIT), pp. 128–132 (2004)

[24] Tweed, D., Fisher, R., Bins, J., List, T.: Efficient Hidden semi-Markov model inference for structured video sequences. In: Proc. 2nd Joint IEEE Int. Workshop on VSPETS (Beijing, China), pp. 247–254 (2005)

[25] Yang, Z., Kitsuregawa, M.: Lapin-spam: An improved algorithm for mining sequential pattern. In: ICDEW (2005)

[26] Yu, S.-Z., Kobashi, H.: An efficient forward-backward algorithm for an explicit duration Hidden Markov model. IEEE Signal Processing Letters 10(1) (2003)

[27] Zaki, J.: Spade: An efficient algorithm for mining frequent sequences. Machine Learning 42(1/2), 31–60 (2001)

[28] Galassi, U., Giordana, A., Julien, C., Saitta, L.: Modeling Temporal Behavior via Structured Hidden Markov Models: An Application to Keystroking Dynamics. In: Proceedings 3rd Indian International Conference on Artificial Intelligence (Pune, India) (2007)

Author Index